edexcel
advancing learning, changing lives

Edexcel GCSE

Mathematics B Modular Foundation

Practice Book

Series Director: Keith Pledger
Series Editor: Graham Cumming

Authors:
Julie Bolter
Gareth Cole
Gill Dyer
Michael Flowers
Karen Hughes
Peter Jolly
Joan Knott
Jean Linsky
Graham Newman
Rob Pepper
Joe Petran
Keith Pledger
Gillian Rich
Rob Summerson
Kevin Tanner
Brian Western

A PEARSON COMPANY

Published by Pearson Education Limited, a company incorporated in England and Wales, having its registered office at Edinburgh Gate, Harlow, Essex, CM20 2JE. Registered company number: 872828

Edexcel is a registered trademark of Edexcel Limited

Text © Julie Bolter, Gareth Cole, Gill Dyer, Michael Flowers, Karen Hughes, Peter Jolly, Joan Knott, Jean Linsky, Graham Newman, Rob Pepper, Joe Petran, Keith Pledger, Gillian Rich, Rob Summerson, Kevin Tanner, Brian Western and Pearson Education Limited 2010

The rights of Julie Bolter, Gareth Cole, Gill Dyer, Michael Flowers, Karen Hughes, Peter Jolly, Joan Knott, Jean Linsky, Graham Newman, Rob Pepper, Joe Petran, Keith Pledger, Gillian Rich, Rob Summerson, Kevin Tanner and Brian Western to be identified as the authors of this Work have been asserted by them in accordance with the Copyright, Designs and Patent Act, 1988.

First published 2010

13 12 11 10
10 9 8 7 6 5 4 3 2 1

British Library Cataloguing in Publication Data
A catalogue record for this book is available from the British Library

ISBN 978-1-84690-097-6

Copyright notice
All rights reserved. No part of this publication may be reproduced in any form or by any means (including photocopying or storing it in any medium by electronic means and whether or not transiently or incidentally to some other use of this publication) without the written permission of the copyright owner, except in accordance with the provisions of the Copyright, Designs and Patents Act 1988 or under the terms of a licence issued by the Copyright Licensing Agency, Saffron House, 6–10 Kirby Street, London EC1N 8TS (www.cla.co.uk). Applications for the copyright owner's written permission should be addressed to the publisher.

Typeset by Tech-Set Ltd, Gateshead
Project management by Wearset Ltd, Boldon, Tyne and Wear
Printed in Great Britain at Scotprint, Haddington

Disclaimer
This material has been published on behalf of Edexcel and offers high-quality support for the delivery of Edexcel qualifications.
This does not mean that the material is essential to achieve any Edexcel qualification, nor does it mean that it is the only suitable material available to support any Edexcel qualification. Edexcel material will not be used verbatim in setting any Edexcel examination or assessment. Any resource lists produced by Edexcel shall include this and other appropriate resources.

Copies of official specifications for all Edexcel qualifications may be found on the Edexcel website: www.edexcel.com

Contents

About this book

All set to make the grade!

Edexcel GCSE Mathematics is specially written to help you get your best grade in the exams.
Unit 1 and Unit 3 allow the use of calculators.
Unit 2 is a non-calculator unit.

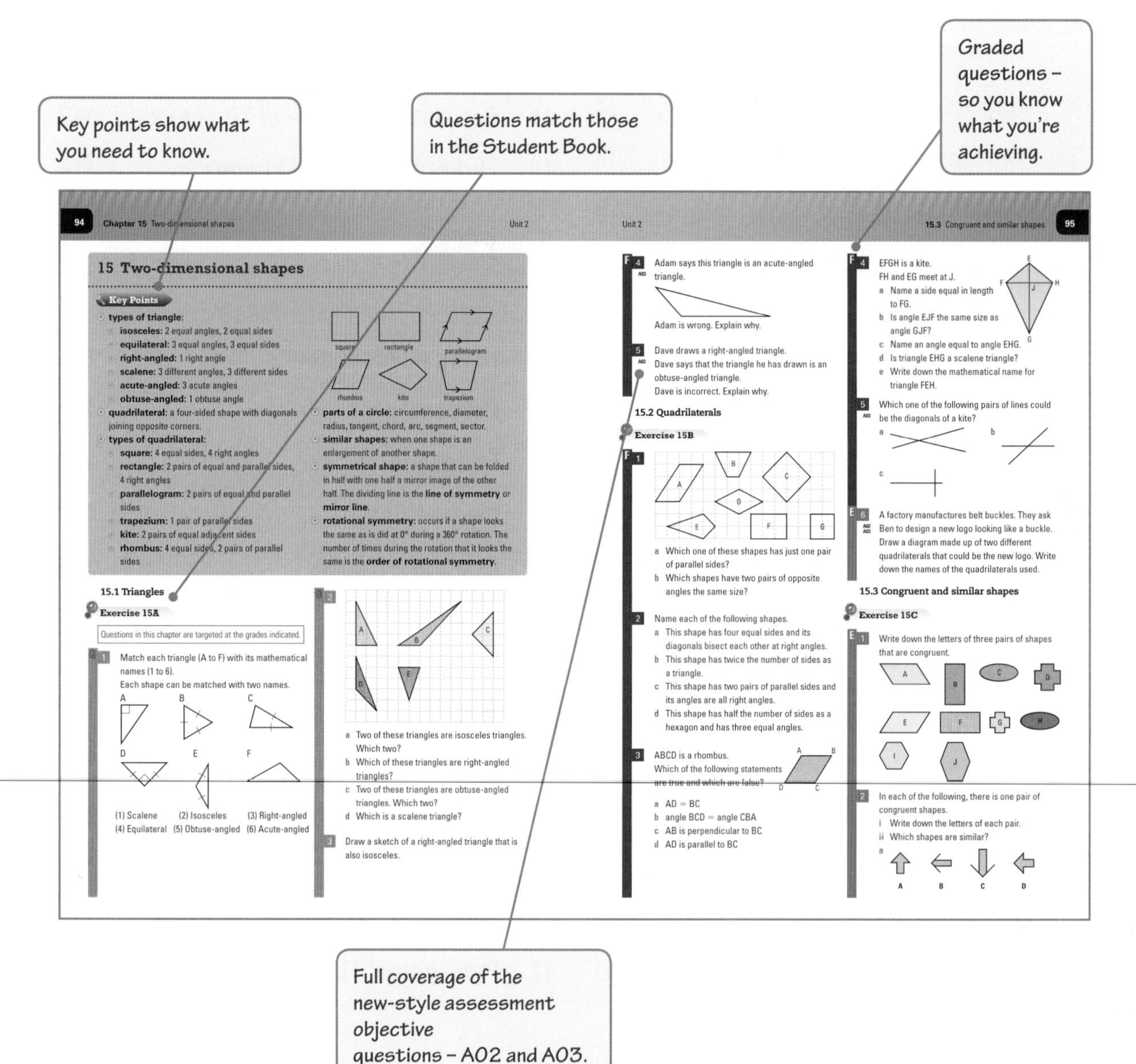

94 Chapter 15 Two-dimensional shapes Unit 2

15 Two-dimensional shapes

Key Points

- types of triangle:
  - isosceles: 2 equal angles, 2 equal sides
  - equilateral: 3 equal angles, 3 equal sides
  - right-angled: 1 right angle
  - scalene: 3 different angles, 3 different sides
  - acute-angled: 3 acute angles
  - obtuse-angled: 1 obtuse angle
- quadrilateral: a four-sided shape with diagonals joining opposite corners.
- types of quadrilateral:
  - square: 4 equal sides, 4 right angles
  - rectangle: 2 pairs of equal and parallel sides, 4 right angles
  - parallelogram: 2 pairs of equal and parallel sides
  - trapezium: 1 pair of parallel sides
  - kite: 2 pairs of equal adjacent sides
  - rhombus: 4 equal sides, 2 pairs of parallel sides
- parts of a circle: circumference, diameter, radius, tangent, chord, arc, segment, sector.
- similar shapes: when one shape is an enlargement of another shape.
- symmetrical shape: a shape that can be folded in half with one half a mirror image of the other half. The dividing line is the line of symmetry or mirror line.
- rotational symmetry: occurs if a shape looks the same as is did at 0° during a 360° rotation. The number of times during the rotation that it looks the same is the order of rotational symmetry.

15.1 Triangles

Exercise 15A

Questions in this chapter are targeted at the grades indicated.

1 Match each triangle (A to F) with its mathematical names (1 to 6).
Each shape can be matched with two names.
(1) Scalene (2) Isosceles (3) Right-angled (4) Equilateral (5) Obtuse-angled (6) Acute-angled

2 a Two of these triangles are isosceles triangles. Which two?
b Which of these triangles are right-angled triangles?
c Two of these triangles are obtuse-angled triangles. Which two?
d Which is a scalene triangle?

3 Draw a sketch of a right-angled triangle that is also isosceles.

4 Adam says this triangle is an acute-angled triangle.
Adam is wrong. Explain why.

5 Dave draws a right-angled triangle.
Dave says that the triangle he has drawn is an obtuse-angled triangle.
Dave is incorrect. Explain why.

15.2 Quadrilaterals

Exercise 15B

1 a Which one of these shapes has just one pair of parallel sides?
b Which shapes have two pairs of opposite angles the same size?

2 Name each of the following shapes.
a This shape has four equal sides and its diagonals bisect each other at right angles.
b This shape has twice the number of sides as a triangle.
c This shape has two pairs of parallel sides and its angles are all right angles.
d This shape has half the number of sides as a hexagon and has three equal angles.

3 ABCD is a rhombus.
Which of the following statements are true and which are false?
a AD = BC
b angle BCD = angle CBA
c AB is perpendicular to BC
d AD is parallel to BC

Unit 2 15.3 Congruent and similar shapes 95

4 EFGH is a kite.
FH and EG meet at J.
a Name a side equal in length to FG.
b Is angle EJF the same size as angle GJF?
c Name an angle equal to angle EHG.
d Is triangle EHG a scalene triangle?
e Write down the mathematical name for triangle FEH.

5 Which one of the following pairs of lines could be the diagonals of a kite?

6 A factory manufactures belt buckles. They ask Ben to design a new logo looking like a buckle.
Draw a diagram made up of two different quadrilaterals that could be the new logo. Write down the names of the quadrilaterals used.

15.3 Congruent and similar shapes

Exercise 15C

1 Write down the letters of three pairs of shapes that are congruent.

2 In each of the following, there is one pair of congruent shapes.
i Write down the letters of each pair.
ii Which shapes are similar?

And

- Functional elements highlighted.
- ResultsPlus features.

Assessment Objectives

There are three types of question that are set in the exam.

Assessment Objective	What it is	What this means	Range % of marks in the exam
A01	**Recall** and use knowledge of the prescribed content.	Standard questions testing your knowledge of each topic.	45-55
A02	**Select** and apply mathematical methods in a range of contexts.	Deciding what method you need to use to get to the correct solution to a contextualised problem.	25-35
A03	**Interpret** and analyse problems and generate strategies to solve them.	Solving problems by deciding how and explaining why.	15-25

The proportion of marks available in the exam varies with each Assessment Objective. Don't miss out, make sure you know how to do A02 and A03 questions!

What does an AO2 question look like?

This just needs you to
(a) read and understand the question and
(b) decide how to get the correct answer.

16 Katie wants to buy a car.
She decides to borrow £3500 from her father. She adds interest of 3.5% to the loan and this total is the amount she must repay her father. How much will Katie pay back to her father in total?

What does an AO3 question look like?

Here you need to read and analyse the question. Then use your mathematical knowledge to solve this problem.

17 Rashida wishes to invest £2000 in a building society account for one year.
The Internet offers two suggestions. Which of these two investments gives Rashida the greatest return?

CHESTMAN BUILDING SOCIETY
£3.50 per month
Plus **1% bonus** at the end of the year

DUNSTAN BUILDING SOCIETY
4% per annum. Paid yearly by cheque

Quality of written communication

There will be marks in the exam for showing your working 'properly' and explaining clearly. In the exam paper, such questions will be marked with a star (*). You need to:

- use the correct mathematical notation and vocabulary, to show that you can communicate effectively
- organise the relevant information logically.

About functional elements

What does a question with functional maths look like?

Functional maths is about being able to apply maths in everyday, real-life situations.

GCSE Tier	Range % of marks in the exam
Foundation	30-40
Higher	20-30

The proportion of functional maths marks in the GCSE exam depends on which tier you are taking. Don't miss out, make sure you know how to do functional maths questions!

In the exercises...

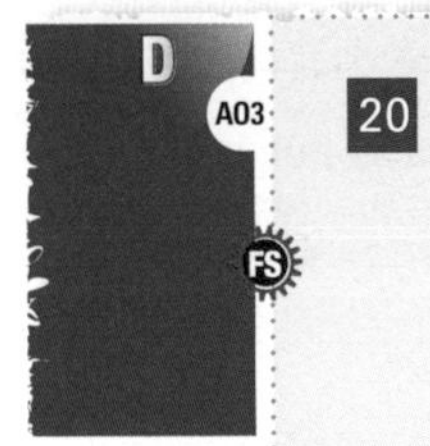

20 The Wildlife Trust are doing a survey into the number of field mice on a farm of size 240 acres. They look at one field of size 6 acres. In this field they count 35 field mice.

a Estimate how many field mice there are on the whole farm.

b Why might this be an unreliable estimate?

> You need to read and understand the question. Follow your plan.
>
> Think what maths you need and plan the order in which you'll work.
>
> Check your calculations and make a comment if required.

ResultsPlus features

ResultsPlus feaures combine exam performance data with examiner insight to give you more information on how to succeed. ResultsPlus tips show students how to avoid errors in solutions to questions.

ResultsPlus
Watch Out!

Some students use the term average – make sure you specify mean, mode or median.

> This warns you about common mistakes and misconceptions that examiners frequently see students make.

ResultsPlus
Examiner's Tip

Make sure the angles add up to 360°.

> This gives exam advice, useful checks, and methods to remember key facts.

1 Collecting and recording data

Key Points

- **statistics:** an area of mathematics concerned with collecting and interpreting data.
- **data:** the information that has been collected.
  - **qualitative:** data described in words
  - **quantitative:** data given as a numerical value
  - **discrete:** data that can only have a whole number value
  - **continuous:** data that can take any numerical value
- **frequency:** the total number of times a certain event happens.
- **class interval:** a group of numbers that data can fall into. Often used when recording continuous data.
- **questionnaire:** a list of questions designed to collect data.
- **an open question:** one that has no suggested answers.
- **a closed question:** one that has a set of answers to choose from.
- **population:** the whole group of people you wish to find something out about.
- **sample:** a select number of people from the population that can be surveyed to represent the total population.
  - the larger the sample the more representative it is of the population
- **biased sample:** an unfair sample that may occur when:
  - the sample does not truly represent the population
  - the sample is too small
- **random sample:** a sample that has given every member of the population an equal chance of being chosen.
- **two-way table:** a table that shows how data falls into two different categories.
- **primary data:** data that has been collected by the person who is going to use it.
- **secondary data:** data that has been collected by somebody else.
- **12-hour clock:** uses am for times between midnight and midday and pm for times between midday and midnight.
- **24-hour clock:** goes from 0 to 24, where 0 and 24 are midnight.
- **speed, time and distance formula:**
  - $\text{speed} = \frac{\text{distance}}{\text{time}}$,
    usually measured in miles per hour (mph), kilometres per hour (km/h), metres per second (m/s)
  - $\text{time} = \frac{\text{distance}}{\text{speed}}$
  - $\text{distance} = \text{speed} \times \text{time}$
  - $\text{average speed} = \frac{\text{total distance travelled}}{\text{total time taken}}$
- **multiplying by 10, 100, 1000:** move all the digits 1, 2 or 3 place values to the left.
- **dividing by 10, 100, 1000:** move all the digits 1, 2 or 3 place values to the right.
- **collecting data by observation:** use a data collection sheet to record a tally of how often a certain event happens.
- **converting metric units of measure:** when changing from a large unit to a smaller unit, multiply by 10, 100 or 1000. When changing from a small unit to a larger unit, divide by 10, 100 or 1000.

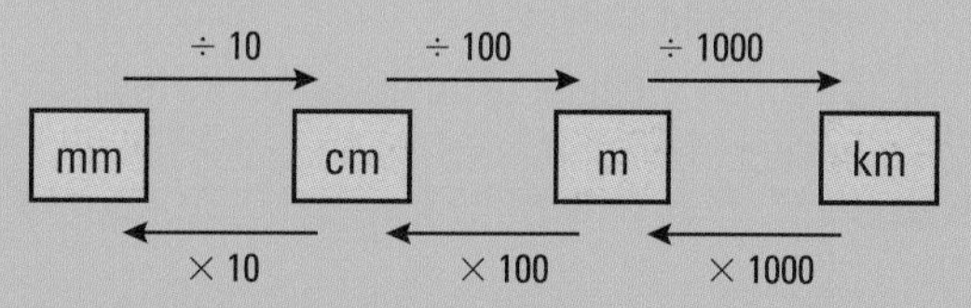

- **rounding numbers:** look at the digit before the place value you are rounding to. If it is less than 5, round down. If it is 5 or more, round up.
- **rounding to a given number of significant figures (s.f.):** count this number of digits from the first non-zero digit. If the next digit is 5 or more then round up.

1.1 Introduction to data

Exercise 1A

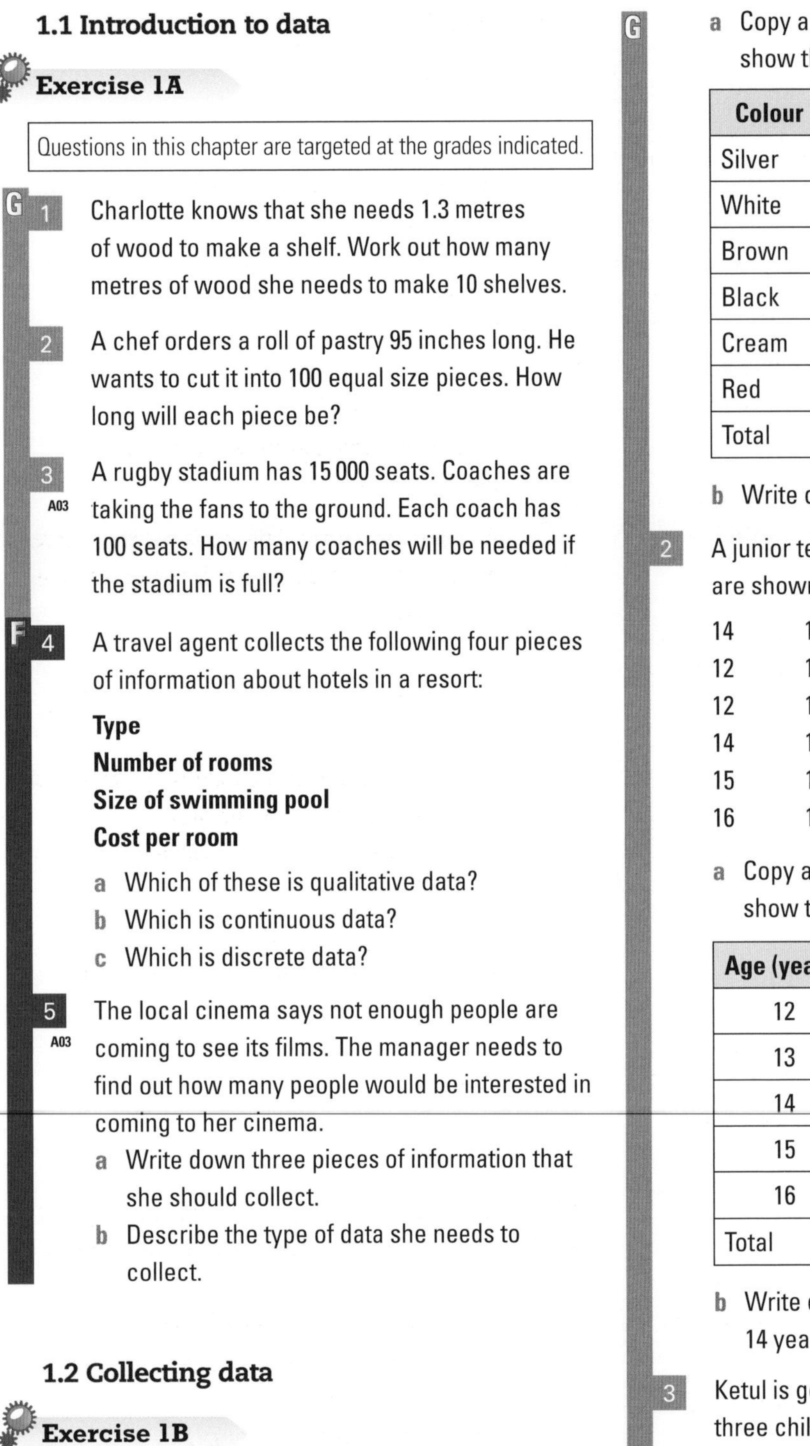

Questions in this chapter are targeted at the grades indicated.

G

1. Charlotte knows that she needs 1.3 metres of wood to make a shelf. Work out how many metres of wood she needs to make 10 shelves.

2. A chef orders a roll of pastry 95 inches long. He wants to cut it into 100 equal size pieces. How long will each piece be?

3. A03 A rugby stadium has 15 000 seats. Coaches are taking the fans to the ground. Each coach has 100 seats. How many coaches will be needed if the stadium is full?

F

4. A travel agent collects the following four pieces of information about hotels in a resort:

   Type
   Number of rooms
   Size of swimming pool
   Cost per room

   a Which of these is qualitative data?
   b Which is continuous data?
   c Which is discrete data?

5. A03 The local cinema says not enough people are coming to see its films. The manager needs to find out how many people would be interested in coming to her cinema.

   a Write down three pieces of information that she should collect.
   b Describe the type of data she needs to collect.

1.2 Collecting data

Exercise 1B

G

1. A kitchen shop has 36 kettles in stock. The colour of each kettle is shown below.

silver	white	brown	silver	black	white
cream	black	silver	white	white	cream
silver	cream	brown	black	cream	silver
silver	white	black	cream	white	white
brown	silver	cream	silver	white	white
silver	silver	silver	white	white	cream

   a Copy and complete the frequency table to show the colours of the kettles.

Colour	Tally	Frequency
Silver		
White		
Brown		
Black		
Cream		
Red		
Total		

   b Write down the most popular colour.

2. A junior tennis club has 36 members. Their ages are shown below.

14	15	16	13	14	16
12	14	16	16	13	15
12	14	16	16	15	15
14	15	13	14	12	14
15	15	15	14	15	16
16	16	16	15	16	13

   a Copy and complete the frequency table to show the members' ages.

Age (years)	Tally	Frequency
12		
13		
14		
15		
16		
Total		

   b Write down the number of members who are 14 years old or less.

3. Ketul is going to make fancy-dress outfits for the three children in his family. The table shows the length of material, in centimetres, that he needs for each outfit.

Child	Joshan	Siana	Chirag
Length of material (cm)	75	107	88

   a Work out how many centimetres of material Ketul needs altogether.
   b Work out how many metres of material Ketul needs altogether.

G 4 AO3 Here is part of a railway timetable.

	Train A	Train B	Train C
Tottenham Hale	17:24	17:54	18:09
Broxbourne	17:39	18:09	18:24
Harlow	17:47	18:18	18:31
Sawbridgeworth	17:54	18:24	18:35
Bishop's Stortford	18:01	18:31	18:42

Simon has to travel from Tottenham Hale and arrive in Sawbridgeworth before 6 pm.

a Which train should he catch?

Barry is happy because he thinks there is a train that arrives in Bishop's Stortford at 8.42 pm.

b Explain why Barry is wrong.

F 5 AO2 A shop manager times the number of minutes 20 customers spend in his store one morning. The times are shown below.

10	12	8	12	15	10
14	9	10	11	20	16
12	15	12	18	10	10
8	10				

a Draw a table or chart to show the amount of time spent in the store.

b How many customers spent more than 10 minutes in the store?

6 AO3 A recipe says that it takes 40 g of icing to cover a cupcake. Jane has prepared a 1.5 kg batch of icing and 40 cupcakes. Does Jane have enough icing to cover all of her cupcakes? Explain your answer.

7 AO2 AO3 Dan likes lifting weights. He needs 10.5 kg of weight for one of his exercises. A sports shop only has 750 g weights for sale.

a Work out how many weights he needs to buy from the shop.

The shop also sells a pack of five 750 g weights for £15.20.
For a second exercise, Dan needs 30 kg of weight.

b How many packs will he need?

c How much would this cost?

1.3 Questionnaires

Exercise 1C

E 1 Adam is doing a survey on cricket.
He writes the following question for a questionnaire.
'How often do you watch cricket on TV each month?'

☐ 0–1 time ☐ 1–2 times

☐ 3–4 times

Write down one reason why this is a poor questionnaire.

2 A market researcher intends to put the following question on a questionnaire for shoppers in a supermarket.
'What is your age?'
Write down one reason why this is a poor question.

3 The local library asks the following question in a survey about libraries in the area.
'Do you agree that the library should be refurbished?'

☐ Yes ☐ Not sure

Write down two problems with this question and the response boxes.

1.4 Sampling

Exercise 1D

F 1 AO3 The local cinema manager is conducting a survey to find out how far people travel to the cinema.
He is going to ask a sample of people.
He decides to ask the first 10 people who come to the first showing of a film.
What is wrong with this sample?

2 AO3 A head teacher decides to conduct a survey to find out how students feel about the school dinners provided.
He asks the students in Class 3Y.
What is wrong with using these students as a sample?

E 3* A03 A TV programme producer wants to find out people's views about her TV series.
She organises a poll where 25 people are telephoned and asked their opinions.
Give reasons why you think this would not give a true picture of people's views.

1.5 Two-way and other tables

Exercise 1E

E 1 A teacher is working out a timetable for Class 11A. Of the 28 students:

eight want to do Maths and ICT
eleven want to do Science and D&T
four want to do ICT and D&T
five want to do Science and Maths.

Copy and complete the two-way table below to show these data.

	ICT	Science	Total
Maths			
D&T			
Total			28

2 The two-way table gives some information about the numbers of different drinks sold at a stall in the market in one hour.

	Large	Small	Total
Tea	10	12	
Coffee	9	5	
Total			

Copy and complete the table.

3 The following two-way table gives information about the numbers of different types of membership at a gym club.

	Junior	Senior	Family	Total
Full week	16	34	25	
Weekends	30	26	28	
Total				

a Copy and complete the table.

b Write down the number of weekend junior members.

D 4 The two-way table below gives information about the meals chosen by people on a flight.

	Meat	Fish	Pasta	Total
Mousse	14	12		36
Ice cream	11		18	40
Fruit tart	6	3		
Total				100

a Copy and complete the table.

b Write down the number of people who chose pasta and fruit tart.

c Write down the total number of people represented by the table.

Exercise 1F

G 1 The following table provides information about the weather in Anytown.

	January	April	July	October
Maximum temperature (degrees C)	5.6	12.4	20.2	13.9
Minimum temperature (degrees C)	2.1	6.5	10.2	5.6
Sunshine (hours)	43.3	185.1	190.9	100.6
Rainfall (mm)	121.9	55.7	60.2	100.9

a Write down the amount of sunshine in April.

b Write down the maximum temperature in October.

c Write down the month that has the most sunshine.

G 2 The table below provides information about some seas and oceans.

Name	Area (sq km)	Average depth (m)	Greatest known depth (m)
Pacific Ocean	155 557 000	4028	11 033
Atlantic Ocean	76 762 000	3926	9219
Indian Ocean	68 556 000	3963	7455
Southern Ocean	20 327 000	4500	7235
Mediterranean Sea	14 056 000	1205	5625
Caribbean Sea	2 718 200	2647	6946
South China Sea	2 319 000	1652	5016
Bering Sea	2 291 900	1547	4773
Gulf of Mexico	1 592 800	1486	3787

a What is the area of the Mediterranean Sea?
b Write down the average depth of the Southern Ocean.
c Which sea has a greatest known depth of 5000 m to the nearest 100 m?
d Write down the ocean that has the largest average depth.
e How many oceans and seas have an area of less than 3 000 000 sq km?

G 3 The following table shows information about the distance and travelling times of various cities from Manchester.

From Manchester to:	Distance (miles)	Car (minutes)	Train (minutes)	Coach (minutes)
Aberdeen	321	390	553	510
Leeds	42	56	55	120
Nottingham	68	107	110	230
Cardiff	188	208	203	430
Portsmouth	237	261	278	620
London	200	233	129	330

a Write down how long it takes to drive from Manchester to Cardiff.
b Write down the place that is 200 miles from Manchester.
c What journey takes the least time if travelling by train from Manchester?
d What journey takes the longest if travelling from Manchester by coach?

4 Here is some data Raj collected on the lengths, in centimetres, of 6 crisps.

6.4 3.48 6.5 0.9 5.56 4.33

Round all these lengths to the nearest whole centimetre.

F 5 Here is some data Claire collected on the distance, in kilometres, of 6 people's journeys to work.

8.56 0.52 3.48 3.51 1.12 5.41

The manager of the company where they work needs to know the distances to one decimal place.

Round all these distances to one decimal place.

2 Processing, representing and interpreting data

Key Points

- **pictogram:** represents qualitative data using symbols or pictures, with a key, to show numbers of items.
- **an angle:** a measure of turn that is formed when two lines meet.
  - **full turn:** 360°
  - **half turn:** 180°
  - **quarter turn:** 90°, also known as a right angle
  - **angles around a point** add up to 360°
- **pie chart:** a circle divided into sectors to show how the total is split up between different categories.
  - area of each sector represents the number of items in that category
  - sector angle $= \dfrac{\text{frequency} \times 360°}{\text{total frequency}}$
  - frequency $= \dfrac{\text{sector angle} \times \text{total frequency}}{360°}$
  - angles at the centre of the pie chart add up to 360°
- **frequency table:** a table that arranges data, giving frequency.
- **bar chart:** a chart used to represent qualitative data and grouped discrete data.
- **comparative or dual bar chart:** a chart that shows two or more bars drawn side by side for each category.
- **composite bar chart:** a chart that shows the size of individual categories split into separate parts.
- **vertical line graph:** a graph used to display ungrouped discrete data.
- **histogram:** a graph used to display grouped continuous data.
- **drawing a frequency polygon:**
  - **discrete data:** draw straight lines to join the tops of the lines on a vertical line chart
  - **continuous data:** draw straight lines to join the midpoints of the tops of the bars in a histogram

2.1 Pictograms

Exercise 2A

Questions in this chapter are targeted at the grades indicated.

G **1** The pictogram shows the number of texts sent from a mobile phone in one week.

Number of texts

Monday	
Tuesday	
Wednesday	
Thursday	
Friday	

Key represents 10 texts

a On which day were the greatest number of texts sent from the mobile phone?

b How many texts were sent from the mobile phone on Thursday?

c How many texts were sent from the mobile phone on Monday?

d How many texts were sent from the mobile phone on Tuesday?

2 The pictogram shows the number of ice cream cones sold in one day from an ice cream stall.

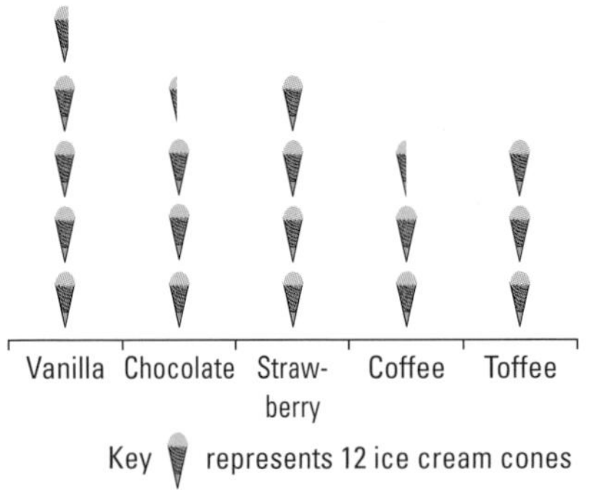

a Write down the flavour that was the least popular.

b How many coffee cones were sold?

c How many cones of chocolate ice cream were sold?

d Work out the total number of ice creams sold.

G 3 The pictogram shows the numbers of three different types of sandwich sold in one day.

Sandwiches sold

Chicken and sweetcorn

Cheese and tomato

Ham and salad

Tuna mayonnaise

Key
represents 20 sandwiches

20 tuna mayonnaise sandwiches were also sold.

a Copy and complete the pictogram.

b How many of each type of sandwich were sold?

F 4 A02 Thirty-six tourists going into a museum in London were asked which language they wanted on their audio guide.

Language	Number of tourists
English	5
French	7
Spanish	6
German	8
Japanese	10

Represent this information in a suitable chart.

5 A02 A nursery equipment shop has stock as shown in the table.

Item	Number in stock
Buggy	12
High chair	6
Car seat	5
Cot	10
Baby bath	8

Represent this information in a suitable chart.

2.2 Pie charts

Exercise 2B

F 1 Work out the size of angle b.

156°

66°

b

F 2 The minute hand of a clock moves from 30 to 50.

a Write down the size of the angle it has gone through.

The hour hand of the clock moves through 150°.

b Write down the number of hours that have passed.

E 3* 60 passengers at a bus station were asked which city they were travelling to.

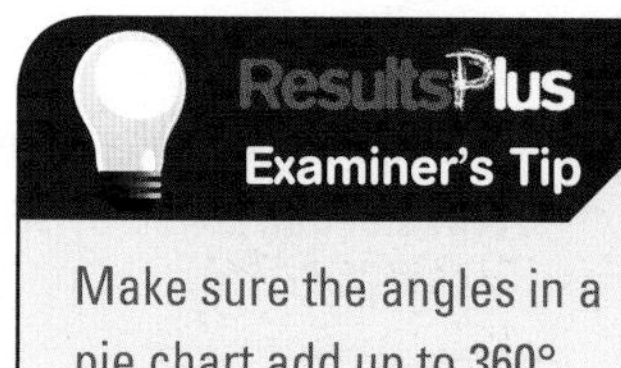

Make sure the angles in a pie chart add up to 360°.

The table shows this information.

City	Reading	Basingstoke	Portsmouth	Southampton
Frequency	8	7	15	30

Work out the sector angle for each city. Draw an accurate pie chart to show this information.

4* Dina asked 40 friends to name their favourite colour. The table shows her results.

Colour	Frequency
Red	12
Orange	8
Yellow	3
Green	5
Blue	7
Purple	5

Work out the sector angle for each colour. Draw an accurate pie chart to show this information.

D 5* A02 This table shows the time it takes 20 people to get an operator at a call centre to answer the phone.

Time (mins)	Frequency
less than 5	2
between 5 and 10	10
between 10 and 15	5
more than 15	3

Draw an accurate pie chart to show this information.

D

6* A02 The top four authors of teenage fiction books were read by the following numbers in a recent library survey.

Author	JK	PP	SM	MB
Frequency	210	85	265	70

Draw an accurate chart to show this information.

7* A02 The table shows the numbers of the different types of car driven by 90 workers in one company.

Type of car	Ford	Nissan	Toyota	Renault
Frequency	28	7	30	25

Draw an accurate chart to show this information.

8* A02 A Maths teacher kept a record of homework not handed in for each school year he taught.

Year	7	8	9	10	11
% homework not handed in	8	12	15	20	17

Draw an accurate chart to show this information.

Exercise 2C

ResultsPlus Examiner's Tip

In an exam 'work out' means calculate the frequency, so don't just measure the angle.

ResultsPlus Examiner's Tip

Always add up the frequencies for each sector to make sure they total to the right number.

E

1 90 boys were asked to name their favourite football team. The pie chart shows the results of this survey.

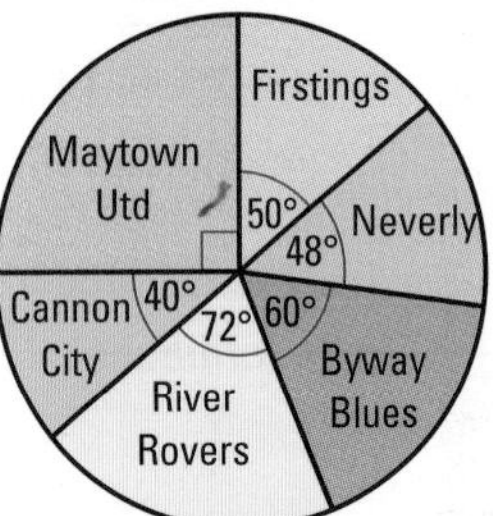

a Which was the most popular football team?

b Which team was the least popular?

c What fraction of the boys say that Cannon City is their favourite team?
Give your fraction in its simplest form.

d How many of the boys prefer Neverly?

2 The pie chart shows the travel methods of 120 students in Year 10.

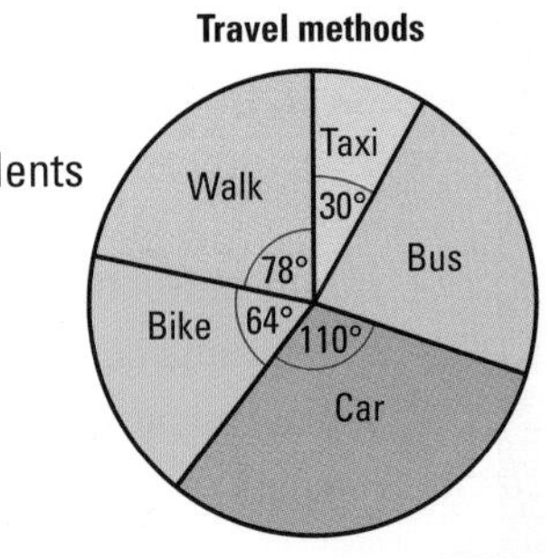

a Which method is the most popular?

b How many degrees represent one person on the pie chart?

c How many students said walking was their chosen method of travel?

d What angle represents the bus on the pie chart?

e How many students said the bus was their chosen method of travel?

3 The pie chart shows information about how Jack spends his time in one 24-hour day. Copy and complete the table. You will need to measure the angles in the pie chart.

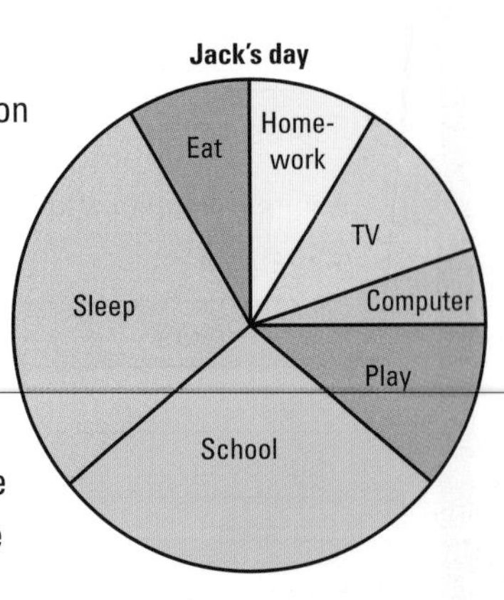

Activity	**Angle (degrees)**	**Number of hours and minutes**
Eat		
Sleep		
School		
Play		
Computer		
Homework		
Watch TV		

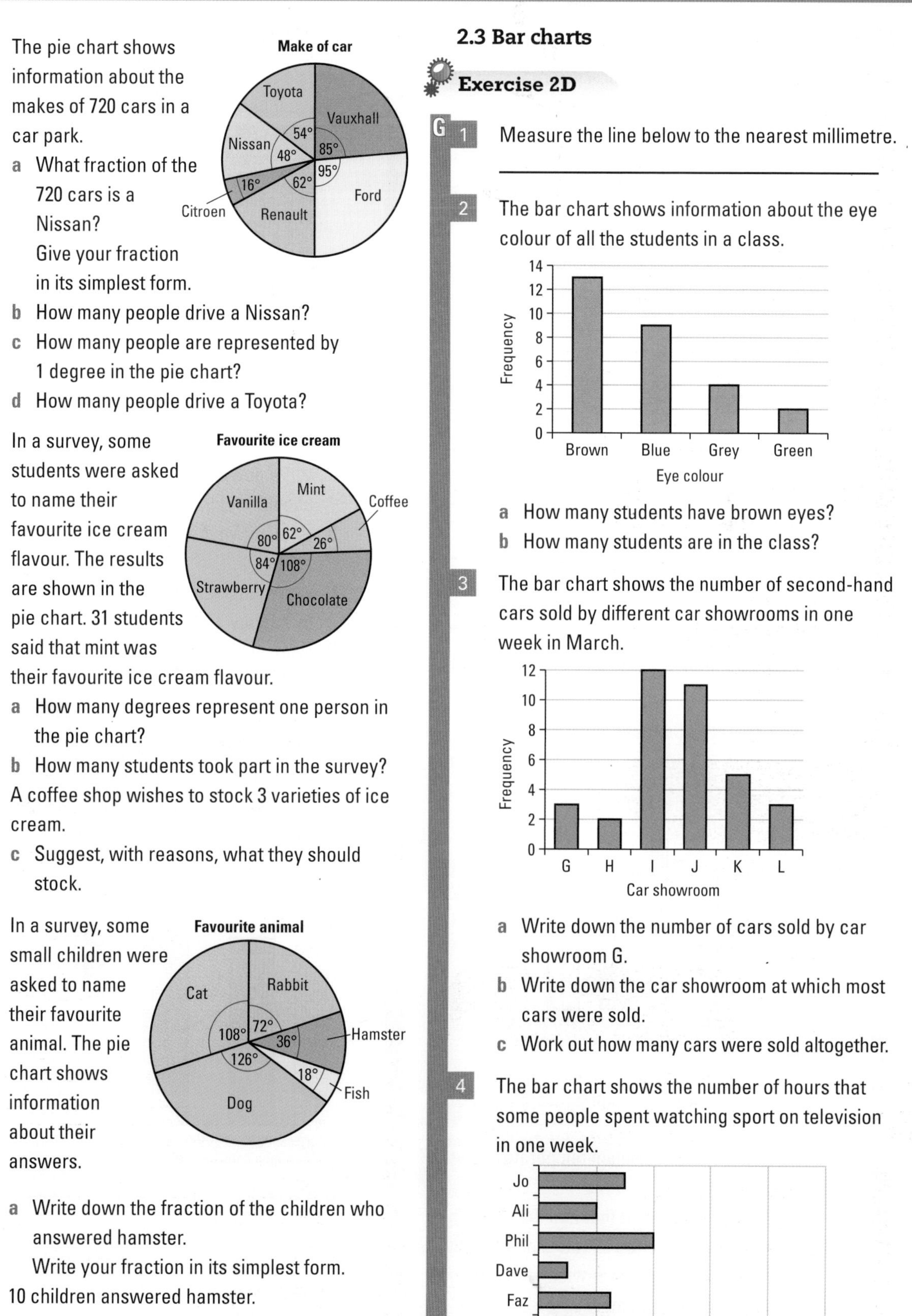

E **4** The pie chart shows information about the makes of 720 cars in a car park.

a What fraction of the 720 cars is a Nissan?
Give your fraction in its simplest form.

b How many people drive a Nissan?

c How many people are represented by 1 degree in the pie chart?

d How many people drive a Toyota?

5 In a survey, some students were asked to name their favourite ice cream flavour. The results are shown in the pie chart. 31 students said that mint was their favourite ice cream flavour.

a How many degrees represent one person in the pie chart?

b How many students took part in the survey?

A coffee shop wishes to stock 3 varieties of ice cream.

c Suggest, with reasons, what they should stock.

6 In a survey, some small children were asked to name their favourite animal. The pie chart shows information about their answers.

a Write down the fraction of the children who answered hamster.
Write your fraction in its simplest form.

10 children answered hamster.

b Work out the number of children that took part in the survey.

2.3 Bar charts

Exercise 2D

G **1** Measure the line below to the nearest millimetre.

2 The bar chart shows information about the eye colour of all the students in a class.

a How many students have brown eyes?

b How many students are in the class?

3 The bar chart shows the number of second-hand cars sold by different car showrooms in one week in March.

a Write down the number of cars sold by car showroom G.

b Write down the car showroom at which most cars were sold.

c Work out how many cars were sold altogether.

4 The bar chart shows the number of hours that some people spent watching sport on television in one week.

G

a Who watched the most hours of sport on television?

b Write down the number of hours of sport on television watched by Dave.

c Write down the two people who watched the same number of hours of sport on television.

d Who watched 4 hours of sport on television?

5 The table gives information about the meals served to Year 11 in the school canteen.
Draw a vertical bar chart to show this information.

Meal	Frequency
Pizza	36
Spaghetti bolognese	45
Fish pie	23
Salad	14
Sausage and chips	32

2.4 Comparative and composite bar charts

Exercise 2E

F

1 The comparative bar chart shows the temperature in a number of resorts in March and September.

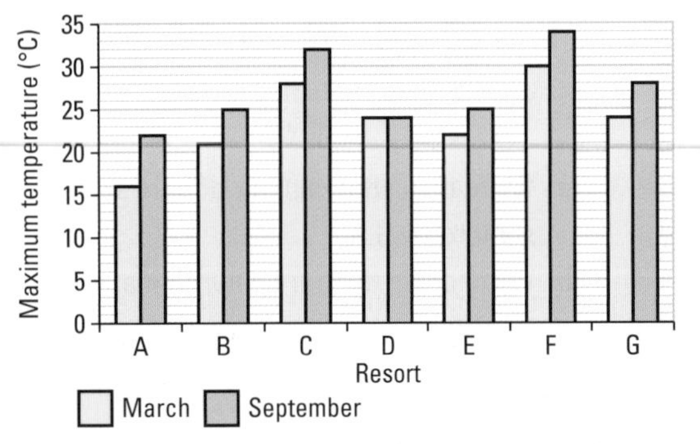

a Write down the maximum temperature in March.

b Write down the maximum temperature in September.

c Write down the resort that had the same maximum temperature in both months.

d Write down the resorts in which the maximum temperature in September was 32°C.

e Write down the resort in which the maximum temperature in March was 21°C.

F

2 Two factories making tools employ male and female workers.
The numbers of males and females are shown on the composite bar chart.

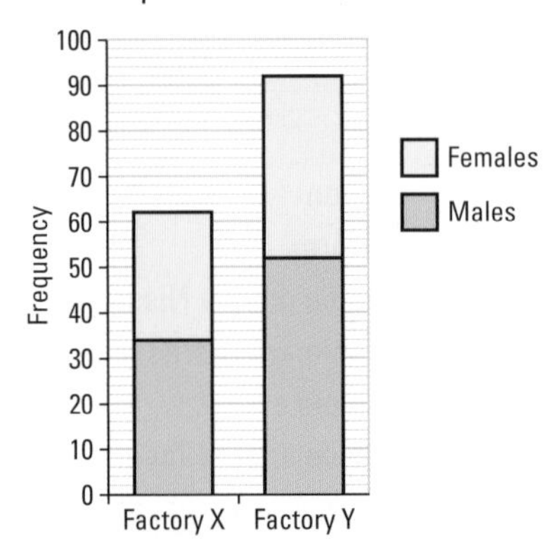

a Write down the factory that employed more people.

b Write down the number of males employed in Factory X.

c Write down the number of people employed by Factory Y.

d Work out how many people were employed by both factories altogether.

One factory has a bigger wage bill than the other.

e Which do you think this is and why?

3 The composite bar chart gives information about the nutritional content of ginger and lemon cookies and oat digestive biscuits.

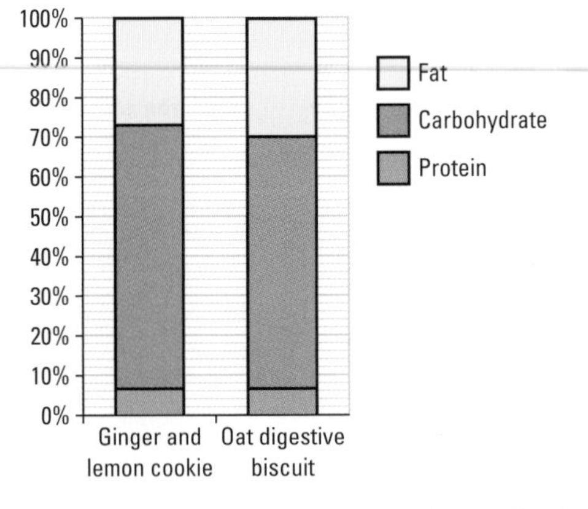

a Write down the name of the biscuit that has more carbohydrates.

b Which type of the nutritional content is the same in both biscuits?

c Work out the percentage of fat in a ginger and lemon cookie.

F **4*** The table gives information about the numbers of males and females in each of the five classes in Year 10.

Class	F	G	H	I	J
Males	12	13	15	13	10
Females	12	14	11	13	12

a Draw a comparative bar chart to represent these data.

b Describe the distribution of males and females in the classes.

2.5 Line diagrams for discrete data and histograms for continuous data

Exercise 2F

G **1** The vertical line graph shows how often each number on a dice came up when it was thrown a number of times.

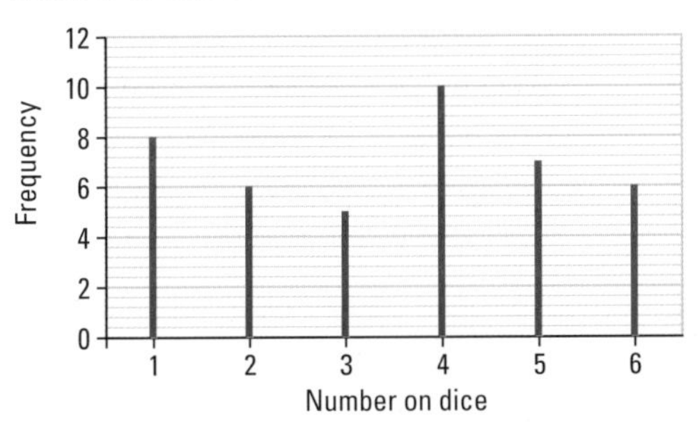

a How many times was the number 5 thrown?

b Which number was thrown seven times?

c How many times was the dice thrown altogether?

2 A survey was done to see the number of people travelling in each car entering a multi-storey car park. The results are shown on the vertical line graph.

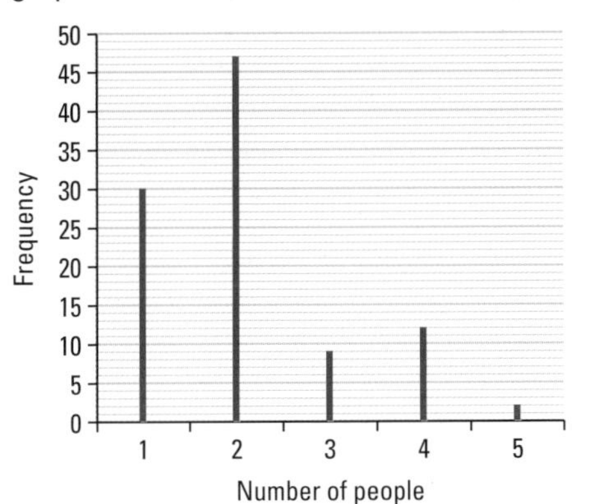

a Write down the number of cars that had three people in them.

b Write down the most common number of people.

c Work out the total number of cars in the survey.

The car park owners wish to encourage car sharing.

They decide to ban all cars carrying fewer than 3 people from using the car park.

d How many people are going to have to change their travel arrangements?

3 The histogram shows information about the training times taken by some 100 m runners before the Olympic Games.

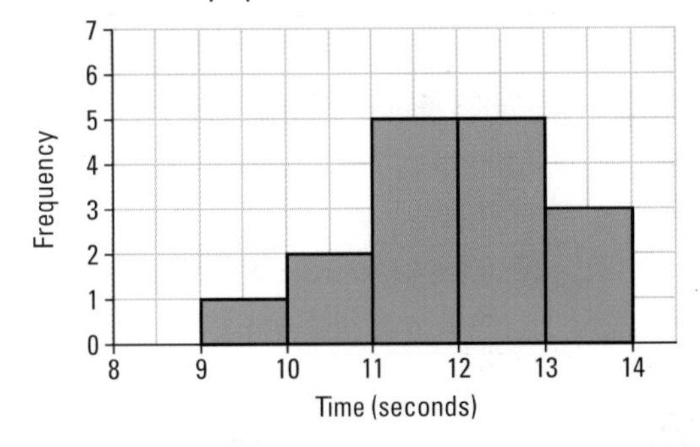

a Write down the reason why there are no gaps between the bars.

b Write down the number of runners that took between 10 and 12 seconds.

c Work out the number of runners that took more than 12 seconds.

d Work out how many runners there were altogether.

4 The speed of cars travelling on a dual carriageway during a one-hour period is measured by a speed camera. The table shows this information.

Speed (s mph)	Frequency
$10 \leqslant s < 20$	2
$20 \leqslant s < 30$	10
$30 \leqslant s < 40$	32
$40 \leqslant s < 50$	28
$50 \leqslant s < 60$	15
$60 \leqslant s < 70$	12

Draw a histogram to show these data.

2.6 Frequency polygons

Exercise 2G

E **1** The table shows information about the shoe sizes of the members of a dance group.

Shoe size	Frequency
3	5
4	6
5	10
6	9
7	4
8	1

a Draw a vertical line graph for these data.

b Use your answer to part **a** to draw a frequency polygon for these data.

D **2** The table shows the distance the members of a sports centre threw a tennis ball in a competition.

Distance (d metres)	Frequency
$25 \leqslant d < 30$	5
$30 \leqslant d < 35$	10
$35 \leqslant d < 40$	16
$40 \leqslant d < 45$	7
$45 \leqslant d < 50$	2

a Draw a histogram for these data.

b Use your answer to part **a** to draw a frequency polygon for these data.

C **3** This term Zac travelled to college by train 20 times and by bus 20 times.

The frequency polygons show information about the amount of time Zac spent on the train and on the bus.

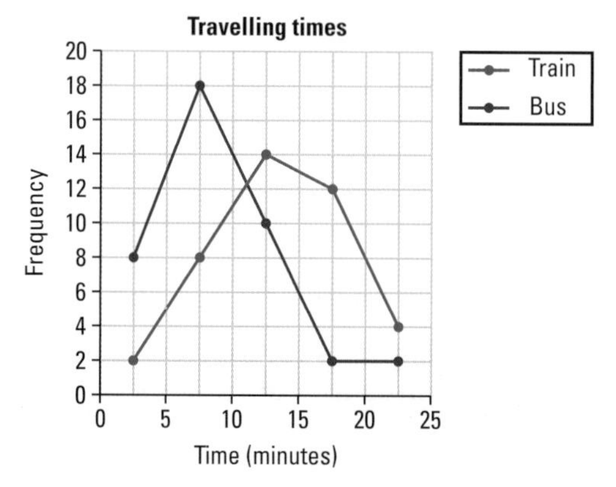

a How many times did Zac spend between 15 and 20 minutes on the bus?

b How many times did Zac spend between 5 and 10 minutes on the train?

c For what fraction of the times Zac went by train did he spend less than 10 minutes on the train? Give your fraction in its simplest form.

A03 **d** On which transport did Zac generally spend the longer time, the train or the bus?
You must give a reason for your answer.

3 Averages and range

Key Points

- **BIDMAS:** the order of number operations. **B**rackets, **I**ndices, **D**ivide, **M**ultiply, **A**dd, **S**ubtract.
- **average:** a single value that is representative of all the values in a set of data.
- **mode:** the most frequent value in a set of data.
- **median:** the middle value, or halfway between two middle values, in an ascending set of data.
- **mean:** the most common average used. mean $= \frac{\text{sum of the values}}{\text{the number of values}}$.
- **range:** describes the spread of the data. range = the highest value – the lowest value.
- **stem and leaf diagram:** a table that displays data to make it easier to find the mode, median and range.
- **frequency distribution:** a list or table for a set of observations which shows all possible events or quantities and the frequencies with which they occur.
- **modal class of grouped data:** the class interval with the highest frequency.
- **median of grouped data:** the class interval in which the median falls.
- **variable:** something that can be changed, and is shown using a letter.
- **term:** a multiple of a variable. It can be a combination of variables and numbers such as x^3, ab, $3y^2$.
- **expression:** a collection of terms and variables.
- **like terms:** terms that use the same variable(s).
- **an equation:** has an equals sign and is used to find a numerical value for a variable.
- **formula:** where one variable is equal to an expression in a different variable(s).
- **adding and subtracting decimals:** keep the decimal points in line so that the place values match.
- **multiplying decimals:** the total number of decimal places in the answer is the same as in the question.
- **dividing decimals:** multiply both numbers by 10, 100, 1000, etc. until you are dividing by a whole number.
- **finding the mean of discrete data in a frequency table:** use the formula
  mean $= \frac{\sum f \times x}{\sum f}$
  where f is the frequency, x is the variable and $\sum$ means 'the sum of'.
- **estimating the mean of grouped data in a frequency table:** use the formula
  mean $= \frac{\sum f \times x}{\sum f}$
  where f is the frequency, x is the class midpoint and $\sum$ means 'the sum of'.
- **using a calculator:**
  - x^2 works out squares
  - x^y or $x^{■}$ works out powers
  - $\sqrt{}$ works out square roots
  - $\sqrt[3]{}$ works out cube roots
- **estimating numbers:** round numbers to a relevant place value depending on the situation.

Average	Use	Advantages	Disadvantages
MODE	for non-numeric data or to find the most popular	can be used with categorical data, not affected by extreme values	there may be more than one mode, if the data set is small there may not be a mode
MEDIAN	describes the middle of a data set with extreme values	not affected by extreme values	actual value may not exist, not used as often as the mean
MEAN	describes the middle of a data set with **no** extreme values	uses all the given data, can be used for further calculations, used the most often	affected by extreme values

3.1 Understanding and using numbers

Exercise 3A

Questions in this chapter are targeted at the grades indicated.

G

1 Laura works in a coffee shop on Sundays. She decides to work out how much money she receives from the first five customers she serves.

This is a list of the amount each customer spends.

£2.10 £1.92 £4.50 £1.10 £3.17

a Work out the total amount Laura receives from these customers.

b In the till Laura has £25.70 in 10p coins. How many 10p coins does she have?

c If she changes the 10p coins into 1p coins, how many 1p coins will she have?

2 Andy has 2.50 metres of wood. He uses 1.35 metres to make a bike ramp. Work out how much wood he has left.

3 The manager of a restaurant gives 4 free starters to each of the 14 tables in his restaurant. How many free starters does he give out?

F

4 Gemma wins £182.84 in a raffle. She decides to share the money equally between four charities. Work out how much each charity gets.

E

5 Find the value of:

a 7^2 b 4^3 c $\sqrt{64}$

d $\sqrt[3]{27}$ e $6^2 - (10 - 7) + 5$

6 A02 A03 Dave has been asked to work out the cost of laying turf in three gardens. He records the data he knows in a table.

Garden	1	2	3
Length (m)	5	7	6.5
Width (m)	6	3.1	4.2

The cost will be £2.50 per square metre.

Using the given data, work out the total cost of laying turf in the three gardens.

3.2 Finding the mode, the median and the mean

Exercise 3B

F

1 Find the mode for each of the following data sets.

a 3 4 2 4 5 1 2 4

b 13 11 15 11 14 13 16 12

c 9 2 3 6 8 7 1 4

d daffodil tulip daffodil daisy lily tulip daffodil

2 Benton did a spelling test every week of a 12-week term. His scores are given below.

14 15 16 15 14 17 17 18 16 17 14 17

Find the mode for these data.

3 Sally made a list of the colours of all her shoes in the wardrobe. Her list is shown below.

navy black white red brown
black brown black cream black black

Write down the mode for these data.

4 The hourly rates of pay for seven workers are listed below.

£7.90 £8.25 £8.25 £8.75
£7.50 £9.10 £8.25

Write down the mode for these data.

Exercise 3C

E

1 Find the median for the following data sets.

a 1 2 2 4 5 6 7 8 9

b 3 3 5 6 8 8 8 9

c 3 3 4 5 5 6

d 4 1 2 5 7 8 4 3 2

e 4 5 6 2 8 9 1 9

2 The maximum daytime temperatures (in °C), in April, in nine towns are recorded below.

10 12 14 12 13 12 15 16 14

Find the median temperature.

3 The CO_2 emissions (in thousands of tonnes) for cars in an English county over ten years are shown below.

34 34 38 40 38 40 42 39 39 40

Find the median CO_2 emissions.

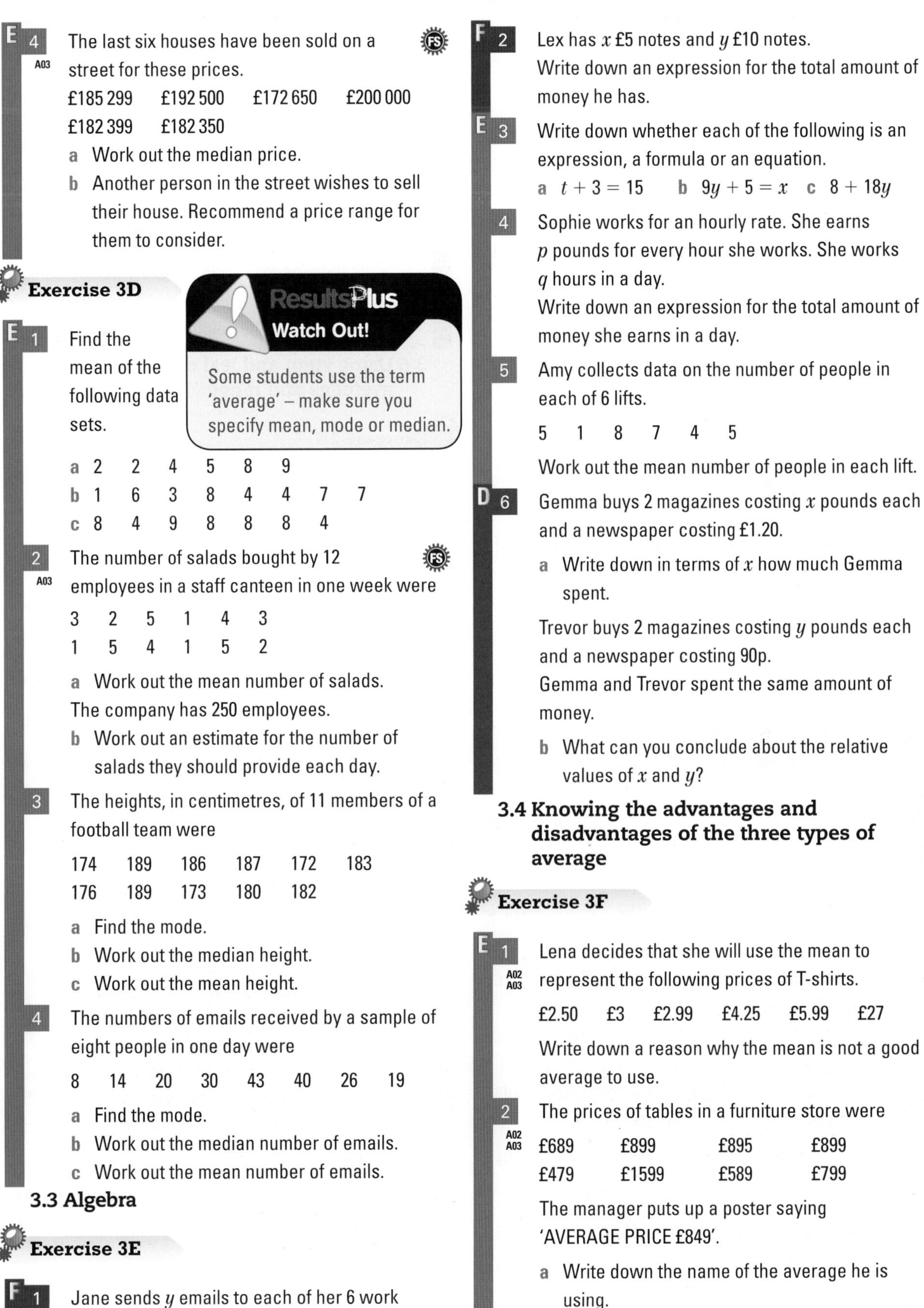

E 4 (A03) The last six houses have been sold on a street for these prices.

£185 299 £192 500 £172 650 £200 000
£182 399 £182 350

a Work out the median price.
b Another person in the street wishes to sell their house. Recommend a price range for them to consider.

Exercise 3D

ResultsPlus
Watch Out!
Some students use the term 'average' – make sure you specify mean, mode or median.

E 1 Find the mean of the following data sets.

a 2 2 4 5 8 9
b 1 6 3 8 4 4 7 7
c 8 4 9 8 8 8 4

2 (A03) The number of salads bought by 12 employees in a staff canteen in one week were

3 2 5 1 4 3
1 5 4 1 5 2

a Work out the mean number of salads.

The company has 250 employees.

b Work out an estimate for the number of salads they should provide each day.

3 The heights, in centimetres, of 11 members of a football team were

174 189 186 187 172 183
176 189 173 180 182

a Find the mode.
b Work out the median height.
c Work out the mean height.

4 The numbers of emails received by a sample of eight people in one day were

8 14 20 30 43 40 26 19

a Find the mode.
b Work out the median number of emails.
c Work out the mean number of emails.

3.3 Algebra

Exercise 3E

F 1 Jane sends y emails to each of her 6 work colleagues.
How many emails does she send?

F 2 Lex has x £5 notes and y £10 notes.
Write down an expression for the total amount of money he has.

E 3 Write down whether each of the following is an expression, a formula or an equation.

a $t + 3 = 15$ b $9y + 5 = x$ c $8 + 18y$

4 Sophie works for an hourly rate. She earns p pounds for every hour she works. She works q hours in a day.
Write down an expression for the total amount of money she earns in a day.

5 Amy collects data on the number of people in each of 6 lifts.

5 1 8 7 4 5

Work out the mean number of people in each lift.

D 6 Gemma buys 2 magazines costing x pounds each and a newspaper costing £1.20.

a Write down in terms of x how much Gemma spent.

Trevor buys 2 magazines costing y pounds each and a newspaper costing 90p.
Gemma and Trevor spent the same amount of money.

b What can you conclude about the relative values of x and y?

3.4 Knowing the advantages and disadvantages of the three types of average

Exercise 3F

E 1 (A02 A03) Lena decides that she will use the mean to represent the following prices of T-shirts.

£2.50 £3 £2.99 £4.25 £5.99 £27

Write down a reason why the mean is not a good average to use.

2 (A02 A03) The prices of tables in a furniture store were

£689 £899 £895 £899
£479 £1599 £589 £799

The manager puts up a poster saying 'AVERAGE PRICE £849'.

a Write down the name of the average he is using.
b Is this a fair average to use?
Give a reason for your answer.

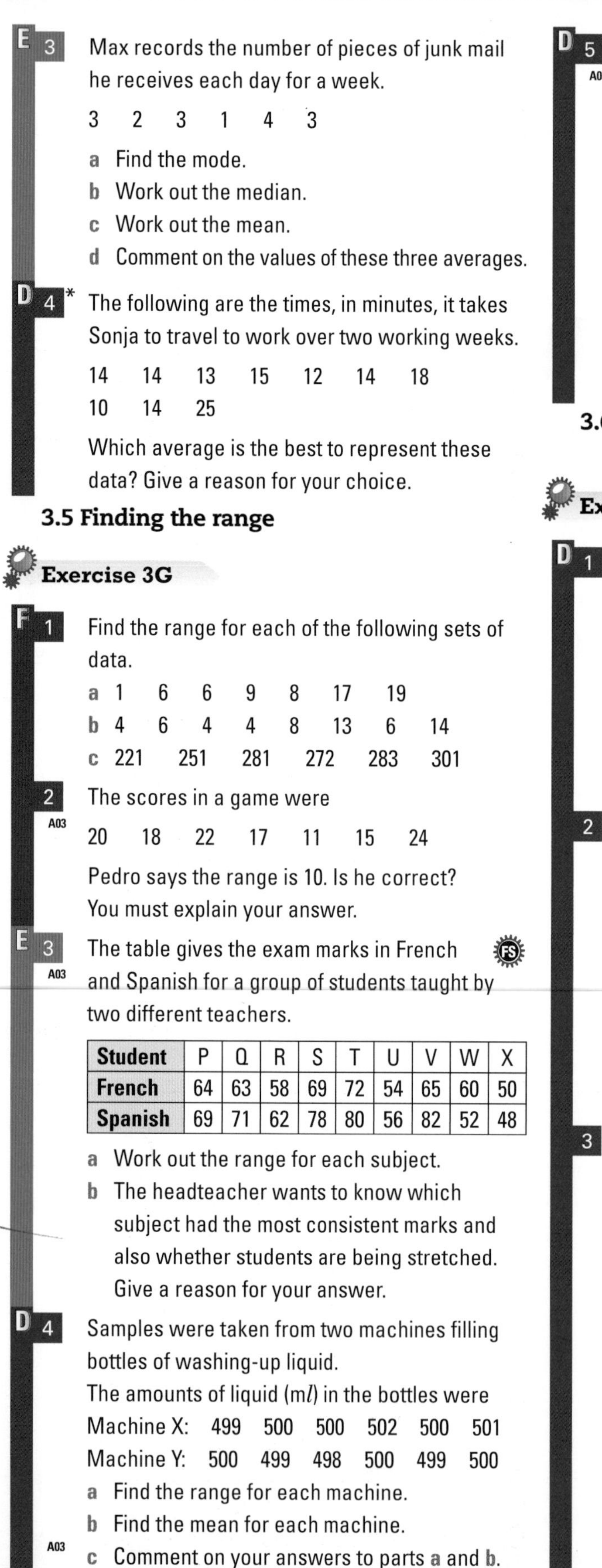

E **3** Max records the number of pieces of junk mail he receives each day for a week.

3 2 3 1 4 3

a Find the mode.

b Work out the median.

c Work out the mean.

d Comment on the values of these three averages.

D **4*** The following are the times, in minutes, it takes Sonja to travel to work over two working weeks.

14 14 13 15 12 14 18
10 14 25

Which average is the best to represent these data? Give a reason for your choice.

3.5 Finding the range

Exercise 3G

F **1** Find the range for each of the following sets of data.

a 1 6 6 9 8 17 19

b 4 6 4 4 8 13 6 14

c 221 251 281 272 283 301

2 (A03) The scores in a game were

20 18 22 17 11 15 24

Pedro says the range is 10. Is he correct? You must explain your answer.

E **3** (A03) The table gives the exam marks in French and Spanish for a group of students taught by two different teachers. (FS)

Student	P	Q	R	S	T	U	V	W	X
French	64	63	58	69	72	54	65	60	50
Spanish	69	71	62	78	80	56	82	52	48

a Work out the range for each subject.

b The headteacher wants to know which subject had the most consistent marks and also whether students are being stretched. Give a reason for your answer.

D **4** Samples were taken from two machines filling bottles of washing-up liquid.
The amounts of liquid (m*l*) in the bottles were

Machine X: 499 500 500 502 500 501
Machine Y: 500 499 498 500 499 500

a Find the range for each machine.

b Find the mean for each machine.

c (A03) Comment on your answers to parts **a** and **b**.

D **5*** (A03) The heights of two sample groups of students from different years were

Sample A: 176 176 177 165 174
174 186 174 174 184
179

Sample B: 168 159 156 159 152
156 158 156 156 160
157

Calculate the mean and range for each year and compare and contrast the frequency distributions of the heights of the two years.

3.6 Using stem and leaf diagrams to find averages and range

Exercise 3H

D **1** Here is an unordered stem and leaf diagram.

0 | 7 6 3 2 5
1 | 4 7 8 4 3
2 | 2 5 0 1 6 3
3 | 3 4 0 3 1 4 5
4 | 7 6 0 9

Key:
1|4 stands for 14

Draw this as an ordered stem and leaf diagram.

2 Fiona manages several pharmacies. She has to drive between them.
Here are the distances, in kilometres, that she drove during May.

34 21 6 14 23 35 20 30
23 35 39 8 20 12 21 17
12 23 26 6

Draw a stem and leaf diagram to represent these data.

3 Here is a stem and leaf diagram showing the numbers of cars sold by a garage group over each of a number of weeks.

0 | 3 7
1 | 2 3 5 6
2 | 0 1 2 5 6 6 6
3 | 1 1 8 8 9 9 9
4 | 0 1 1 2 3

Key:
2|5 stands for 25 cars

a Write down the number of weeks represented in this diagram.

b Write down the mode for these data.

c Find the median number of cars.

d Work out the range of these data.

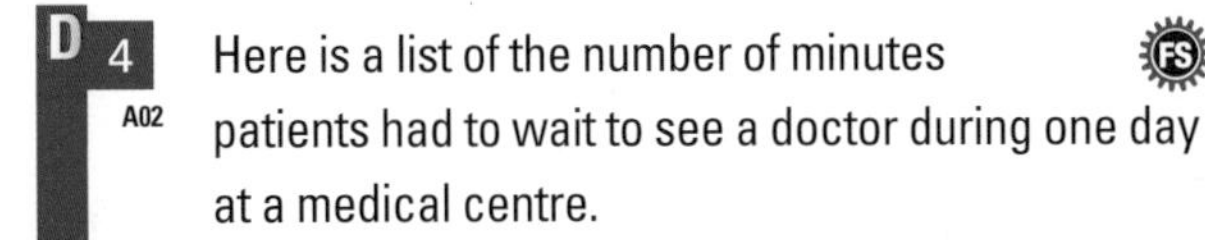

D 4 (A02) Here is a list of the number of minutes patients had to wait to see a doctor during one day at a medical centre.

12	2	5	5	20	8	5	5
12	20	5	10	5	10	10	15
18	25	15	10	8	5	12	15

Draw a suitable diagram that the doctor could use to

a display these data

b show a measure of spread

c show a measure of average.

3.7 Using frequency tables to find averages for discrete data

Exercise 3I

D 1 (A03) A council is introducing parking permits for its residents. They asked a sample of households how many cars they had.

The results are shown in the frequency table.

Number of cars	Frequency
0	3
1	16
2	24
3	7
4	0

a Write down the mode of these data.

b Find the median number of cars.

c Work out the mean number of cars.

d Recommend the number of extra parking permits they should provide. You must explain your answer.

2 A sample of a sweet red pepper crop was taken and each pepper was weighed.

The weights to the nearest 5 g are shown in the frequency table.

Weight of pepper (g)	Frequency
65	1
70	6
75	12
80	5
85	1

a Write down the mode of these data.

b Find the median weight of the peppers.

c Work out the mean weight of the peppers.

D 3 In an experiment with different flowers the number of petals per flower was recorded.

The results are shown in the frequency table.

Number of petals per flower	Frequency
3	5
4	3
5	12
6	10
7	8
8	7
9	5

a Write down the mode of these data.

b Find the median number of petals per flower.

c Work out the mean number of petals per flower.

3.8 Working with grouped data

Exercise 3J

E 1 A group of students were asked the number of books they read in a term.

The results are shown in the frequency table.

Class interval	0 to 2	3 to 5	6 to 8	9 to 11
Frequency	1	7	12	5

a Write down the modal class.

b Find the class into which the median falls.

2 A group of students did a French vocabulary test.

The results are shown in the frequency table.

Class interval	1 to 5	6 to 10	11 to 15	16 to 20
Frequency	1	7	16	6

a Write down the modal class.

b Find the class into which the median falls.

3 The frequency table gives the height, h in mm, of 60 seedlings grown in a quality control investigation.

Class interval	Frequency
$0 \leqslant h < 2.5$	5
$2.5 \leqslant h < 5$	6
$5 \leqslant h < 7.5$	18
$7.5 \leqslant h < 10$	15
$10 \leqslant h < 12.5$	16

a Write down the modal class.

b Find the class into which the median falls.

3.9 Estimating the mean of grouped data

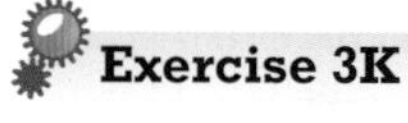

Exercise 3K

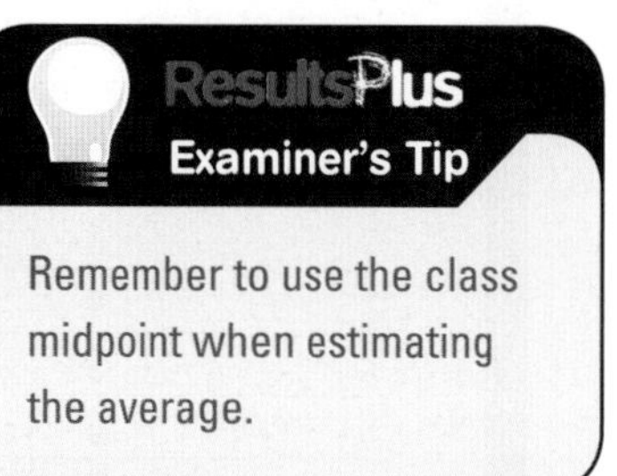

C

1 (A02) A primary school recorded how many times pupils brought fruit in their lunch boxes.

a Copy and complete the table.

Class interval	Frequency (f)	Class midpoint (x)	Frequency × midpoint (fx)
1 to 3	5		
4 to 6	10		
7 to 9	18		
10 to 12	12		

b Find an estimate for the mean number of times, to the nearest whole number.

2 (A02) The heating system in a college is unreliable. The caretaker records the number of times an engineer is called out over a 14-week term. The results are shown in the frequency table.

Class interval	Frequency (f)
0 to 1	6
2 to 3	5
4 to 5	2
6 to 7	1

Find an estimate for the mean of the number of engineer's visits, to the nearest whole number.

3 (A02) Bella made a note of the length of answerphone messages left on her home phone over one week.
The results are shown in the frequency table.

Class interval (t mins)	Frequency (f)
$0 \leqslant t < 5$	18
$5 \leqslant t < 10$	8
$10 \leqslant t < 15$	2
$15 \leqslant t < 20$	1
$20 \leqslant t < 25$	1

Find an estimate for the mean of the length of message, to the nearest whole number.

3.10 Using calculators

Exercise 3L

E

1 Find the values of:

a 4.2^2 b 3.9^3 c $\sqrt{4160.25}$ d $\sqrt[3]{21\,952}$

2 The costs of seven books are recorded as follows:

£5.99, £10.49, £15.05, £9.99, £12.50, £21.00, £11.21

Find the mean cost.

3 A teacher counts up the number of pupils who have a name beginning with T in each year in the school. The data is as follows:

26 31 38 11 22 47 29

a Find the total number of pupils with a name beginning with T.

b Check your calculation by estimating the total number of pupils with a name beginning with T.

D

4 a Estimate the value of $\frac{6.2^2 \times 2}{5.1}$

b Find the value of $\frac{6.2^2 \times 2}{5.1}$ correct to 2 decimal places.

4 Line diagrams and scatter graphs

Key Points

- **coordinate:** a point on a grid (x, y).
- **x-coordinate:** the number of units horizontally, given first.
- **y-coordinate:** the number of units vertically, given second.
- **line graph:** a graph drawn from two pieces of data. Points are plotted on the graph and joined with straight lines.
- **scatter graph:** a graph used to show whether there is any relationship between two variables.
- **correlation:** a relationship between pairs of variables.

Positive correlation

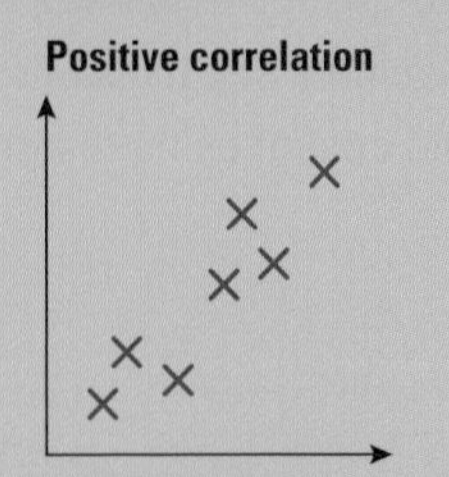

As one value increases the other one increases.

Negative correlation

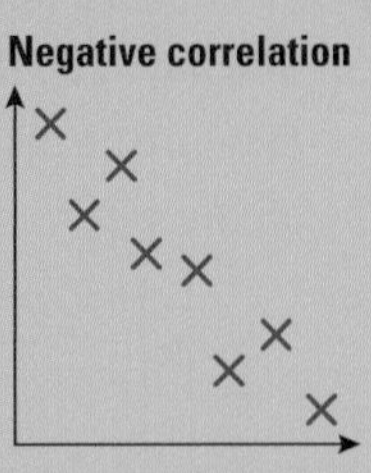

As one value increases the other decreases.

No correlation

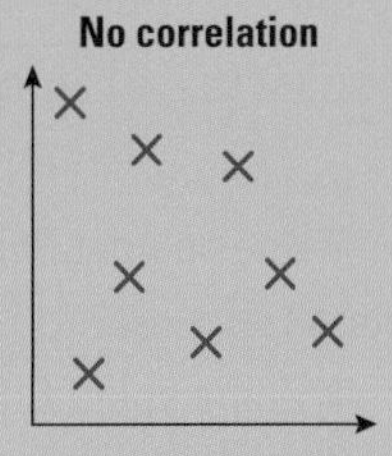

The points are random and widely spaced.

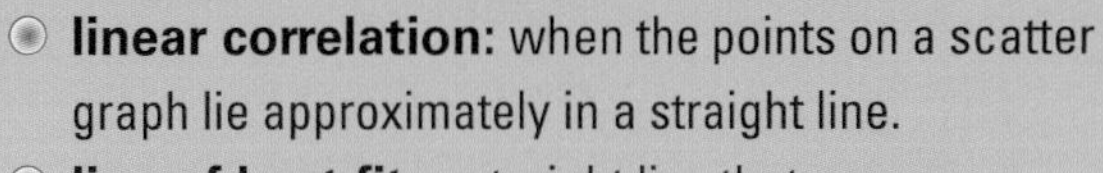

- **linear correlation:** when the points on a scatter graph lie approximately in a straight line.
- **line of best fit:** a straight line that passes as near as possible to the points on a scatter graph so as to best represent the trend of the graph.
  - there should be approximately the same number of points on either side of the line
  - further results can be estimated using the line of best fit
- **plotting the graph of a formula:** find pairs of corresponding values and use them as coordinates for the graph.
- **finding the steepness of a straight line:** use the formula

$$\text{steepness} = \frac{\text{increase in } y}{\text{increase in } x}$$

- **finding the equation of a straight line:** find the steepness of the line and the point it crosses the y-axis. Put these into the form $y = mx + c$, where m is the steepness and c is where the line crosses the y-axis.

4.1 Plotting points on a graph

Exercise 4A

Questions in this chapter are targeted at the grades indicated.

E **1** This graph opposite shows the number of pounds you get for a number of euros.

a Estimate the number of pounds you get for 4.2 euros.

b Estimate the number of euros you get for £7.60.

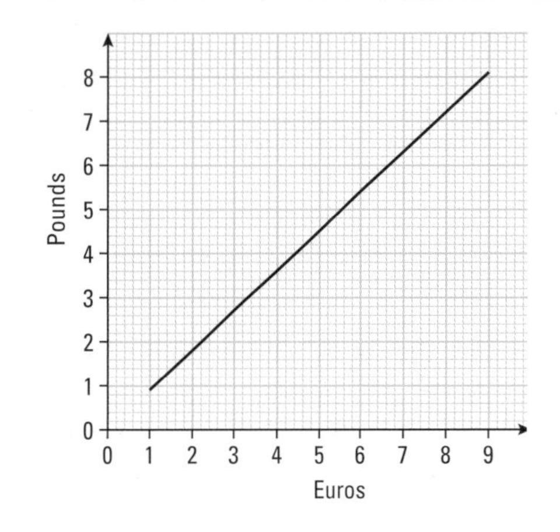

E 2

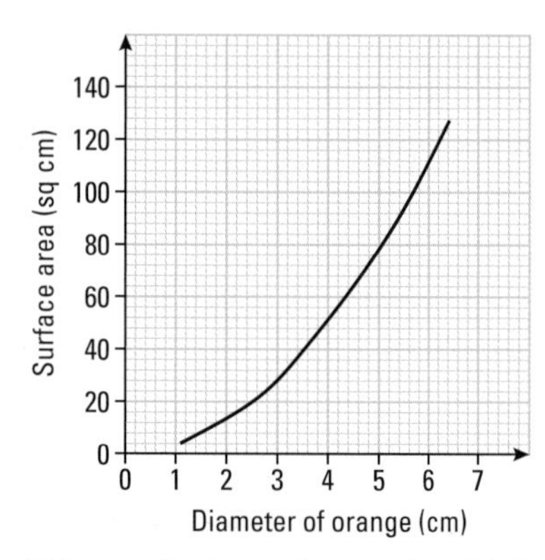

This graph shows the relationship between the diameter of an orange and its surface area.

a Estimate the surface area if the diameter is 5.2 cm.

b Estimate the diameter if the surface area is 104 cm^2.

D 3 a Copy and complete the following table.

x	0	1	2	3	4	5
$y = 3x$	0	3	6			

b Using values 0 to 6 on the x-axis and 0 to 16 on the y-axis, plot the graph of the line $y = 3x$ for values of x between 0 and 5.

c Use your graph to find the value of y when x is 3.5.

d Use your graph to find the value of y when x is 11.

4.2 Straight-line graphs

Exercise 4B

D 1 a Draw axes for 0 to 10 on the x-axis and 0 to 20 on the y-axis.

b Plot the points shown in the table on these axes.

x	0	2	4	6	8	10
y	3	6	9	12	15	18

c Join the points with a straight line.

d Work out the gradient of the line.

2 The gradient of a line is $\frac{5}{2}$.
For this line, write down the increase in y for every one increase in x.

C 3 Find the gradient of the line.

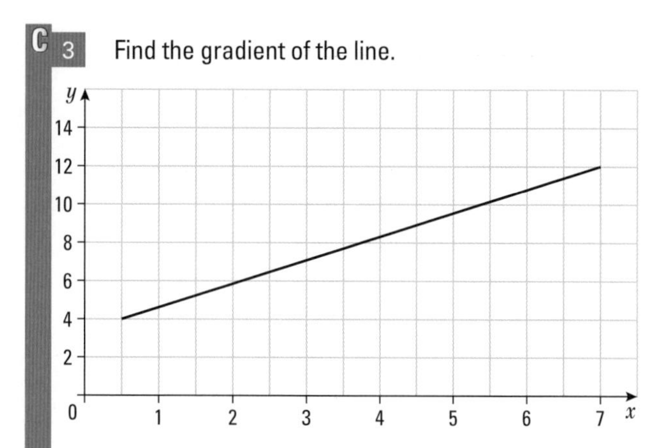

4.3 Drawing and using line graphs

Exercise 4C

E 1 Zaina has a pay-as-you-go mobile phone. She pays 30p for each minute she uses her phone. She displays the cost of using the phone on the graph.

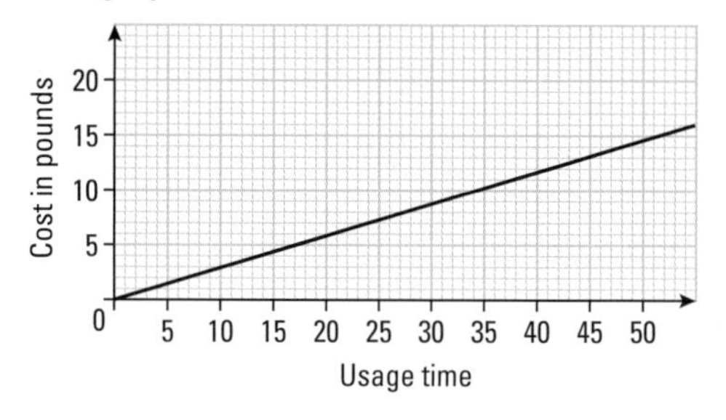

a How much does it cost Zaina to use her phone for 43 minutes?

b One month Zaina spent £10.50 on using her phone. For how many minutes did Zaina use her phone that month?

2 This graph converts temperatures between Fahrenheit and Celsius.

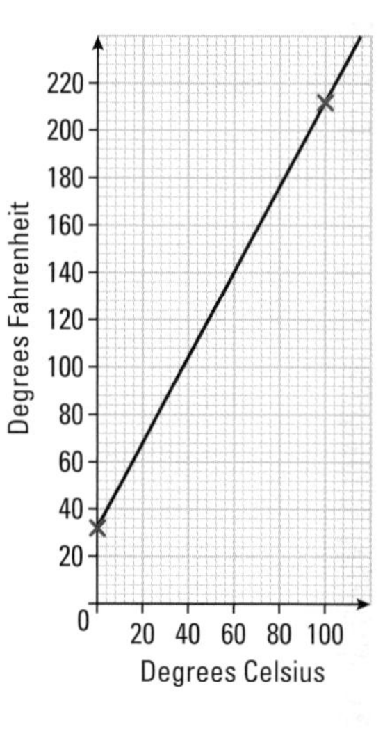

a Use the graph to convert these temperatures to degrees Fahrenheit.
  i 40°C
  ii 60°C
  iii 72°C

b Use the graph to convert these temperatures to degrees Celsius.
  i 70°F ii 100°F iii 128°F

E 3* A woman sells burgers from a burger van. She kept a tally of the number of burgers sold during the eight hours she worked one day. During the day she sold a total of 700 burgers. The table shows how many burgers she had sold, in total, by the end of each hour.

Hours	Burgers sold
0	0
1	12
2	28
3	160
4	250
5	320
6	350
7	500
8	700

a Draw a line graph for these data.

b Write down the hour in which she sold the most burgers. Give a possible reason.

4.4 Drawing and using scatter graphs

Exercise 4D

D 1 This scatter graph illustrates the mileage and age of 15 cars.

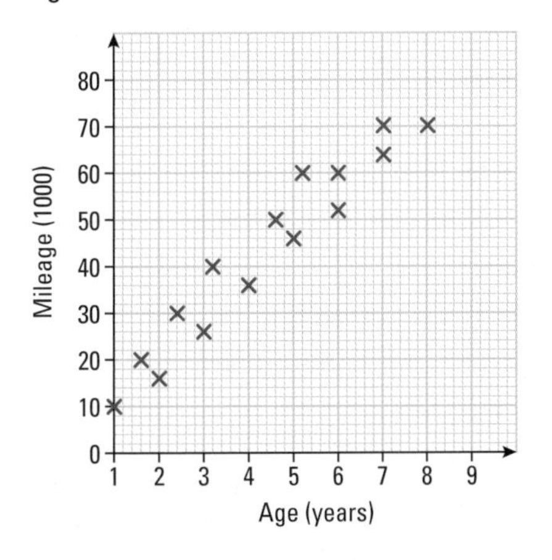

Describe the relationship between the age and mileage of a car.

D 2 The heights and weights of 10 children in a gym class are shown in the table.

Height (cm)	Weight (kg)
153	43
162	58
142	35
149	40
144	39
150	42
170	63
175	70
155	52
160	54

a Using the scales shown on the diagram, draw a scatter graph for these data.

b Describe the relationship between height and weight.

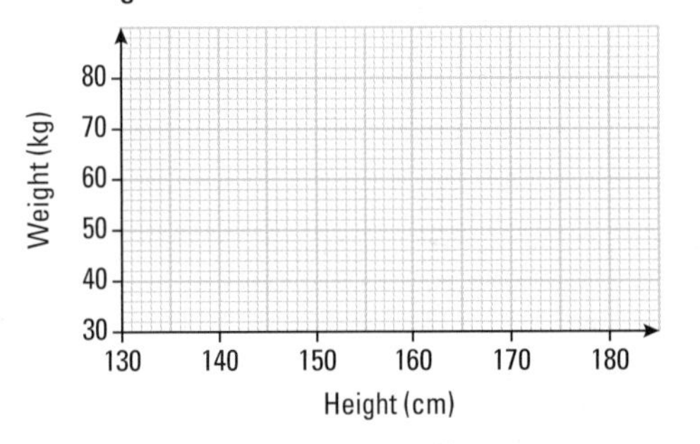

3 A03 The data table shows the weight and top speeds of nine cars.

Weight (kg)	Top speed (mph)
850	95
950	95
1050	105
1150	105
1250	110
1450	115
1550	120
1650	125
1750	125

a Using the scales shown on the diagram, draw a scatter graph for these data.

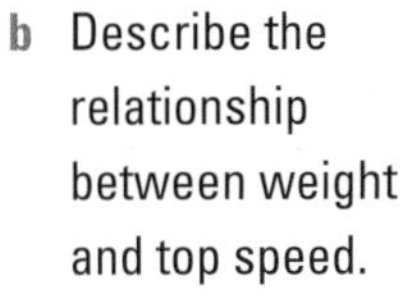

b Describe the relationship between weight and top speed.

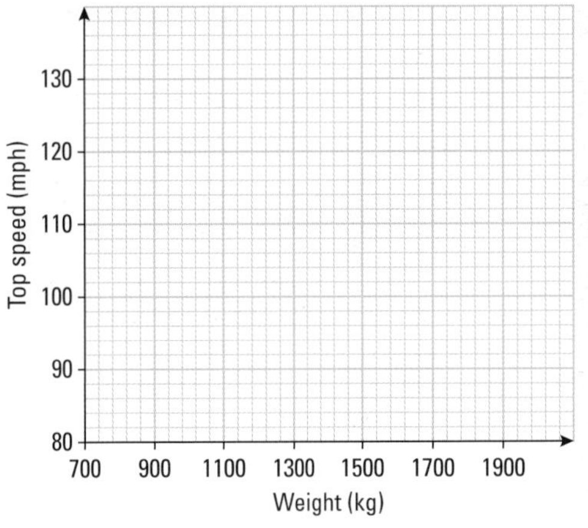

D 4* The data show the number of pedestrian crossings and the number of pedestrian accidents in each of 12 areas over a period of six months.

Number of crossings	Number of accidents
10	6
8	11
16	3
6	9
12	3
10	8
9	5
16	4
18	1
12	4
5	15
9	8

Draw a scatter graph for these data.
Describe the relationship between the number of pedestrian crossings and the number of pedestrian accidents.

4.5 Recognising correlation

Exercise 4E

D 1 The scatter graph shows the marks achieved by a group of 12 students in a French exam and in an art exam.

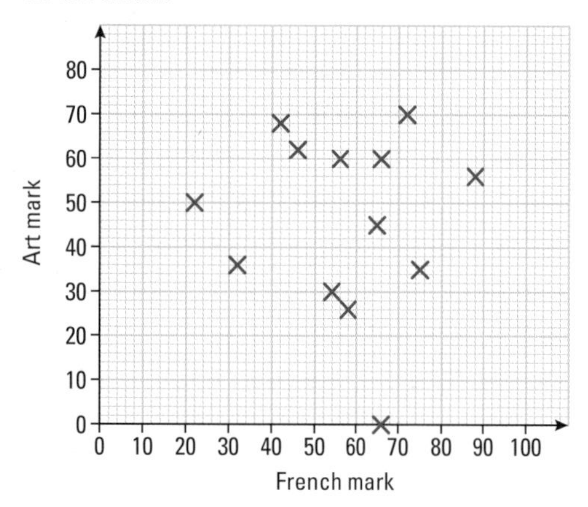

Describe the correlation shown by this scatter graph.

D 2 A03 A garage sells second-hand vans. The scatter graph shows the engine sizes and costs of 12 vans of the same model.

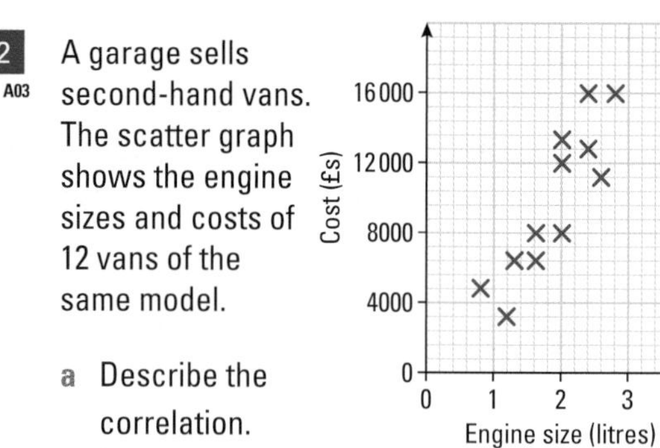

a Describe the correlation.

b Describe the relationship between engine size and cost.

3 A03 Copy and complete the following table. Tick the type of correlation for each set of data.

Variables	Positive correlation	Negative correlation	No correlation
Distance travelled by car and Fuel in the tank			
Intelligence and Height of people			
Size of pond and Number of ducks			
Age and Cycling speed of adults			
Temperature and Shoe size of people			
Age of cars and Miles driven			
Temperature of day and Number of cold drinks sold			

C 4* A03 Joe sells ice cream from a stall on the sea front. He thinks that he will sell more ice creams when the weather is hot. For 10 days he records the number of ice creams sold and gets the maximum temperature for the day from a website.
The data he collects are shown in the table.

Maximum temperature (°C)	Number of ice creams sold
20	58
18	48
22	60
25	70
27	73
24	67
28	80
30	87
23	62
19	45

a Draw a scatter graph for these data.
b Describe the correlation and decide whether his claim is correct, with reasons.

4.6 Lines of best fit

Exercise 4F

D 1 This scatter graph shows the hours per week students watch TV and their test marks.

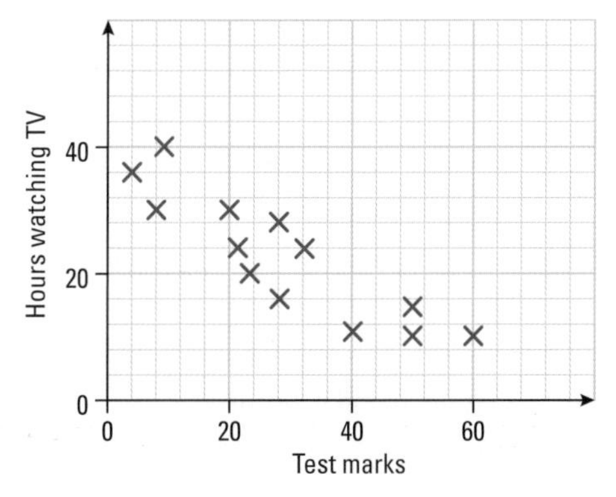

a Describe the correlation.
b Copy the scatter diagram.
c Draw the line of best fit on your diagram.
A03 d Describe the relationship between hours spent watching TV and test marks.

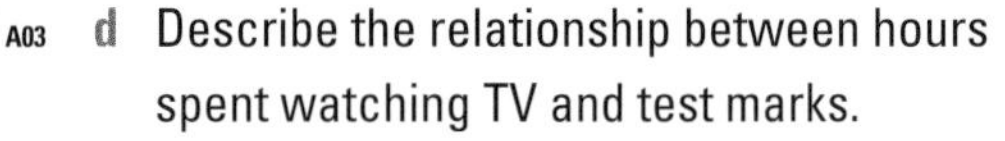

C 2 An estate agent wants to find out if its advertising boards have an effect on its sales. It looks at the number of boards and sales in eight different areas.

Number of boards	Number of sales
24	18
30	26
25	22
33	29
20	15
24	22
27	25
30	24

a Copy the scatter graph and plot the remaining points.

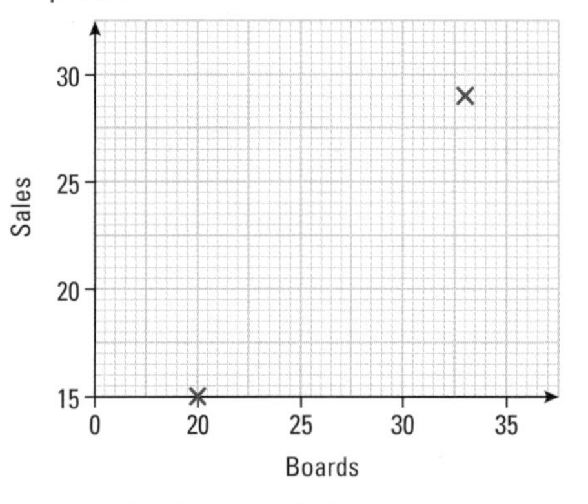

b Describe the correlation.
c Draw the line of best fit on your diagram.
A03 d Describe the relationship between the number of advertising boards and house sales.

3 A Primary Care Trust checks the number of patients being seen in its clinics in one week. The data are shown in the table.

Number of clinics	Number of patients
8	65
7	60
5	55
5	45
6	50
6	55
6	60
4	45

C

a Copy the diagram and complete the scatter graph for these data.

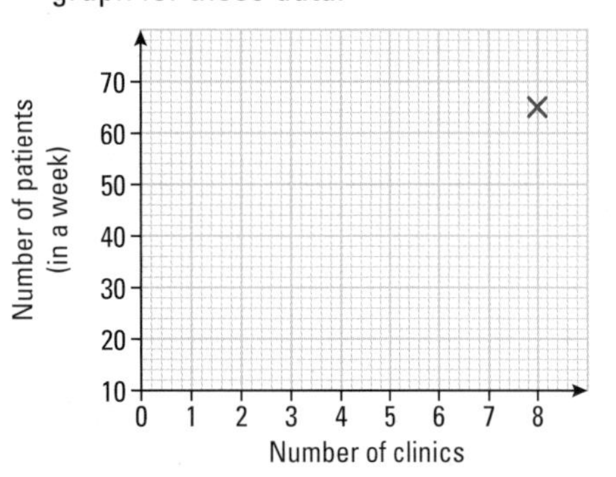

b Describe the correlation.

c Draw a line of best fit on your diagram.

A03 **d** Describe the relationship between the number of clinics and the number of patients in a week.

4.7 Using lines of best fit to make predictions

Exercise 4G

ResultsPlus Examiner's Tip

Always draw in the lines – you may get method marks for this even if they are in the wrong place.

C **1** A03 A workshop makes model airplanes. FS
The scatter diagram shows some information about the numbers of models made and the cost of making them.

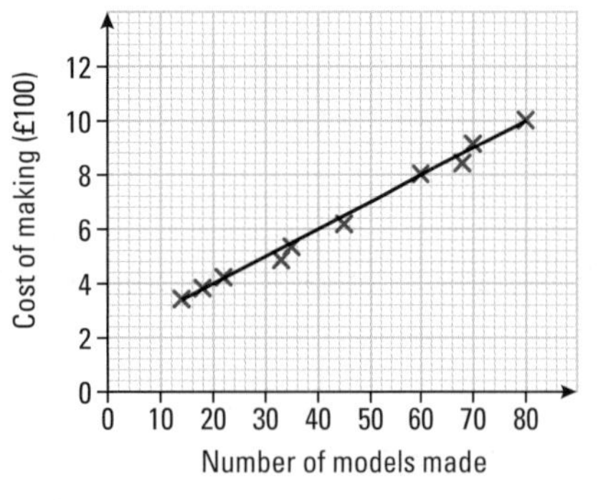

a Describe the correlation.

b The manager wants to estimate:

i the cost of making 40 models

ii the number that can be made for £700.

Use the graph to make these estimates for the manager.

C **2** The scatter graph shows the midday temperature and the number of units of gas used by a house on each of 20 days.

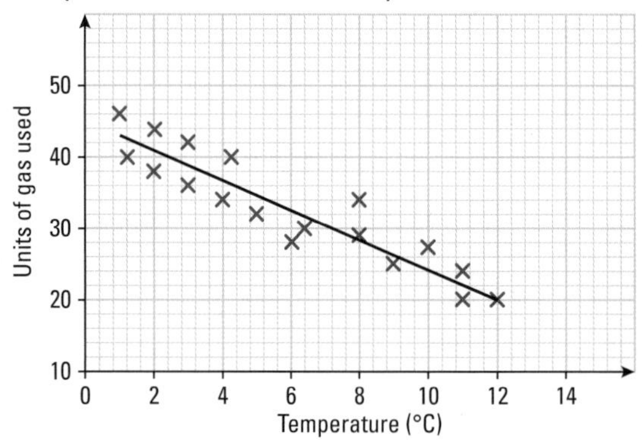

a Describe the correlation.

b Use the graph and the line of best fit to find an estimate for:

i the number of units of gas used when the midday temperature was 13°C

ii the midday temperature when 32 units of gas were used.

3 A03 The following scatter graph shows the ages and systolic blood pressure of a group of 10 women.

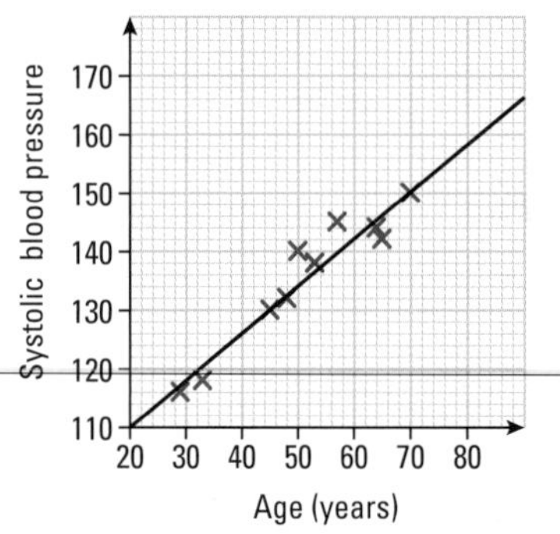

a Estimate the systolic blood pressure of a woman aged 60.

b Predict the age of a woman whose systolic blood pressure is 135.

c Describe the relationship between age and systolic blood pressure.

5 Probability

Key Points

- **fraction:** part of a unit.
- **numerator:** the top part of a fraction.
- **denominator:** the bottom part of a fraction.
- **per cent:** out of 100.
- **percentage (%):** the number of parts per hundred.
- **probability:** how likely it is an event will occur.
  - $0 \leqslant \text{probability} \leqslant 1$
  - an event that is certain to happen has a probability of 1
  - an event that is impossible has a probability of 0
  - can be written as a fraction, decimal or percentage
- **mutually exclusive outcomes:** events that cannot happen at the same time. The sum of the probabilities of all the mutually exclusive outcomes is 1.
- **sample space diagram:** a table that shows all the possible outcomes in an experiment with two variables. Can be used to find a theoretical probability.
- **relative frequency:** the frequency of an observed value divided by the total number of observations.
- **two-way table:** a table that shows how data falls into two different categories.
  - the sum of the row totals = the sum of the column totals
- **ratio:** a comparison of a part to a part, written in the form $a:b$.
- **converting fractions to decimals:** divide the numerator by the denominator using long division.
- **finding percentages of quantities:** write the percentage as a fraction, and then multiply the fraction by the quantity.
- **writing one quantity as a percentage of another quantity:** write the first quantity as a fraction of the second quantity, then convert the fraction to a percentage.
- **finding the probability of equally likely outcomes:** use the formula

$$\text{probability} = \frac{\text{number of successful outcomes}}{\text{total number of possible outcomes}}$$

- **finding the probability of an event not happening:** calculate 1 – the probability the event does happen.
- **calculating the estimated probability:** use the formula

$$\text{estimated probability} = \frac{\text{number of successful trials}}{\text{total number of trials}}$$

- **predicting the number of outcomes:** multiply the probability by the total number of trials.

5.1 The probability scale

Exercise 5A

Questions in this chapter are targeted at the grades indicated.

F

1 How likely is each of the following events? In each case, represent your answer on a probability scale.

a The sun will set tomorrow.

b A pet dog will reach its 100th birthday.

c The next baby born will be a boy.

d You will use the internet to download music.

e It will snow on New Year's Day in Brighton.

F

2 How likely do you think it is that the next Wimbledon Lawn Tennis Men's Champion will come from:

a China b the UK c the EU?

Represent your answers on the same probability scale.

3 a Convert the following to decimals: $\frac{3}{4}$, $\frac{3}{5}$, 60%, 48%.

b Convert the following to fractions: 0.7, 0.15, 45%, 28%.

c Convert the following to percentages: 0.85, 0.43, $\frac{1}{5}$, $\frac{3}{10}$.

4 Write the following in order, starting with the smallest: $\frac{2}{5}$, 0.35, $\frac{2}{10}$, 52%.

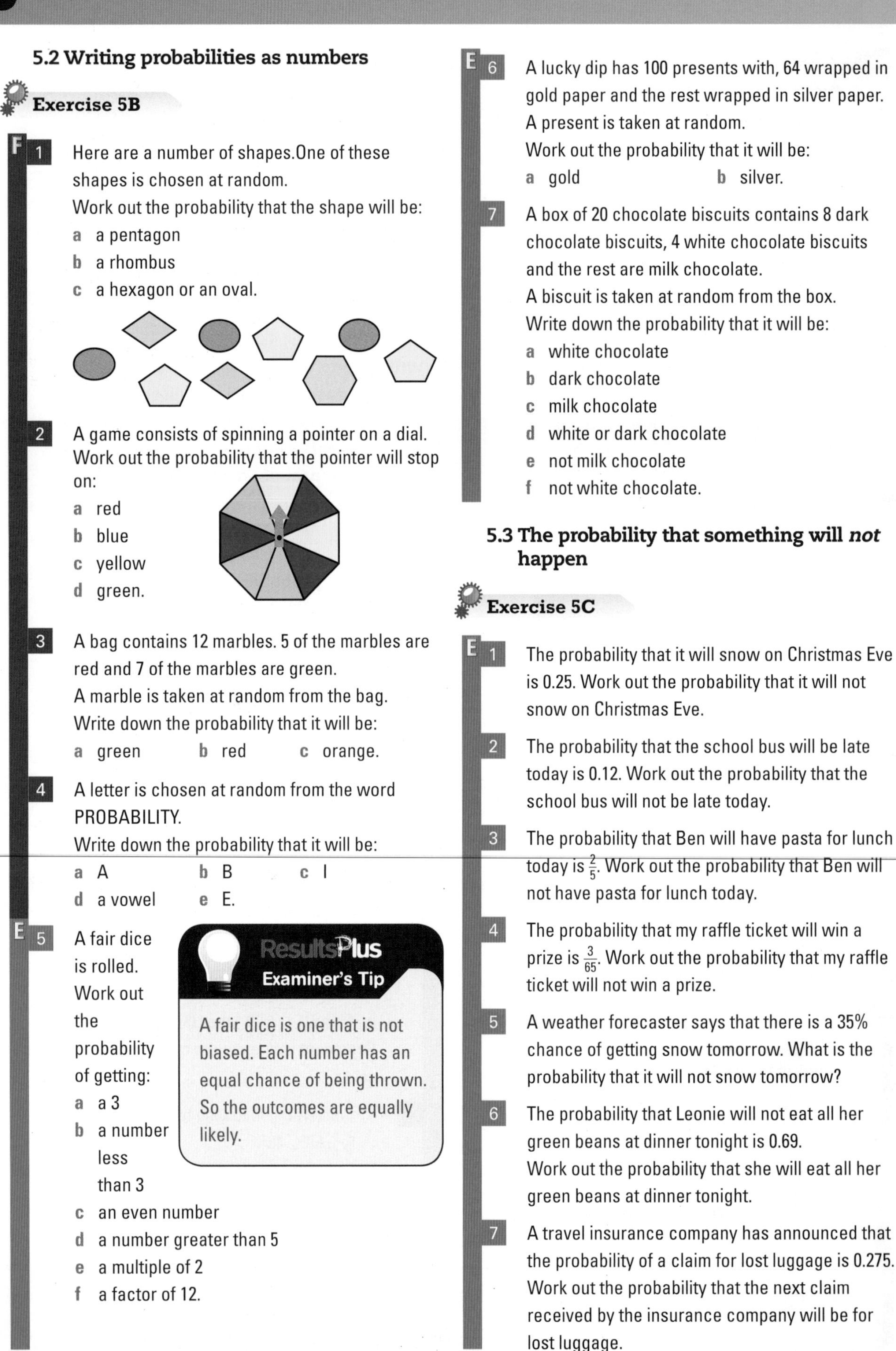

5.2 Writing probabilities as numbers

Exercise 5B

F

1 Here are a number of shapes.One of these shapes is chosen at random.
Work out the probability that the shape will be:
a a pentagon
b a rhombus
c a hexagon or an oval.

2 A game consists of spinning a pointer on a dial. Work out the probability that the pointer will stop on:
a red
b blue
c yellow
d green.

3 A bag contains 12 marbles. 5 of the marbles are red and 7 of the marbles are green.
A marble is taken at random from the bag.
Write down the probability that it will be:
a green b red c orange.

4 A letter is chosen at random from the word PROBABILITY.
Write down the probability that it will be:
a A b B c I
d a vowel e E.

E

5 A fair dice is rolled. Work out the probability of getting:
a a 3
b a number less than 3
c an even number
d a number greater than 5
e a multiple of 2
f a factor of 12.

ResultsPlus
Examiner's Tip
A fair dice is one that is not biased. Each number has an equal chance of being thrown. So the outcomes are equally likely.

E

6 A lucky dip has 100 presents with, 64 wrapped in gold paper and the rest wrapped in silver paper.
A present is taken at random.
Work out the probability that it will be:
a gold b silver.

7 A box of 20 chocolate biscuits contains 8 dark chocolate biscuits, 4 white chocolate biscuits and the rest are milk chocolate.
A biscuit is taken at random from the box.
Write down the probability that it will be:
a white chocolate
b dark chocolate
c milk chocolate
d white or dark chocolate
e not milk chocolate
f not white chocolate.

5.3 The probability that something will *not* happen

Exercise 5C

E

1 The probability that it will snow on Christmas Eve is 0.25. Work out the probability that it will not snow on Christmas Eve.

2 The probability that the school bus will be late today is 0.12. Work out the probability that the school bus will not be late today.

3 The probability that Ben will have pasta for lunch today is $\frac{2}{5}$. Work out the probability that Ben will not have pasta for lunch today.

4 The probability that my raffle ticket will win a prize is $\frac{3}{65}$. Work out the probability that my raffle ticket will not win a prize.

5 A weather forecaster says that there is a 35% chance of getting snow tomorrow. What is the probability that it will not snow tomorrow?

6 The probability that Leonie will not eat all her green beans at dinner tonight is 0.69.
Work out the probability that she will eat all her green beans at dinner tonight.

7 A travel insurance company has announced that the probability of a claim for lost luggage is 0.275.
Work out the probability that the next claim received by the insurance company will be for lost luggage.

E 8 Eastway Express says that the probability of their trains being late in the morning is 0.19. Sam says that the probability that his train will not be late is 0.91. Why is he wrong?

Exercise 5D

D 1 Rob's post is usually delivered at 10.30 am.
This table gives the probability that his post will be on time and the probability that his post will be late.

	early	on time	late
Probability		0.7	0.1

Work out the probability that Rob's post will be delivered before 10.30 am.

2 A biased dice is rolled.
This table gives the probability that it will land on each of the numbers 1, 2, 4, 5 and 6.

	1	2	3	4	5	6
Probability	0.1	0.1		0.2	0.3	0.1

Work out the probability that the dice will land on 3.

3 Isaac has some cards with letters A, B, C, D marked on them. One of these cards is taken at random. This table gives the probability that the card will be B or C or D.

Letter	A	B	C	D
Probability		0.21	0.29	0.32

Work out the probability that the card will:
a be A b not be D c be B or C.

4 A box contains a number of counters. Each counter is either purple or orange or white or black. A counter is taken at random from the box.
The probability that the counter will be purple is 0.35.
The probability that the counter will be orange is 0.24.
The probability that the counter will be white is 0.1.
Work out the probability that the counter will be black.

D 5 When Sara spins this 3-sided spinner, the probability that it will land on 3 is $\frac{5}{9}$.
The probability that it will land on 2 is $\frac{1}{9}$.
Work out the probability that it will land on 1.

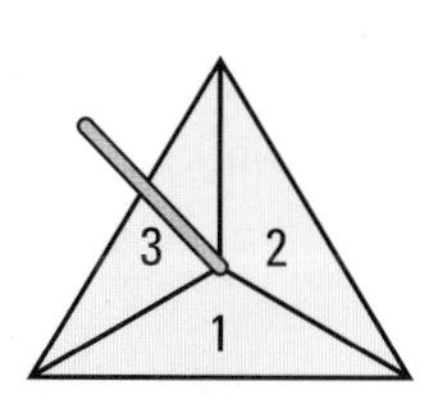

6 When High House hockey team play a match they can win, lose or draw the game.
The probability that they will win the game is $\frac{2}{5}$.
The probability that they will lose the game is $\frac{1}{10}$.
Work out the probability that they will draw the game.

5.4 Using fractions, decimals and percentages in problems

Exercise 5E

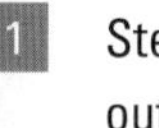

E 1 Steve gets a mark of 49 out of 60 in a test. Work out what this is as a percentage.

D 2 A guitar has a price tag of £400. In the sale there is a 30% reduction. What is the sale price?

3 A shoe manufacturer charges a shop owner £50 for a pair of shoes. The shop owner then adds 35% to this price before he puts it up for sale in his shop. Work out the shop price.

4 Mrs Stokes borrows £5400 from the bank for one year in order to buy a new car. The bank charges her interest of 14% per year. How much does she have to pay back to the bank?

C 5* A local clothes shop has a sale on. There is to be 30% off everything. On the last day they advertise that there will be a further 10% off.
Tim says, 'These jeans cost £60, so I will pay £37.80.'
Cat says, 'No, the cost of the jeans will be £36.'
Explain why Tim and Cat reached different prices.

5.5 Sample space diagrams

Exercise 5F

E 1 Two ordinary dice are rolled. Show all the possible outcomes.
A02 A03

D 2 Use the sample space diagram in question 1 to work out the probability of getting:

a a total score of 6
b the same number on both dice
c a total score greater than 10.

3 The ace, king, queen and jack of clubs and the ace, king, queen and jack of diamonds are put into two piles. The sample space diagram shows all the possible outcomes when a card is taken from each pile.

Clubs					
	J	AJ	KJ	QJ	JJ
	Q	AQ	KQ	QQ	JQ
	K	AK	KK	QK	JK
	A	AA	KA	QA	JA
		A	K	Q	J
		Diamonds			

Work out the probability that:
a both cards will be kings
b at least one card will be a queen
c only one of the cards will be a queen
d the cards will make a matching pair
e one card will be a club
f neither card will be a jack
g both cards will be clubs.

4 A coin and a 3-sided spinner are spun.

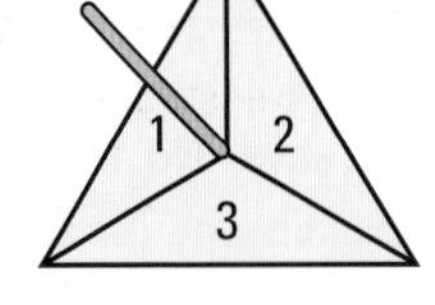

a Draw a sample space diagram to show all the possible outcomes.

Work out the probability that:
b an even number is shown with a head
c an odd number is shown with a tail.

D 5 A02 Mark has his socks and underpants in a drawer. The colours of the socks are either blue or black or brown or grey. The colours of his underpants are the same except for brown. Mark takes at random some socks and underpants from the drawer.
a Draw a sample space diagram to show all the possible combinations of colours for the socks and underpants.
b Work out the probability that Mark takes socks and underpants of:
i the same colour
ii different colours.

6 A03 Samuel is going to spin a 3-sided spinner and a 4-sided spinner and add the scores. The spinners are fair. What is the probability of getting an even number? Explain why.

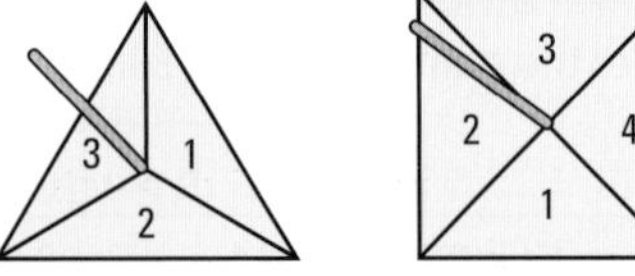

5.6 Relative frequency

Exercise 5G

E 1 Roll a 4-sided spinner 60 times and record your results in a frequency table like this.

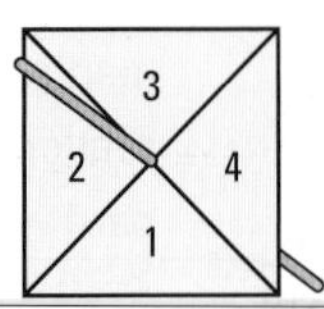

Number	Tally	Frequency
1		
2		
3		
4		
	Total	60

a Use the results in your table to work out the estimated probability of getting:
i a 4 ii an even number
iii a number less than 3.
b Write down the theoretical probability of getting:
i a 4 ii an even number
iii a number less than 3.
c Do you think your spinner is fair? Give a reason for your answer.

E 2 Shuffle a pack of cards and choose a card at random. Record it and replace it. Repeat this 100 times.

a Use your results to write down the estimated probability of getting an ace.

b How could you improve on your answer to part **a**?

3 Make a 6-sided dice of your own out of card. Test the dice to see whether it is fair or not.

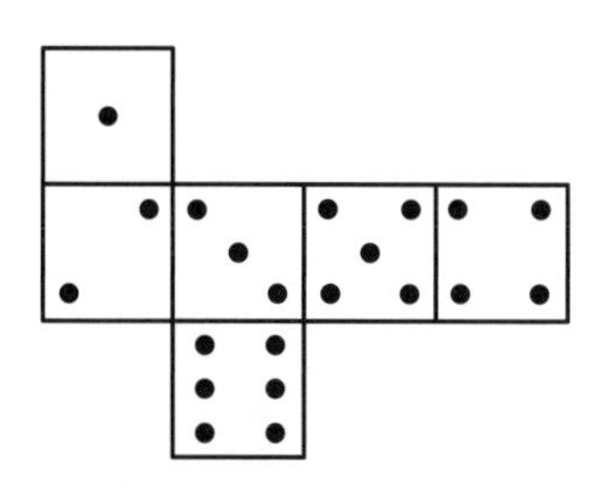

Net of a 6-sided dice

4 There are 1000 letter tiles in a bag. A letter is chosen at random from the bag.

A02 A03

a Work out an estimate of the probability of getting the letter m.

b What affects this estimate?

5.7 Two-way tables

Exercise 5H

D 1 Keiran counted the number of blackbirds and the number of robins in the garden in front of his block of flats in March and April.
The following two-way table provides some of the information from his results.

	Blackbirds	Robins	Total
March	4	2	
April			
Total		6	16

Copy and complete the table.

D 2 The table below gives some information about PE activities chosen by some Year 10 students.

	Swimming	Gymnastics	Cross-country	Total
Boys	20		6	33
Girls	9			
Total		25		65

a Copy and complete the table.

b One of the students is picked at random.
Work out the probability that this student is:

i a boy

ii a boy who does gymnastics

iii a girl who swims

iv a girl.

3 45 students each went to one activity at the sports centre on Saturday night.
The following two-way table shows some information about which activity they chose.

	Swimming	Squash	Bowling	Total
Boys	6			22
Girls			6	
Total	15	16		45

a Copy and complete the table.

b One of the students is picked at random.
Write down the probability that this student:

i will be a girl

ii played squash

iii will be a boy who went bowling.

4 Packed lunches were ordered for a school trip. Some students each had one drink and one snack in the school canteen.
The table below gives some information about what the students ordered.

	Orange juice	Apple juice	Water	Total
Cheese sandwiches	8			25
Sausage rolls	3		2	
Fish bites			4	12
Total	17	9	19	

D

a Copy and complete the table.

b One of the students is picked at random. Use your table to write down the probability that this student had:

i apple juice

ii fish bites

iii water and sausage rolls

iv orange juice and cheese sandwiches.

c Jake says that the probability of picking someone who had apple juice and fish bites is the same as picking someone who had water and sausage rolls. Is he right? Give a reason for your answer.

5.8 Predicting outcomes

Exercise 5I

E 1 Hakim is going to roll an ordinary dice 84 times. Work out an estimate for the number of times it will land on:

a 5 b an odd number c 3 or 4.

2 An ordinary coin is spun 300 times. How many times do you expect it to land on a tail?

3 Kate spins an ordinary 4-sided spinner (numbered 1 to 4) 240 times. How many times can she expect it to land on 4?

4 The table gives information about the probability that Tamsin will win, draw or lose a game of Fizz.

	Win	Draw	Lose
Probability	0.34	0.3	0.36

Tamsin is going to play 50 games of Fizz. Find an estimate for the number of games she will:

a lose

b draw.

5 A bag contains 6 green marbles and 4 yellow marbles. A marble is taken at random from the bag and its colour is recorded. The marble is now put back into the bag and another marble is taken at random from the bag. This is repeated 50 times. Find an estimate for the total number of:

a green marbles taken from the bag

b yellow marbles taken from the bag.

E 6 The probability that an insurance company will get a claim for a leaking roof is 0.36. If the insurance company gets 270 claims during the next month, find an estimate for the number of these claims that will not be for a leaking roof, to the nearest whole number.

C 7* The diagram shows part of Hannah's design for a game. In her game a player pays p pence to spin a star. When the star stops spinning the player wins the amount shown by the arrow.

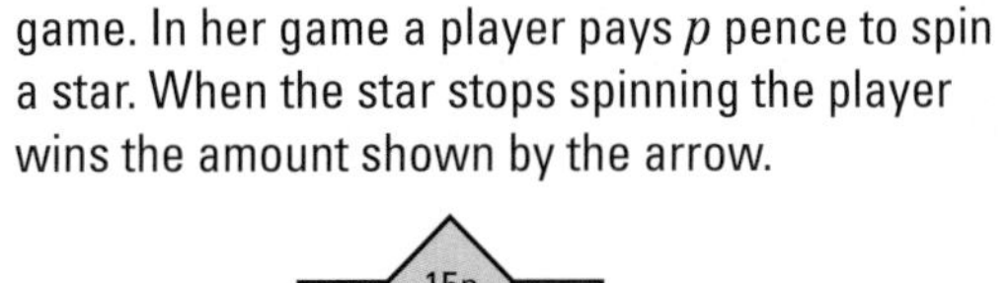

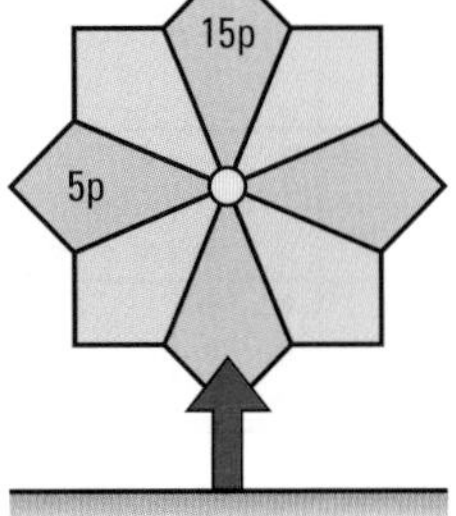

Hannah wants to gain an average of 10p each time the game is played. Show how this can be done by adding six more numbers to the star and finding a suitable value for p.

5.9 Ratio and proportion

Exercise 5J

E 1 A team played all 38 matches in the football season. They won 13, lost 15 and drew the rest. Write down the matches won, lost and drawn as a ratio.

2 In Ketan's DVD collection he has 15 horror movies, 35 thrillers and 20 comedies. Write down the ratio of the number of horror movies to thrillers to comedies in its simplest form.

D 3 There are 180 skiers on a ski slope. The ratio of snowboarders to skiers is 2 : 3. Write down the number of snowboarders.

4 Laura is making boiled eggs with soldiers for a family breakfast. The ratio of soldiers to eggs is 6 : 1. She has boiled 10 eggs. How many soldiers will she need?

D

5 The cost of four burgers at a barbecue is £18.40. Work out the cost of seven burgers.

6 Richard is 15 years old. Rob is 13 years old. They are to share £140 in the ratio of their ages. Work out how much each will get.

7 Here is the list of ingredients to make 5 shortbread biscuits:
35 g butter   15 g sugar   60 g flour.
Imran is going to make 35 shortbread biscuits. How much of each ingredient will he need?

C

8 In a bag of money the ratio of £5 notes to £10 notes to £20 notes is 3 : 5 : 1.

a What fraction of the money is in £10 notes?

Alex picks a note out of the bag.

b What is the probability that it is a £20 note?

1 Number

Key Points

- **digits or figures:** 0 to 9, used for writing all numbers.
- **place value:** the position of a digit within a number. Helps when reading, writing or ordering numbers.
- **rounding:** reducing the accuracy of a number by replacing the right-hand digits with zeros.
- **number line:** an ordered line of numbers. Helps when increasing or decreasing a number.
- **negative numbers:** numbers less than zero, that appear to the left of zero on a number line. The further from zero, the smaller the number.
- **multiplying by 10, 100, 1000:** move all the digits 1, 2 or 3 place values to the left.
- **dividing by 10, 100, 1000:** move all the digits 1, 2 or 3 place values to the right.
- **rounding numbers:** look at the digit before the place value you are rounding to. If it is less than 5, round down. If it is 5 or more, round up.
- **adding a negative number:** the same as subtracting a positive number.
- **subtracting a negative number:** the same as adding a positive number.
- **multiplying or dividing negative numbers:** if the signs are the same, the result is positive. If the signs are different, the result is negative.

1.1 Understanding digits and place value

Exercise 1A

Questions in this chapter are targeted at the grades indicated.

G 1 Draw a place value diagram and write in:

a a four-digit number with a 3 in the hundreds column

b a two-digit number with a 5 in the tens column

c a five-digit number with a 0 in the thousands column

d a three-digit number with an 8 in the units column

e a three-digit number with a 1 in the tens column

f a five-digit number with a 7 in the hundreds column

g a four-digit number with a 6 in every column except the thousands column

h a five-digit number with a 4 in the thousands column and the units column.

G 2 For each teacher, write down five different numbers that they could be thinking about.

3 Write down the value of the 6 in each of these numbers.

a 263 b 6543 c 632

d 25 460 e 63 421

1.2 Reading, writing and ordering whole numbers

Exercise 1B

G 1 Write these numbers in figures.

a Two hundred and fifty-two

b Seven thousand one hundred and sixteen

c Four thousand six hundred and two

d Twenty-one thousand eight hundred and thirty

G 2 Write these numbers in words.

a 732 b 123 c 2971
d 5206 e 3015

3 Write each set of numbers in order of size, starting with the smallest.

a 381, 532, 301, 37, 397
b 4051, 4105, 709, 4501, 4321
c 60, 600, 6066, 6000, 6006, 5996
d 46 762, 49 234, 46 745, 46 123

4 Write the numbers that have been highlighted in blue in figures.

a The numbers of people employed by a local hospital are:
Doctors: two hundred and three
Office staff: four hundred and twenty
Nurses: one thousand and forty-two

b The distances (in miles) driven by three people in a year are:
Ann: eleven thousand seven hundred and fifty
Craig: eight thousand five hundred and sixteen
Sara: six thousand three hundred and two

5 This table gives the populations of five member states of the European Union in 2009.
Write the numbers in words.

	Country	Population
a	Austria	8 298 923
b	Belgium	10 274 595
c	Italy	59 715 625
d	Malta	407 810
e	Spain	44 474 631

6 The table gives the prices of some second-hand cars.

Car	Price
Golf	£6990
A3	£10 695
Mazda 3	£5835
Clio	£4549
Picasso	£12 205

a Write down the price of each car in words.
b Rewrite the list in price order, starting with the most expensive.

1.3 The number line

Exercise 1C

G 1 Draw a number line from 0 to 30.
Mark these numbers on your number line.

a 5 b 13 c 25
d 30 e 21

2 Use a number line from 0 to 25 to

a increase 13 by 6 b decrease 17 by 5
c increase 17 by 4 d decrease 23 by 19
e increase 8 by 17 f decrease 21 by 14.

3 For each of these moves, write down the difference between them, and whether it is an increase or decrease.

a 3 to 7 b 12 to 19 c 9 to 2
d 19 to 14 e 4 to 28 f 18 to 3

1.4 Adding and subtracting

Exercise 1D

G 1 Find the total of 16 and 27.

2 Work out 48 plus 32.

3 Work out 163 + 59.

4 In four maths tests, Sophie scored 51 marks, 56 marks, 78 marks and 67 marks.
How many marks did she score altogether?

5 Find the sum of all the two-digit numbers with 2 in the tens column.

6 In a photography exhibition, five photographers exhibited 14 photos, 29 photos, 6 photos, 18 photos and 23 photos, respectively.
Find the total number of photographs in the exhibition.

7 The number of passengers on a ferry was 34 downstairs and 46 upstairs.
How many passengers were on the ferry altogether?

8 On her MP3 player, Akhil had 82 pop songs, 57 rock songs and 73 dance songs.
How many songs did she have in total?

9 Work out 84 + 29 + 172 + 69 + 83.

10 On six days in May, 68, 34, 57, 113, 83 and 12 people went whitewater rafting on a river.
How many people went rafting in total?

Exercise 1E

G

1 Work out 722 − 507.

2 How much is 6170 − 3082?

3 Take 2009 from 4020.

4 In a car boot sale Simon sells 16 of his 38 DVDs. How many does he have left?

5 When playing darts, Alex scored 121 with his first three darts, Lorna scored 96 with her first three darts and Dita scored 85 with her first three darts.
   a How many more than Dita did Lorna score?
   b What is the difference between Alex's and Lorna's scores?
   c How many less than Alex did Dita score?

6 A03 Phil, Naomi and Dave are delivering leaflets to houses on 3 separate streets. They are each given 120 leaflets. Phil delivers 84, Naomi delivers 93 and Dave delivers 109.
   a How many leaflets do they deliver altogether?
   b How many leaflets do they each have left?
   c How many more leaflets does Dave deliver than i Phil and ii Naomi?

1.5 Multiplying and dividing

Exercise 1F

G

1 Multiply each of these numbers by
   i 10 ii 100 iii 1000
   a 37 b 7 c 103
   d 40 e 2050

2 Work out
   a 53 × 20 b 62 × 200
   c 211 × 30 d 34 × 500
   e 225 × 40 f 412 × 3000

3 Work out
   a 56 × 7 b 62 × 3
   c 134 × 6 d 235 × 8
   e 327 × 5 f 298 × 4

4 Work out
   a 23 × 13 b 74 × 18
   c 64 × 32 d 513 × 13
   e 339 × 35 f 431 × 57
   g 178 × 47 h 49 × 58

G

5 Find the product of 47 and 14.

6 How many is 123 multiplied by 13?

7 A03 Jacob cycles there and back to the station, 9 miles each way, on Monday, Wednesday and Friday. On Tuesday and Thursday he drives to and from the city centre 23 miles away. In a working week, how far does he travel
   a by bicycle
   b by car
   c altogether?

8 Southton Rugby Club hires 32 coaches to take supporters to an away match.
Each coach can take 52 passengers. How many supporters can be taken to the match?

9 Each box of paperclips contains 45 paperclips. How many paperclips are there in 34 boxes?

10 Ashen buys oranges which are packed in cases of 36.
He buys 9 cases. How many oranges does he buy?

Exercise 1G

G

1 Work out
   a 4210 ÷ 10 b 2600 ÷ 100
   c 7000 ÷ 10 d 51 000 ÷ 1000

2 Work out
   a 84 ÷ 2 b 96 ÷ 3 c 68 ÷ 4
   d 84 ÷ 6 e 560 ÷ 4 f 425 ÷ 5
   g 96 ÷ 4 h 784 ÷ 8 i 261 ÷ 9
   j 511 ÷ 7 k 612 ÷ 2 l 2010 ÷ 3

F

3 Work out
   a 288 ÷ 16 b 696 ÷ 12 c 576 ÷ 32
   d 945 ÷ 21 e 972 ÷ 36 f 2010 ÷ 30
   g 1440 ÷ 24 h 1500 ÷ 20

4 a Work out 465 ÷ 15.
   b How many 50s make 950?
   c Work out 629 ÷ 17.
   d Work out 900 divided by 30.
   e Divide 128 by 8.

5 Five people paid £3600 in total for a holiday.
They each paid the same amount.
How much did each person pay?

F 6 a There are 96 football teams in 4 leagues. Each league has the same number of teams in it. How many teams are in each league?

A03 b In the 4 junior leagues, there are 60 teams. Two of the leagues have 14 teams. How many teams are there in the other two leagues that have the same number?

7 A single school bus can take 36 passengers.
A03 Years 7, 8 and 9 are going to take part in a sports day at the local stadium. How many bus trips have to be made for the following Years?

a Year 7 (252 students)
b Year 8 (324 students)
c Year 9 (288 students)

8 Cartons containing cereal packets are delivered
A03 to a supermarket. They can hold 18 × 1 kg packets, 22 × 750 g packets or 30 × 500 g packets of cereal. The supermarket orders

a 216 × 1 kg packets of cereal
b 330 × 750 g packets of cereal
c 390 × 500 g packets of cereal.

How many cartons will be delivered?

1.6 Rounding

Exercise 1H

G 1 Round these numbers to the nearest ten.

a 48 b 52 c 275
d 183 e 896 f 1308

2 Round these numbers to the nearest hundred.

a 447 b 785 c 1307
d 2093 e 7888 f 39 355

3 Round these numbers to the nearest thousand.

a 3126 b 47 171 c 82 525
d 123 705 e 818 282 f 1 123 405

F 4

	Length (ft)	Cruising speed (mph)	Takeoff weight (lb)
Airbus A380	240	634	1 200 010
Boeing 787-3	171	587	374 950
Gulfstream G200	62	528	35 455
Bombardier CS300	124	541	131 790
Citation X	72	607	36 395

For each of these aircraft, round:

a the length to the nearest ten feet
b the takeoff weight to the nearest hundred pounds
c the cruising speed to the nearest ten mph.

1.7 Negative numbers

Exercise 1I

G 1 For each list of numbers write down

i the highest and the lowest number
ii the numbers in order, starting with the lowest.

a 7, −12, −1, 2, 6 b −5, 0, −7, −11, 2
c −1, 8, 15, −13, −4 d −11, 0, −18, −19, −3

2 Write down the two missing numbers in each sequence.

a 7, 5, 3, __, __, −3, −5
b −14, −10, −6, −2, __, __, 10
c 8, 5, 2, −1, __, __, −10
d 19, 14, 9, 4, __, __, −11
e 25, 18, 11, 4, __, __, −17
f −11, −8, −5, −2, __, __, 7

F 3 Use the number line to find the number that is:

9, 8, 7, 6, 5, 4, 3, 2, 1, 0, −1, −2, −3, −4, −5, −6, −7

a 6 more than 3
b 5 more than −7
c 8 less than 7
d 3 less than −4
e 4 less than 0
f 10 more than −5
g 7 more than −7
h 3 less than −3
i 10 less than 6
j 2 less than −2.

4 What number is:

a 40 more than −70
b 30 less than −20
c 90 greater than −50
d 90 smaller than 50
e 230 smaller than −30
f 70 bigger than 100
g 170 bigger than −300
h 300 bigger than −300
i 160 more than −20
j 100 less than −100?

F

5 The table gives the highest and lowest temperatures recorded in five cities during one month.

	Highest temperature	Lowest temperature
Washington	1°C	−8°C
Marrakesh	19°C	−12°C
Cape Town	26°C	19°C
Amsterdam	−2°C	−3°C
Anchorage	−9°C	−11°C

a Which city recorded the lowest temperature?
b Which city recorded the biggest difference between its highest and lowest temperatures?
c Which city recorded the smallest difference between its highest and lowest temperatures?

6 The temperature of the fridge compartment of a fridge-freezer is set at 3°C.
The freezer compartment is set at −20°C.
What is the difference between these temperature settings?

+10°C
+5°C
0°C
−5°C
−10°C
−15°C
−20°C

This thermometer is showing a temperature of −20°C.

7 The temperature of a shop freezer should be set at −24°C. It is set to −17°C by mistake.
What is the difference between these temperature settings?

1.8 Working with negative numbers

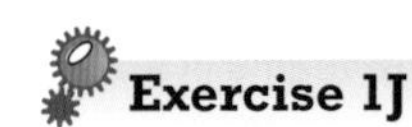

Exercise 1J

Use the number line going from −10°C to +10°C to help you with these questions.

ResultsPlus
Examiner's Tip
A number line can help you when working with negative numbers.

F

1 Find the difference between each pair of temperatures.
a −3°C, 4°C
b −6°C, −1°C
c 2°C, 7°C
d −8°C, 4°C
e 7°C, −5°C
f 2°C, 9°C
g −3°C, −9°C
h −7°C, 5°C

2 Find the new temperature after:
a a 3° rise from −4°C
b a 7° fall from 3°C
c 7°C falls by 15°
d −4°C rises by 8°
e −5°C rises by 9°
f 5°C falls by 10°
g −3°C falls by 5°.

+10°
+9°
+8°
+7°
+6°
+5°
+4°
+3°
+2°
+1°
0°
−1°
−2°
−3°
−4°
−5°
−6°
−7°
−8°
−9°
−10°

1.9 Calculating with negative numbers

Exercise 1K

F

1 Work out
a −5 + −2
b 8 − +6
c 7 − −3
d 6 + +5
e −6 − −5
f −3 + +5
g 5 + −7
h −2 − +8

2 A03 The record lowest 72-hole score in a golf tournament is −26 (26 under par). The winning golfer in a recent tournament scored −17.
What is the difference between the scores?

3 The temperature outside is recorded as −20°C one night. The following day it rises by 7°C.
What is the temperature during the day?

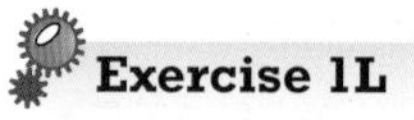

Exercise 1L

Work out these calculations.

F

1
a $+3 \times -2$
b $+24 \div -6$
c $+2 \div +1$
d $+2 \times +8$
e $-9 \div +3$
f $-4 \times +4$

2
a $-9 \times +12$
b $-32 \div -4$
c $-16 \div -4$
d $-2 \times +9$
e $+25 \div -5$
f -3×-6

3
a $-5 \times +3$
b $-16 \div -2$
c $-4 \times +6$
d $-9 \div -3$
e $+4 \div +2$
f $-6 \times +8$

4
a -7×-3
b $-30 \div +6$
c $-8 \div +4$
d -3×-11
e $+4 \times -8$
f $+64 \div +8$

5
a $-50 \div -10$
b $-7 \times +9$
c $+4 \times +6$
d -4×-7
e $-12 \div +3$
f $-7 \times +5$

2 Factors, multiples and primes

Key Points

- **even numbers:** numbers that divide exactly by 2. Any number that ends in 0, 2, 4, 6 or 8.
- **odd numbers:** numbers that do not divide exactly by 2. Any number that ends in 1, 3, 5, 7 or 9.
- **factors of a number:** whole numbers that divide exactly into the number. They always include 1 and the number itself.
- **multiples of a number:** the results of multiplying the number by positive whole numbers.
- **common multiple:** a number that is a multiple of two or more numbers.
- **prime number:** a whole number greater than 1 whose only factors are itself and 1.
- **prime factor:** a factor that is also a prime number. A number can be written as a product of its prime factors.
- **common factor:** a number that is a factor of two or more numbers.
- **lowest common multiple (LCM):** the lowest multiple that is common to two or more numbers.
- **highest common factor (HCF):** the highest factor common to two or more numbers.
- **square number:** a number that is the result of squaring a whole number.
- **cube number:** a number that is the result of cubing a whole number.
- **finding the square of a number:** multiply the number by itself.
- **finding the cube of a number:** multiply the number by itself and then multiply the result by the original number.

2.1 Factors, multiples and prime numbers

Exercise 2A

Questions in this chapter are targeted at the grades indicated.

G

1 Write down all the even numbers from this list.
32, 28, 47, 855, 2220, 84 548, 300 000

2 Write down all the odd numbers from this list.
205, 437, 8328, 63 447, 100 000

3 Write down the next two even numbers after:
a 16 b 32 c 298.

4 Write down the odd number that comes before:
a 9 b 43 c 100.

5 Write down all the even four-digit numbers that can be made using only the numbers on the cards below.

2 3 8 9

6 (A03) The first two odd numbers and the first two even numbers are written on cards.
a What numbers are written on the cards?
b What is the smallest even number that can be made using the four cards?
c What is the largest odd number that can be made using the four cards?

F

7 (A03) A number of migrating birds at a wildlife centre are ringed, so that their journeys can be recorded when they return the next spring. Their numbers are listed in the order in which they return.
82, 35, 59, 63, 22, 77, 81, 90, 93, 2, 4, 3, 71, 73, 8, 14, 28, 83, 30, 54, 13, 25, 66, 1, 12
a How many birds returned?
The birds with even numbers are put in one enclosure and the birds with odd numbers are put in another enclosure.
b Arrange the ring numbers to show into which enclosure they are placed.

Exercise 2B

G

1 Write down all the factors of the following numbers.
a 12 b 22
c 16 d 11
e 33 f 72

Students sometimes get confused between factors and multiples – remember multiples are from multiplying.

2 List the first five multiples of the following numbers.
a 4 b 9 c 15 d 11 e 20

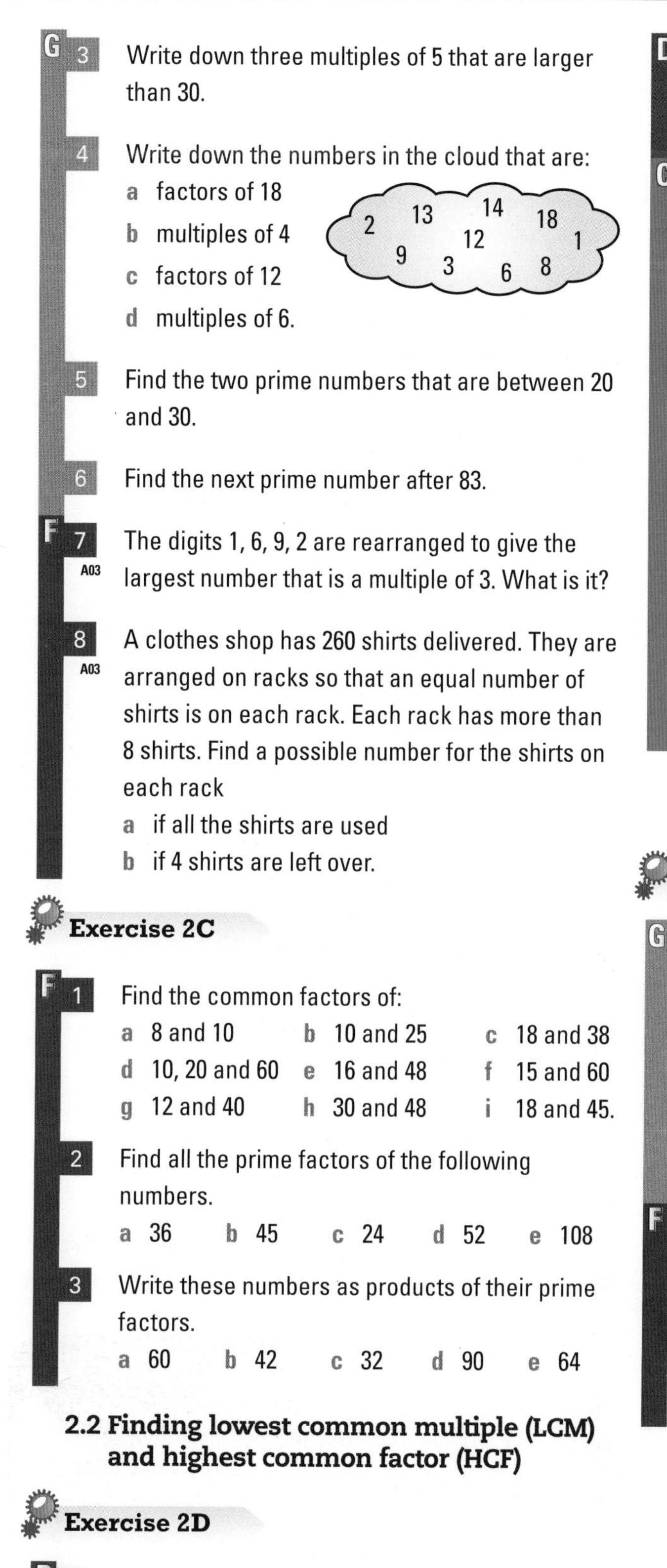

G **3** Write down three multiples of 5 that are larger than 30.

4 Write down the numbers in the cloud that are:

a factors of 18
b multiples of 4
c factors of 12
d multiples of 6.

5 Find the two prime numbers that are between 20 and 30.

6 Find the next prime number after 83.

F **7** (A03) The digits 1, 6, 9, 2 are rearranged to give the largest number that is a multiple of 3. What is it?

8 (A03) A clothes shop has 260 shirts delivered. They are arranged on racks so that an equal number of shirts is on each rack. Each rack has more than 8 shirts. Find a possible number for the shirts on each rack

a if all the shirts are used
b if 4 shirts are left over.

Exercise 2C

F **1** Find the common factors of:

a 8 and 10 b 10 and 25 c 18 and 38
d 10, 20 and 60 e 16 and 48 f 15 and 60
g 12 and 40 h 30 and 48 i 18 and 45.

2 Find all the prime factors of the following numbers.

a 36 b 45 c 24 d 52 e 108

3 Write these numbers as products of their prime factors.

a 60 b 42 c 32 d 90 e 64

2.2 Finding lowest common multiple (LCM) and highest common factor (HCF)

Exercise 2D

D **1** Find the highest common factor of:

a 4 and 10 b 9 and 15 c 18 and 42
d 14 and 32 e 21 and 49.

D **2** Find the lowest common multiple of:

a 3 and 5 b 4 and 8 c 10 and 15
d 36 and 48 e 25 and 85.

C **3** Find the LCM and HCF of:

a 6 and 8 b 120 and 160
c 24 and 64 d 81 and 150
e 36 and 96 f 20 and 30.

4 (A02) Two sensors in a car flash at different intervals. One flashes every 20 seconds and the other every 36 seconds. The driver sees them flash at the same time.
How long will it be before this happens again?

5 (A02) Two robots are set to move in a circle. One takes 20 seconds to complete one circle. The other robot takes 15 seconds. They are set off at the same time.
How long will it be before they are both back at their start positions at the same time?

2.3 Finding square numbers and cube numbers

Exercise 2E

G **1** Work out the square of the following numbers.

a 4 b 5
c 30 d 7
e 8

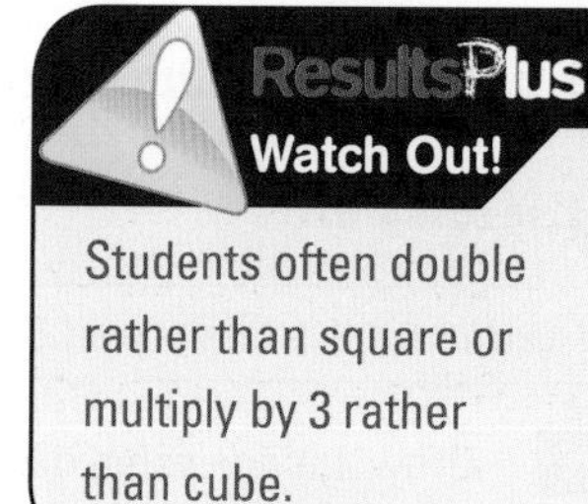

F **2** Work out the cube of the following numbers.

a 3 b 4 c 6 d 10 e 11

3 Work out

a 6 squared b the cube of 2
c 10^2 d 5^3

3 Decimals and rounding

Key Points

- **square root (√):** the opposite of the square of a number e.g. If $3 \times 3 = 9$, $\sqrt{9} = 3$.
- **cube root ($\sqrt[3]{}$):** the opposite of the cube of a number e.g. If $3 \times 3 \times 3 = 27$, $\sqrt[3]{27} = 3$.
- **ordering decimals:** compare the whole number parts, then the digits in the tenths place, then the digits in the hundredths place, and so on.
- **adding and subtracting decimals:** keep the decimal points in line so that the place values match.
- **multiplying decimals:** the total number of decimal places in the answer is the same as that in the question.
- **dividing decimals:** multiply both numbers by 10, 100, 1000, etc. until you have a whole number to divide by.
- **rounding decimals:**
  - **to the nearest whole number:** if the digit in the tenths column is greater than or equal to 5, round the whole number up.
  - **to 1 decimal place:** if the digit in the second decimal place is greater than or equal to 5, round the first decimal place up. If not, leave off this digit and any that follow.
  - **to a given number of decimal places (d.p.):** count this number of decimal places from the decimal point. If the next digit is greater than or equal to 5, round up. If not, leave off this digit and any that follow.
- **rounding to a given number of significant figures (s.f.):** count this number of digits from the first non-zero digit. If the next digit is 5 or more, then round up.
- **estimating an answer:** round each number to 1 significant figure and then calculate.

3.1 Understanding place value

Exercise 3A

Questions in this chapter are targeted at the grades indicated.

G

1 Draw a place value diagram and write in these numbers.

a 32.5 b 3.25 c 346.4 d 3.641
e 0.346 f 5.002 g 1.05 h 0.068

2 What is the place value of the underlined digit in each number?

a $42\underline{5}.3$ b $41.\underline{5}3$ c $6.\underline{4}62$
d $2.7\underline{4}8$ e $\underline{2}3.14$ f $19.66\underline{9}3$
g $\underline{3}24.159$ h $5.712\underline{9}$ i $7.1\underline{0}4$
j $8.\underline{3}27$ k $52.60\underline{4}$ l $0.0\underline{2}33$

3.2 Writing decimal numbers in order of size

Exercise 3B

G

1 The table gives the price of packets of raisins in different shops.

Shop	Price
Stall	£1.49
Corner	£1.42
Market	£1.05
Main	£1.65
Store	£1.45
Super	£1.40

Write the list of prices in order. Start with the lowest price.

2 Rearrange these decimal numbers in order of size. Start with the largest.

a 0.52, 0.54, 0.6, 0.49, 0.62
b 4.3, 4.21, 4.57, 3.21, 3.04
c 0.31, 0.045, 0.04, 0.42, 0.05
d 5.0, 7.25, 7.04, 5.37, 5.68
e 1.07, 1.11, 1.1099, 2.06, 1.0097
f 4.07, 3.06, 4.4, 3.2, 3.15, 3.1078

G 3 The fastest 400 m times, in seconds, of six athletes were:

43.451	43.150	42.145
44.510	43.105	43.045

Write down the athletes' times in order. Start with the fastest.

3.3 Adding and subtracting decimals

Exercise 3C

Work these out, showing all your working.

F
1. 2.9 + 4.6
2. 8 + 0.75
3. 60.7 + 24.2
4. 75.2 + 0.37
5. 0.9 + 0.5
6. 18.1 + 2.526
7. 8.8 + 8.8
8. 20 + 2.001
9. 0.006 + 1.908
10. 116 + 1.16
11. 3.6 + 12.7 + 7.84
12. 18.03 + 3.9 + 3.66
13. 0.712 + 3.61 + 14.9
14. 6.8 + 6.33 + 7.077
15. 8 + 4.823 + 0.333
16. 42.34 + 6.53 + 2.261
17. 7.03 + 14 + 0.0267
18. 2.05 + 51.3 + 124.056

Exercise 3D

F 1 Work out these money calculations, showing all your working.

a £29.90 − £23.70 b £6.84 − £3.70
c £23.50 − £8.46 d £200.70 − £5.40
e £0.58 − £0.27 f £2 − £0.65
g £26.90 − £20.71 h £23.64 − £10.50
i £5.50 − £1.60 j £6.84 − £4.77
k £26.50 − £9.48 l £24 − £0.75

2 Work out these calculations, showing all your working.

a 7.125 − 5.9 b 12.01 − 5.361
c 8.29 − 7.036 d 104.06 − 45.48

3.4 Multiplying decimals

Exercise 3E

Work these out, showing all your working.

G 1 Find the cost of:

a 6 books at £3.25 each
b 4 bags of fruit at £2.37 each
c 5 drinks at £0.65 each
d 2.5 kilos of apples at £0.80 per kilo.

G 2 Work out

a 0.054×100 b 0.54×100
c 5.4×100 d 0.0302×100
e 0.302×100 f 3.02×100

What do you notice about your answers to question 2?

F 3 Work out

a 6.7×4 b 0.67×4
c 0.67×0.4 d 5.25×5
e 5.25×0.5 f 0.525×0.5
g 52.5×0.05 h 5.25×0.005
i 0.525×0.005

4 Work out

a 4.62×10 b 46.2×10
c 0.462×10 d 32.65×10
e 326.5×10 f $0.032\,65 \times 10$

Look carefully at your answers to question 4. What do you notice?

5 Work out

a 34.6×8 kg b 2.15×0.05 seconds
c 0.13×0.13 m d 0.4×0.4 miles
e $1.6 \times 0.5\,l$ f 0.04×0.03 hours

6 A book costs £5.35. Work out the cost of buying:

a 25 copies b 36 copies
c 55 copies.

7 It costs £8.57 for one person to enter the theme park. How much does it cost for:

a 15 people b 25 people
c 43 people?

8 A bucket holds 4.55 litres of water. How much water is contained in:

a 24 buckets b 36 buckets
c 45 buckets?

3.5 Squares and square roots, cubes and cube roots

Exercise 3F

E 1 Work out the square of the following numbers.

a 2.1 b 3.2 c 4.3
d 3.02 e 0.6

2 Work out the cube of the following numbers.

a 1.2 b 2.2 c 3.3
d 0.4 e 0.7

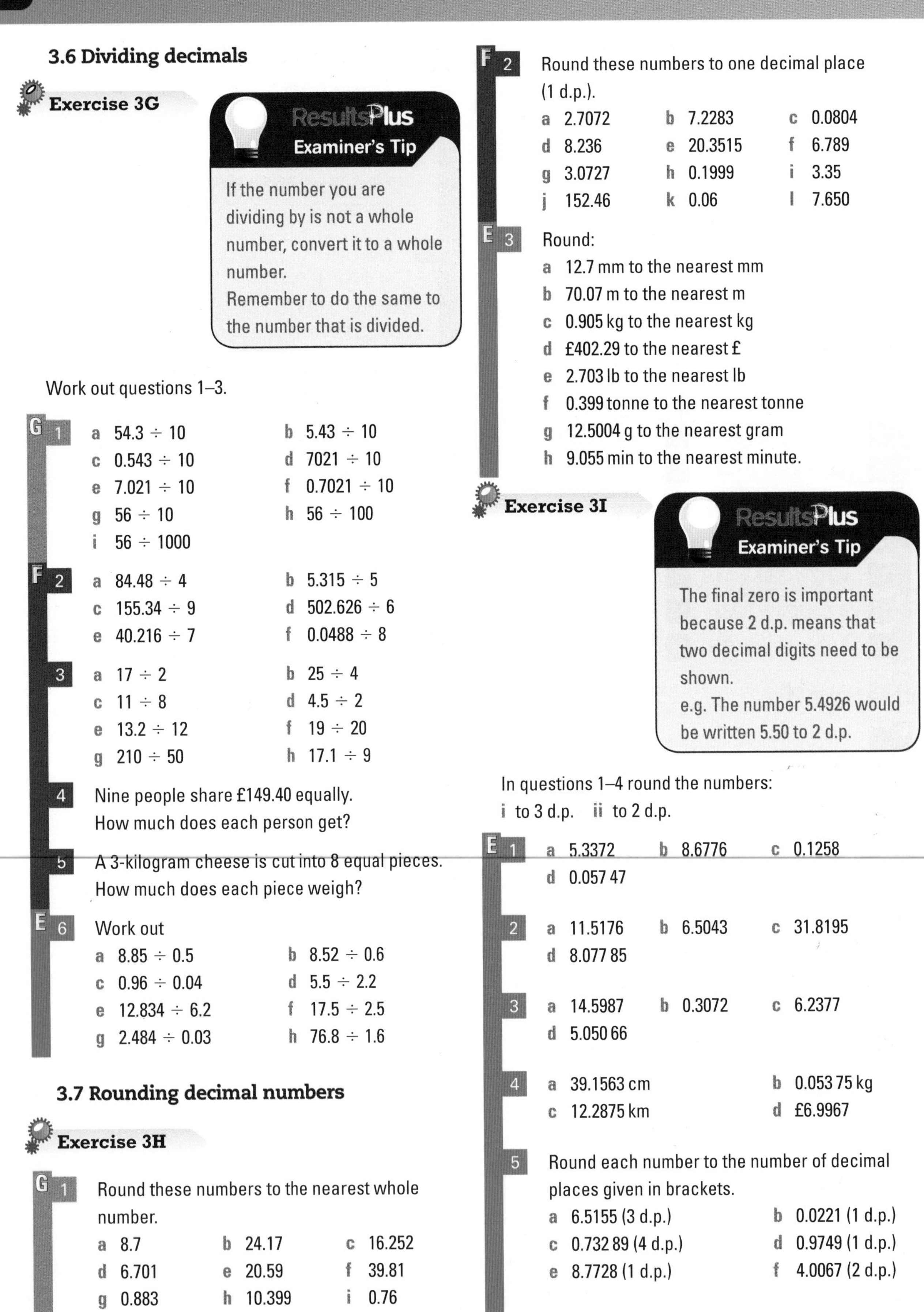

3.6 Dividing decimals

Exercise 3G

ResultsPlus Examiner's Tip

If the number you are dividing by is not a whole number, convert it to a whole number.
Remember to do the same to the number that is divided.

Work out questions 1–3.

G 1
a 54.3 ÷ 10 b 5.43 ÷ 10
c 0.543 ÷ 10 d 7021 ÷ 10
e 7.021 ÷ 10 f 0.7021 ÷ 10
g 56 ÷ 10 h 56 ÷ 100
i 56 ÷ 1000

F 2
a 84.48 ÷ 4 b 5.315 ÷ 5
c 155.34 ÷ 9 d 502.626 ÷ 6
e 40.216 ÷ 7 f 0.0488 ÷ 8

3
a 17 ÷ 2 b 25 ÷ 4
c 11 ÷ 8 d 4.5 ÷ 2
e 13.2 ÷ 12 f 19 ÷ 20
g 210 ÷ 50 h 17.1 ÷ 9

4 Nine people share £149.40 equally.
How much does each person get?

5 A 3-kilogram cheese is cut into 8 equal pieces.
How much does each piece weigh?

E 6 Work out
a 8.85 ÷ 0.5 b 8.52 ÷ 0.6
c 0.96 ÷ 0.04 d 5.5 ÷ 2.2
e 12.834 ÷ 6.2 f 17.5 ÷ 2.5
g 2.484 ÷ 0.03 h 76.8 ÷ 1.6

3.7 Rounding decimal numbers

Exercise 3H

G 1 Round these numbers to the nearest whole number.
a 8.7 b 24.17 c 16.252
d 6.701 e 20.59 f 39.81
g 0.883 h 10.399 i 0.76
j 200.08 k 18.55 l 2.99

F 2 Round these numbers to one decimal place (1 d.p.).
a 2.7072 b 7.2283 c 0.0804
d 8.236 e 20.3515 f 6.789
g 3.0727 h 0.1999 i 3.35
j 152.46 k 0.06 l 7.650

E 3 Round:
a 12.7 mm to the nearest mm
b 70.07 m to the nearest m
c 0.905 kg to the nearest kg
d £402.29 to the nearest £
e 2.703 lb to the nearest lb
f 0.399 tonne to the nearest tonne
g 12.5004 g to the nearest gram
h 9.055 min to the nearest minute.

Exercise 3I

ResultsPlus Examiner's Tip

The final zero is important because 2 d.p. means that two decimal digits need to be shown.
e.g. The number 5.4926 would be written 5.50 to 2 d.p.

In questions 1–4 round the numbers:
i to 3 d.p. ii to 2 d.p.

E 1
a 5.3372 b 8.6776 c 0.1258
d 0.057 47

2
a 11.5176 b 6.5043 c 31.8195
d 8.077 85

3
a 14.5987 b 0.3072 c 6.2377
d 5.050 66

4
a 39.1563 cm b 0.053 75 kg
c 12.2875 km d £6.9967

5 Round each number to the number of decimal places given in brackets.
a 6.5155 (3 d.p.) b 0.0221 (1 d.p.)
c 0.732 89 (4 d.p.) d 0.9749 (1 d.p.)
e 8.7728 (1 d.p.) f 4.0067 (2 d.p.)

3.8 Rounding to 1 significant figure

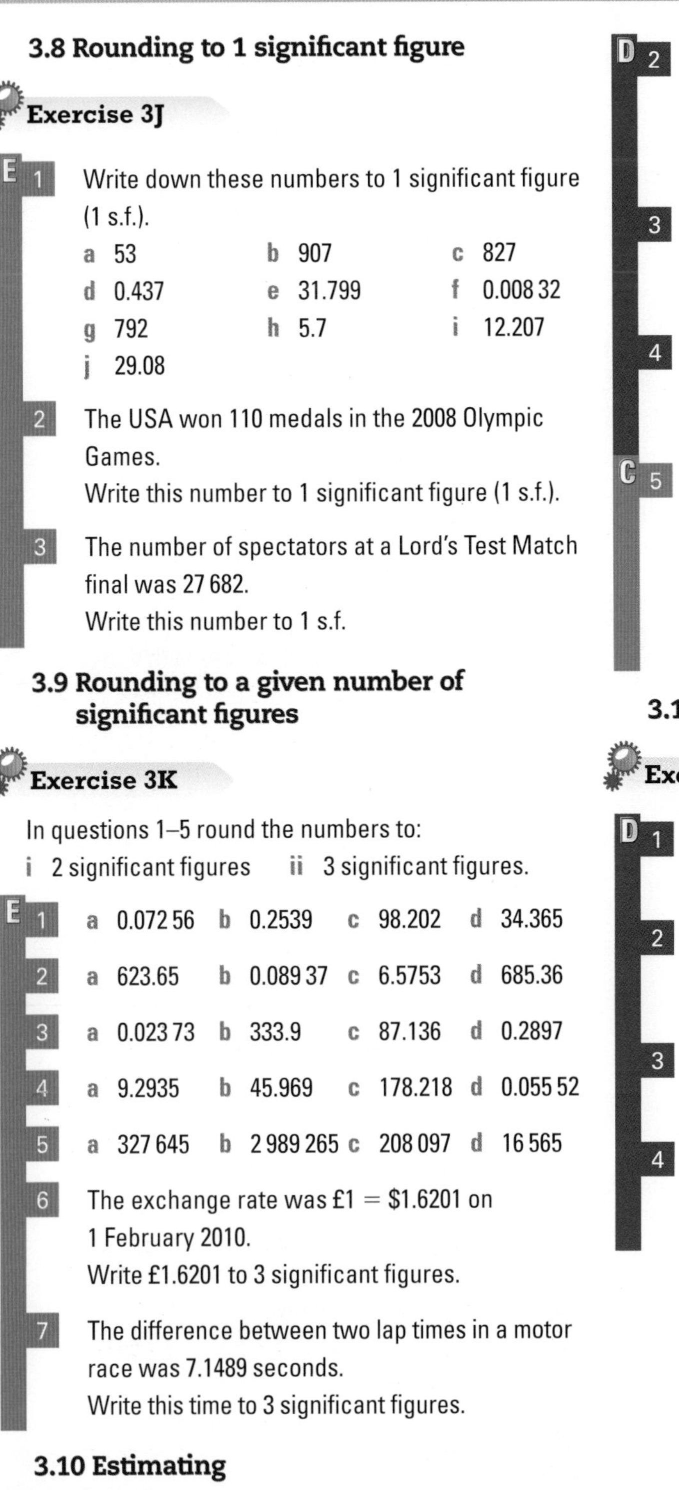

Exercise 3J

E

1 Write down these numbers to 1 significant figure (1 s.f.).

a 53 b 907 c 827
d 0.437 e 31.799 f 0.008 32
g 792 h 5.7 i 12.207
j 29.08

2 The USA won 110 medals in the 2008 Olympic Games.
Write this number to 1 significant figure (1 s.f.).

3 The number of spectators at a Lord's Test Match final was 27 682.
Write this number to 1 s.f.

3.9 Rounding to a given number of significant figures

Exercise 3K

In questions 1–5 round the numbers to:
i 2 significant figures ii 3 significant figures.

E

1 a 0.072 56 b 0.2539 c 98.202 d 34.365

2 a 623.65 b 0.089 37 c 6.5753 d 685.36

3 a 0.023 73 b 333.9 c 87.136 d 0.2897

4 a 9.2935 b 45.969 c 178.218 d 0.055 52

5 a 327 645 b 2 989 265 c 208 097 d 16 565

6 The exchange rate was £1 = \$1.6201 on 1 February 2010.
Write £1.6201 to 3 significant figures.

7 The difference between two lap times in a motor race was 7.1489 seconds.
Write this time to 3 significant figures.

3.10 Estimating

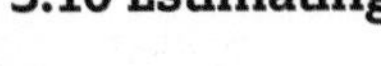

Exercise 3L

D

1 Showing your rounding, work out estimates for:

a $75 \times \frac{67}{41}$ b $\frac{306 \times 411}{54}$ c $\frac{8 \times 41 \times 87}{404}$

d $\frac{400}{13 \times 39}$ e $\frac{598}{12 \times 61}$ f $\frac{104 \times 76}{23 \times 42}$

D

2 An athletics stand has 52 rows of seats.
Each row has 104 seats.
Work out an estimate for the total number of seats.

3 A box contains 48 apples.
Work out an estimate for the total number of apples in 64 boxes.

4 Lucy is buying 28 floor tiles.
Each floor tile costs £5.25.
Work out an estimate for her total cost.

C

5 Work out estimates for each of the following calculations.

a $17.2 \times \frac{0.23}{4.3}$ b $5.76 \times \frac{27.9}{0.87}$

c $\frac{883}{23.4} \times 0.467$

3.11 Manipulating decimals

Exercise 3M

D

1 Given that $4.6 \times 8.2 = 37.72$, work out

a 46×82 b 460×820 c 0.46×82

2 Given that $\frac{24.6}{1.25} = 19.68$, work out

a $\frac{246}{1.25}$ b $\frac{2.46}{1.25}$ c $\frac{0.246}{1.25}$

3 Given that $15.2 \times 5.5 = 83.6$, work out

a 152×5.5 b 1.52×0.55 c 0.152×55

4 Given that $\frac{25.65}{4.75} = 5.4$, work out

a $\frac{25.65}{47.5}$ b $\frac{2.565}{4.75}$ c $\frac{256.5}{4.75}$

4 Fractions

Key Points

- **fraction:** part of a unit.
- **numerator:** the top part of a fraction.
- **denominator:** the bottom part of a fraction.
- **equivalent fractions:** fractions that represent the same quantity but use different denominators, e.g. $\frac{1}{2} = \frac{2}{4}$.
- **improper fraction:** a fraction where the numerator is greater than the denominator.
- **mixed number:** a number with a whole number part and a proper fraction part.
- **terminating decimal:** a fraction that has an exact decimal equivalent.
- **recurring decimal:** a fraction that has a decimal equivalent that repeats itself. A dot goes over the figure that repeats itself.
  $\frac{1}{3} = 0.333333\ldots = 0.\dot{3}$
  $\frac{5}{11} = 0.454545\ldots = 0.\dot{4}\dot{5}$
  $\frac{1}{7} = 0.142\,857\,142\,857\ldots = 0.\dot{1}42\,85\dot{7}$
- **finding an equivalent fraction:** multiply the numerator and the denominator by the same whole number.
- **writing a fraction in its simplest form:** divide the numerator and the denominator by the same whole number.
- **ordering fractions:** find equivalent fractions with the same denominator, and order by numerator.
- **finding a common denominator:** find a common multiple of the denominators.
- **multiplying fractions:** multiply the numerators together and multiply the denominators together.
- **dividing fractions:** turn the dividing fraction upside down and then multiply.
- **adding and subtracting fractions:** use equivalent fractions to get fractions with a common denominator. Add or subtract the numerators.
- **converting fractions to decimals:** divide the numerator by the denominator using long division.
- **converting decimals to fractions:** look at the place value of the smallest digit. E.g. The smallest digit of 0.709 is 9 thousandths, so 0.709 is 709 thousandths or $\frac{709}{1000}$.

4.1 Understanding fractions

Exercise 4A

Questions in this chapter are targeted at the grades indicated.

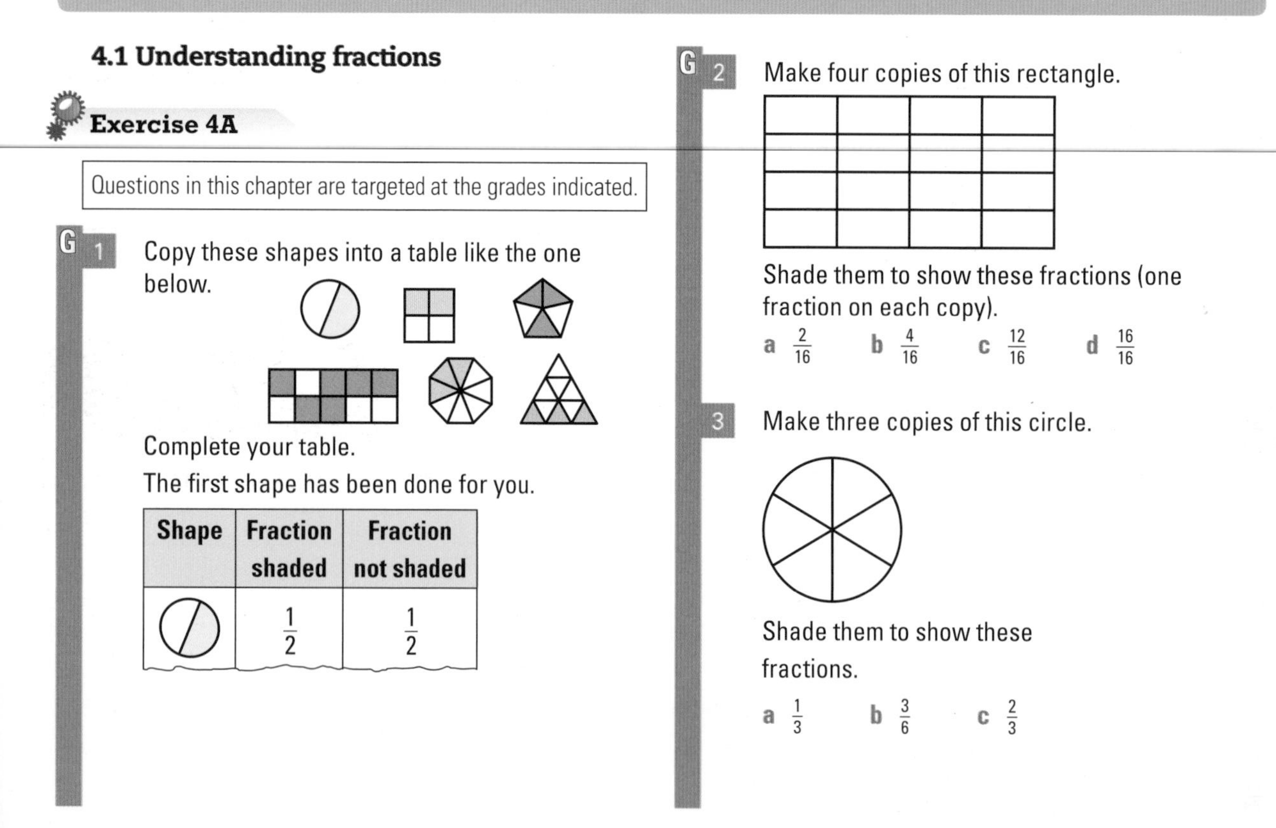

G 1 Copy these shapes into a table like the one below.

Complete your table.
The first shape has been done for you.

Shape	Fraction shaded	Fraction not shaded
(circle)	$\frac{1}{2}$	$\frac{1}{2}$

G 2 Make four copies of this rectangle.

Shade them to show these fractions (one fraction on each copy).

a $\frac{2}{16}$ **b** $\frac{4}{16}$ **c** $\frac{12}{16}$ **d** $\frac{16}{16}$

3 Make three copies of this circle.

Shade them to show these fractions.

a $\frac{1}{3}$ **b** $\frac{3}{6}$ **c** $\frac{2}{3}$

G 4 There are 32 people on a martial arts course. Thirteen are female and 19 are male.
What fraction of the people are

a male b female?

5 32 competitors took part in a swimming competition on a Saturday, and 43 other competitors took part on Sunday.

a How many swimmers were there altogether?

b What fraction of the swimmers competed on Sunday?

c What fraction of the swimmers competed on Saturday?

4.2 Equivalent fractions

Exercise 4B

G 1 For each of these diagrams write down at least two equivalent fractions that describe the shaded fraction.

a 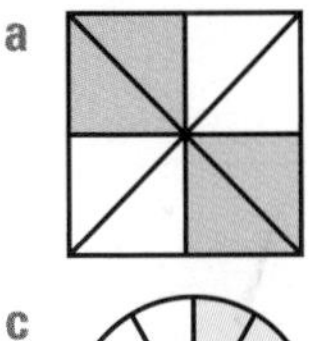

b 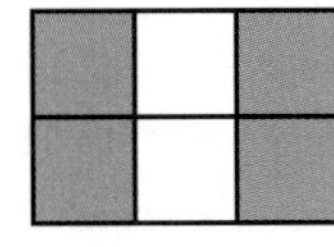

c 

d

2 Copy and complete each set of equivalent fractions.

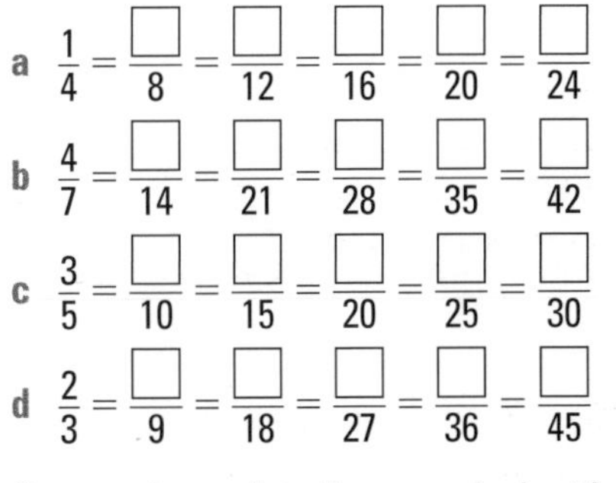

a $\frac{1}{4} = \frac{\square}{8} = \frac{\square}{12} = \frac{\square}{16} = \frac{\square}{20} = \frac{\square}{24}$

b $\frac{4}{7} = \frac{\square}{14} = \frac{\square}{21} = \frac{\square}{28} = \frac{\square}{35} = \frac{\square}{42}$

c $\frac{3}{5} = \frac{\square}{10} = \frac{\square}{15} = \frac{\square}{20} = \frac{\square}{25} = \frac{\square}{30}$

d $\frac{2}{3} = \frac{\square}{9} = \frac{\square}{18} = \frac{\square}{27} = \frac{\square}{36} = \frac{\square}{45}$

3 Copy and complete these equivalent fractions.

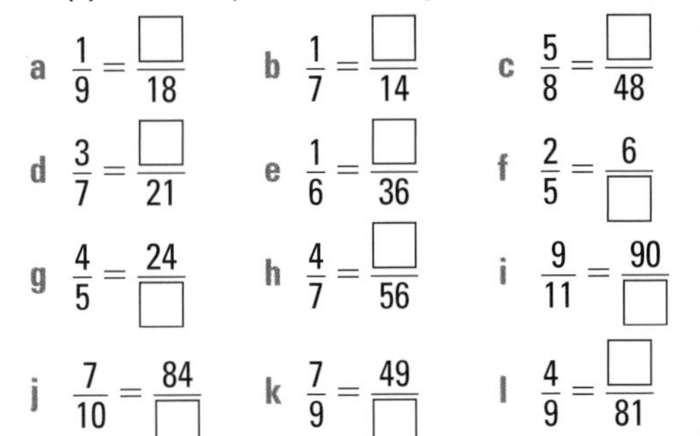

a $\frac{1}{9} = \frac{\square}{18}$ b $\frac{1}{7} = \frac{\square}{14}$ c $\frac{5}{8} = \frac{\square}{48}$

d $\frac{3}{7} = \frac{\square}{21}$ e $\frac{1}{6} = \frac{\square}{36}$ f $\frac{2}{5} = \frac{6}{\square}$

g $\frac{4}{5} = \frac{24}{\square}$ h $\frac{4}{7} = \frac{\square}{56}$ i $\frac{9}{11} = \frac{90}{\square}$

j $\frac{7}{10} = \frac{84}{\square}$ k $\frac{7}{9} = \frac{49}{\square}$ l $\frac{4}{9} = \frac{\square}{81}$

G 4 a Find a fraction equivalent to $\frac{1}{4}$ and a fraction equivalent to $\frac{1}{5}$ so that the bottom numbers of the two new fractions are the same.

b Repeat part **a** for

i $\frac{2}{3}$ and $\frac{3}{5}$ ii $\frac{1}{10}$ and $\frac{1}{8}$ iii $\frac{1}{2}$ and $\frac{2}{3}$

iv $\frac{1}{3}$ and $\frac{3}{4}$ v $\frac{2}{3}$ and $\frac{3}{8}$ vi $\frac{1}{4}$ and $\frac{3}{5}$

4.3 Ordering fractions

Exercise 4C

G 1 By writing equivalent fractions, find the smaller fraction in each pair.

a $\frac{3}{5}$ or $\frac{1}{4}$ b $\frac{2}{4}$ or $\frac{1}{5}$

c $\frac{2}{3}$ or $\frac{1}{2}$ d $\frac{1}{5}$ or $\frac{3}{10}$

2 Which is larger?

a $\frac{2}{5}$ or $\frac{3}{7}$ b $\frac{1}{10}$ or $\frac{1}{8}$

c $\frac{3}{4}$ or $\frac{5}{6}$ d $\frac{1}{2}$ or $\frac{2}{5}$

e $\frac{2}{3}$ or $\frac{3}{8}$ f $\frac{3}{4}$ or $\frac{4}{5}$

3 Write these fractions in order of size. Put the smallest one first.

a $\frac{2}{5}, \frac{3}{4}, \frac{1}{3}$ b $\frac{3}{5}, \frac{5}{6}, \frac{8}{15}$

c $\frac{3}{4}, \frac{3}{5}, \frac{1}{2}$ d $\frac{2}{7}, \frac{5}{14}, \frac{1}{2}, \frac{5}{7}$

4 Put these fractions in order of size, starting with the largest.

$\frac{3}{5}$ $\frac{1}{2}$ $\frac{5}{8}$ $\frac{3}{4}$ $\frac{7}{10}$

4.4 Improper fractions and mixed numbers

Exercise 4D

G 1 Convert these improper fractions to mixed numbers.

ResultsPlus
Examiner's Tip
Remember: a fraction is a division waiting to be done.

a $\frac{3}{2}$

b $\frac{5}{4}$ c $\frac{8}{7}$

d $\frac{13}{8}$ e $\frac{11}{8}$ f $\frac{13}{5}$ g $\frac{21}{10}$

h $\frac{23}{5}$ i $\frac{18}{7}$ j $\frac{14}{5}$ k $\frac{19}{3}$

l $\frac{17}{9}$ m $\frac{37}{4}$ n $\frac{29}{5}$ o $\frac{25}{9}$

p $\frac{19}{10}$

G **2** Convert these mixed numbers to improper fractions.

a $2\frac{1}{2}$ b $6\frac{1}{2}$ c $3\frac{3}{4}$ d $2\frac{2}{3}$
e $8\frac{1}{4}$ f $5\frac{2}{5}$ g $6\frac{7}{10}$ h $7\frac{1}{5}$
i $4\frac{3}{4}$ j $3\frac{1}{4}$ k $2\frac{9}{10}$ l $10\frac{1}{3}$
m $3\frac{5}{6}$ n $4\frac{3}{8}$ o $7\frac{5}{8}$ p $4\frac{9}{100}$

4.5 Multiplying fractions

Exercise 4E

G **1** Work out
a $\frac{1}{2}$ of £6 b $\frac{1}{5}$ of £35 c $\frac{1}{3}$ of £27
d $\frac{1}{6}$ of £48 e $\frac{1}{4}$ of 24 cm f $\frac{3}{4}$ of 24 cm
g $\frac{1}{10}$ of 320 kg h $\frac{3}{5}$ of 40 kg

2 A machine takes $5\frac{1}{2}$ minutes to produce a tennis racquet.
How long would the machine take to produce 12 tennis racquets at the same rate?

E **3** Work out
a $\frac{1}{2}\times\frac{3}{5}$ b $\frac{1}{8}\times\frac{3}{4}$ c $\frac{3}{5}\times\frac{4}{5}$
d $\frac{5}{8}\times\frac{3}{4}$ e $\frac{7}{12}\times\frac{1}{3}$ f $\frac{9}{10}\times\frac{3}{4}$
g $\frac{3}{10}\times\frac{1}{5}$ h $\frac{2}{3}\times\frac{1}{3}$ i $\frac{1}{2}\times\frac{5}{8}$
j $\frac{4}{5}\times\frac{1}{3}$ k $\frac{3}{7}\times\frac{2}{3}$ l $\frac{2}{3}\times\frac{4}{5}$
m $\frac{3}{7}\times\frac{1}{5}$ n $\frac{2}{3}\times\frac{4}{7}$ o $\frac{3}{2}\times\frac{1}{4}$
p $\frac{3}{5}\times\frac{2}{3}$

4 Work out
a $\frac{1}{3}\times\frac{3}{5}$ b $\frac{1}{4}\times\frac{4}{7}$ c $\frac{5}{6}\times\frac{2}{5}$
d $\frac{3}{5}\times\frac{2}{10}$ e $\frac{5}{12}\times\frac{3}{4}$ f $\frac{7}{15}\times\frac{3}{14}$
g $\frac{6}{9}\times\frac{3}{10}$ h $\frac{3}{4}\times\frac{12}{15}$ i $\frac{1}{9}\times\frac{6}{7}$
j $\frac{6}{7}\times\frac{5}{18}$ k $\frac{1}{2}\times\frac{6}{10}$ l $\frac{2}{3}\times\frac{1}{8}$
m $\frac{3}{7}\times\frac{2}{9}$ n $\frac{9}{5}\times\frac{1}{3}$ o $5\times\frac{7}{20}$
p $\frac{6}{10}\times\frac{13}{18}$

5 Work out
a $\frac{1}{2}\times 5$ b $\frac{2}{3}\times 7$ c $8\times\frac{4}{5}$
d $6\times\frac{3}{4}$ e $\frac{7}{10}\times 30$ f $6\times\frac{2}{3}$
g $12\times\frac{2}{5}$ h $\frac{5}{6}\times 10$ i $\frac{2}{5}\times 7$

D **6** Work out
a $\frac{2}{3}\times 2\frac{1}{5}$ b $2\frac{1}{2}\times\frac{1}{8}$ c $1\frac{1}{4}\times 2\frac{1}{4}$
d $3\frac{1}{4}\times\frac{1}{5}$ e $\frac{2}{5}\times 4\frac{1}{4}$ f $\frac{5}{6}\times 1\frac{1}{4}$

C **7** Work out
a $2\frac{1}{3}\times 2\frac{6}{10}$ b $3\frac{1}{2}\times 4\frac{1}{2}$ c $2\frac{2}{3}\times 2\frac{5}{8}$
d $1\frac{4}{5}\times 3\frac{1}{3}$ e $3\frac{3}{4}\times 2\frac{2}{5}$ f $2\frac{1}{2}\times\frac{1}{8}$
g $1\frac{2}{5}\times 1\frac{1}{2}$ h $9\times 2\frac{2}{3}$ i $2\frac{1}{7}\times 2\frac{2}{5}$

4.6 Dividing fractions

Exercise 4F

E **1** Work out
a $\frac{1}{4}\div\frac{1}{3}$
b $\frac{1}{3}\div\frac{1}{4}$
c $\frac{3}{4}\div\frac{1}{8}$
d $\frac{1}{2}\div\frac{7}{12}$
e $\frac{2}{3}\div\frac{1}{4}$
f $\frac{5}{6}\div\frac{1}{3}$
g $\frac{5}{6}\div\frac{1}{4}$ h $\frac{7}{10}\div\frac{4}{20}$ i $\frac{2}{4}\div\frac{1}{2}$
j $\frac{3}{5}\div\frac{3}{4}$ k $\frac{5}{8}\div\frac{2}{3}$ l $\frac{1}{4}\div\frac{1}{2}$

> **ResultsPlus**
> **Watch Out!**
> Sometimes students turn the first fraction upside down by mistake. Make sure you turn the second fraction upside down.

2 Work out
a $8\div\frac{1}{4}$ b $12\div\frac{3}{5}$ c $6\div\frac{3}{7}$
d $6\div\frac{7}{8}$ e $4\div\frac{4}{7}$ f $3\div\frac{7}{12}$
g $4\div\frac{1}{3}$ h $5\div\frac{1}{4}$

3 Work out
a $\frac{3}{4}\div 2$ b $\frac{5}{6}\div 3$ c $\frac{3}{5}\div 3$
d $\frac{4}{5}\div 6$ e $\frac{2}{3}\div 8$ f $\frac{8}{9}\div 2$

D
g $2\frac{5}{6}\div 8$ h $2\frac{1}{2}\div 10$ i $3\frac{1}{4}\div 5$
j $1\frac{1}{3}\div 8$ k $4\frac{2}{3}\div 2$ l $5\frac{1}{4}\div 7$

C **4** Work out
a $2\frac{1}{2}\div 3\frac{1}{4}$ b $2\frac{1}{4}\div 3\frac{1}{2}$ c $2\frac{3}{4}\div 3\frac{1}{4}$
d $3\frac{5}{8}\div 1\frac{1}{6}$ e $7\frac{2}{3}\div 3\frac{1}{3}$ f $4\frac{1}{2}\div 2\frac{3}{4}$
g $2\frac{7}{10}\div 1\frac{7}{10}$ h $4\frac{7}{12}\div 1\frac{2}{3}$

4.7 Adding and subtracting fractions

Exercise 4G

E

1 Work out

a $\frac{1}{8}+\frac{5}{8}$
b $\frac{4}{9}+\frac{1}{9}$
c $\frac{1}{12}+\frac{11}{12}$
d $\frac{7}{18}+\frac{9}{18}$ e $\frac{1}{8}+\frac{5}{8}$ f $\frac{5}{7}+\frac{1}{7}$
g $\frac{1}{5}+\frac{4}{5}$ h $\frac{3}{10}+\frac{9}{10}$ i $\frac{7}{9}+1\frac{4}{9}$
j $\frac{5}{6}+2\frac{5}{6}$ k $\frac{3}{4}+\frac{1}{4}+\frac{1}{4}$ l $\frac{5}{8}+\frac{7}{8}+\frac{1}{8}$

ResultsPlus
Watch Out!
Make sure you don't add the denominators.

D

2 Work out

a $\frac{1}{2}+\frac{3}{4}$ b $\frac{1}{4}+\frac{5}{8}$ c $\frac{1}{2}+\frac{5}{8}$ d $\frac{1}{3}+\frac{1}{6}$
e $\frac{5}{6}+\frac{2}{3}$ f $\frac{1}{5}+\frac{3}{10}$ g $\frac{7}{12}+\frac{1}{4}$ h $\frac{3}{4}+\frac{9}{20}$

3 Work out

a $\frac{1}{2}+\frac{5}{8}$ b $\frac{3}{4}+\frac{3}{10}$ c $\frac{4}{9}+\frac{7}{12}$ d $\frac{5}{8}+\frac{9}{10}$
e $\frac{3}{10}+\frac{4}{15}$ f $\frac{5}{6}+\frac{1}{4}$ g $\frac{5}{8}+\frac{7}{12}$ h $\frac{1}{6}+\frac{5}{9}$
i $\frac{5}{8}+\frac{3}{4}$ j $\frac{1}{6}+\frac{3}{8}$ k $\frac{3}{8}+\frac{9}{16}$

4 Work out

a $\frac{1}{2}+\frac{2}{3}$ b $\frac{1}{5}+\frac{1}{6}$ c $\frac{3}{8}+\frac{1}{5}$ d $\frac{3}{4}+\frac{2}{9}$
e $\frac{5}{6}+\frac{1}{7}$ f $\frac{7}{10}+\frac{2}{7}$ g $\frac{2}{3}+\frac{3}{10}$ h $\frac{3}{5}+\frac{3}{7}$
i $\frac{2}{5}+\frac{3}{8}$ j $\frac{2}{5}+\frac{1}{6}$ k $\frac{2}{3}+\frac{3}{7}$

C

5 Work out

a $3\frac{1}{2}+2\frac{1}{8}$ b $1\frac{3}{4}+3\frac{7}{8}$ c $4\frac{3}{4}+2\frac{5}{16}$
d $\frac{3}{4}+2\frac{5}{8}$ e $3\frac{9}{16}+1\frac{5}{8}$ f $4\frac{3}{10}+1\frac{2}{3}$
g $2\frac{1}{6}+\frac{2}{7}$ h $2\frac{5}{6}+3\frac{1}{7}$ i $1\frac{2}{5}+2\frac{7}{15}$
j $3\frac{2}{3}+1\frac{2}{9}$

6 Jo cycled $2\frac{1}{4}$ miles to one village and then a further $4\frac{2}{3}$ miles to her home.
What was the total distance Jo travelled?

7 Work out

a $3\frac{1}{4}+1\frac{1}{2}$ b $3\frac{1}{2}+\frac{2}{3}$ c $4\frac{1}{4}+2\frac{7}{8}$
d $2\frac{1}{3}+5\frac{3}{4}$ e $2\frac{5}{16}+1\frac{7}{8}$ f $4\frac{11}{12}+\frac{3}{4}$
g $\frac{5}{6}+5\frac{1}{3}$ h $1\frac{2}{3}+4\frac{3}{5}$

Exercise 4H

E

1 Work out

a $\frac{5}{11}-\frac{1}{11}$ b $\frac{7}{9}-\frac{3}{9}$ c $\frac{5}{8}-\frac{1}{8}$ d $\frac{7}{12}-\frac{1}{12}$
e $\frac{2}{3}-\frac{1}{3}$ f $\frac{7}{8}-\frac{1}{8}$ g $\frac{13}{16}-\frac{7}{16}$ h $\frac{6}{7}-\frac{4}{7}$

2 $\frac{3}{8}$ of the students at Weald College wear contact lenses. What fraction of the students do not wear them?

D

3 Work out

a $\frac{3}{4}-\frac{1}{2}$ b $\frac{7}{8}-\frac{1}{4}$ c $\frac{7}{8}-\frac{1}{2}$ d $\frac{3}{4}-\frac{3}{8}$
e $\frac{5}{6}-\frac{2}{3}$ f $\frac{9}{12}-\frac{1}{3}$ g $\frac{9}{10}-\frac{1}{5}$ h $\frac{1}{4}-\frac{3}{20}$
i $\frac{1}{2}-\frac{1}{8}$ j $\frac{5}{8}-\frac{1}{2}$ k $\frac{11}{12}-\frac{1}{4}$

4 Work out

a $\frac{2}{3}-\frac{1}{5}$ b $\frac{5}{8}-\frac{1}{5}$ c $\frac{1}{3}-\frac{1}{6}$ d $\frac{2}{5}-\frac{1}{6}$
e $\frac{4}{5}-\frac{1}{3}$ f $\frac{3}{5}-\frac{3}{6}$ g $\frac{9}{10}-\frac{1}{3}$ h $\frac{7}{10}-\frac{1}{4}$
i $3\frac{1}{4}-\frac{1}{10}$ j $4\frac{1}{2}-\frac{1}{3}$

5 In a school, $\frac{9}{16}$ of the students are girls.
What fraction of students are boys?

C

6 Work out

a $4\frac{5}{8}-1\frac{1}{4}$ b $6\frac{1}{2}-3\frac{1}{4}$ c $9\frac{1}{2}-5\frac{3}{10}$
d $4-2\frac{3}{10}$ e $4\frac{4}{5}-1\frac{9}{10}$ f $2\frac{2}{3}-\frac{11}{12}$
g $5\frac{3}{4}-3\frac{19}{20}$ h $4\frac{7}{8}-2\frac{2}{3}$ i $5\frac{7}{9}-2\frac{1}{3}$
j $5\frac{4}{5}-\frac{3}{8}$ k $7\frac{4}{7}-1\frac{2}{5}$

4.8 Converting between fractions and decimals

Exercise 4I

G

1 Convert these fractions into decimals. Show your working.

a $\frac{2}{5}$ b $\frac{3}{4}$ c $\frac{9}{10}$ d $\frac{9}{20}$ e $\frac{3}{25}$
f $\frac{7}{50}$ g $\frac{5}{8}$ h $\frac{7}{20}$ i $\frac{11}{25}$ j $\frac{7}{16}$
k $\frac{3}{8}$ l $\frac{23}{50}$ m $\frac{7}{100}$ n $\frac{11}{200}$ o $\frac{1}{3}$
p $\frac{17}{20}$

2 Convert these decimals into fractions.

a 0.5 b 0.57 c 0.95 d 0.157
e 0.295 f 0.9 g 0.79 h 0.005
i 0.00005 j 0.0073 k 0.73 l 0.073
m 0.59 n 0.0049 o 0.029 p 0.051

G **3** Write these fractions as decimals.

a $\frac{3}{5}$ b $\frac{1}{4}$ c $1\frac{3}{8}$ d $\frac{29}{100}$

e $2\frac{3}{5}$ f $\frac{11}{25}$ g $\frac{3}{8}$ h $2\frac{17}{40}$

i $\frac{9}{50}$ j $5\frac{3}{16}$ k $8\frac{3}{20}$ l $3\frac{5}{16}$

m $\frac{9}{1000}$ n $3\frac{7}{25}$ o $13\frac{15}{16}$ p $4\frac{7}{20}$

F **4** Write these decimals as fractions in their simplest form.

a 0.84 b 0.75 c 1.9 d 2.406

e 6.003 f 2.075 g 0.046 h 5.875

i 3.25 j 20.202 k 0.635 l 4.512

m 0.8175 n 14.14 o 19.1875 p 50.065

5 Percentages

Key Points

- **per cent:** out of 100.
- **percentage (%):** a quantity out of 100. Can also be written as a decimal or a fraction.
- **knowing basic percentage equivalents:** know these percentages:

Percentage	1%	10%	25%	50%	75%
Decimal	0.01	0.1	0.25	0.5	0.75
Fraction	$\frac{1}{100}$	$\frac{1}{10}$	$\frac{1}{4}$	$\frac{1}{2}$	$\frac{3}{4}$

- **finding percentages of quantities using fractions:** write the percentage as a fraction, and then multiply the fraction by the quantity.
- **finding percentages of quantities using decimals:** write the percentage as a decimal, and then multiply the decimal by the quantity.
- **increasing a quantity by a percentage:** work out the increase and add it to the original quantity.
- **decreasing a quantity by a percentage:** work out the decrease and subtract it from the original quantity.
- **using the multiplier method:** work out the multiplier for an increase or decrease. Then multiply the original amount by the multiplier to find the new amount.

5.1 Converting between percentages, fractions and decimals and ordering them

Exercise 5A

Questions in this chapter are targeted at the grades indicated.

G **1** What percentage of each shape is shaded?

a b c d e f

2 For each shape in question 1 write down the percentage of the shape that is not shaded.

3 a Copy this shape and shade 40% of it.

b Copy this shape and shade 80% of it.

G **4** 65% of the houses in a street have a driveway.
What percentage of the houses do not have a driveway?

5 Emily has some flowers. 55% of the flowers are pink.
What percentage of the flowers are not pink?

6 76% of teachers in a school are female.
What percentage of the teachers are male?

7 On a wall the tiles are white or brown or blue. 15% of the tiles are brown. 30% of the tiles are blue.
What percentage of the tiles are white?

Exercise 5B

G **1** Write these percentages as decimals.

a 25% b 35% c 64% d 85%
e 39% f 40% g 9% h 4%

2 Write 225% as a decimal.

F **3** A shop reduced its prices by 14.5%.
Write 14.5% as a decimal.

4 A savings account has an interest rate of 3.4%.
Write 3.4% as a decimal.

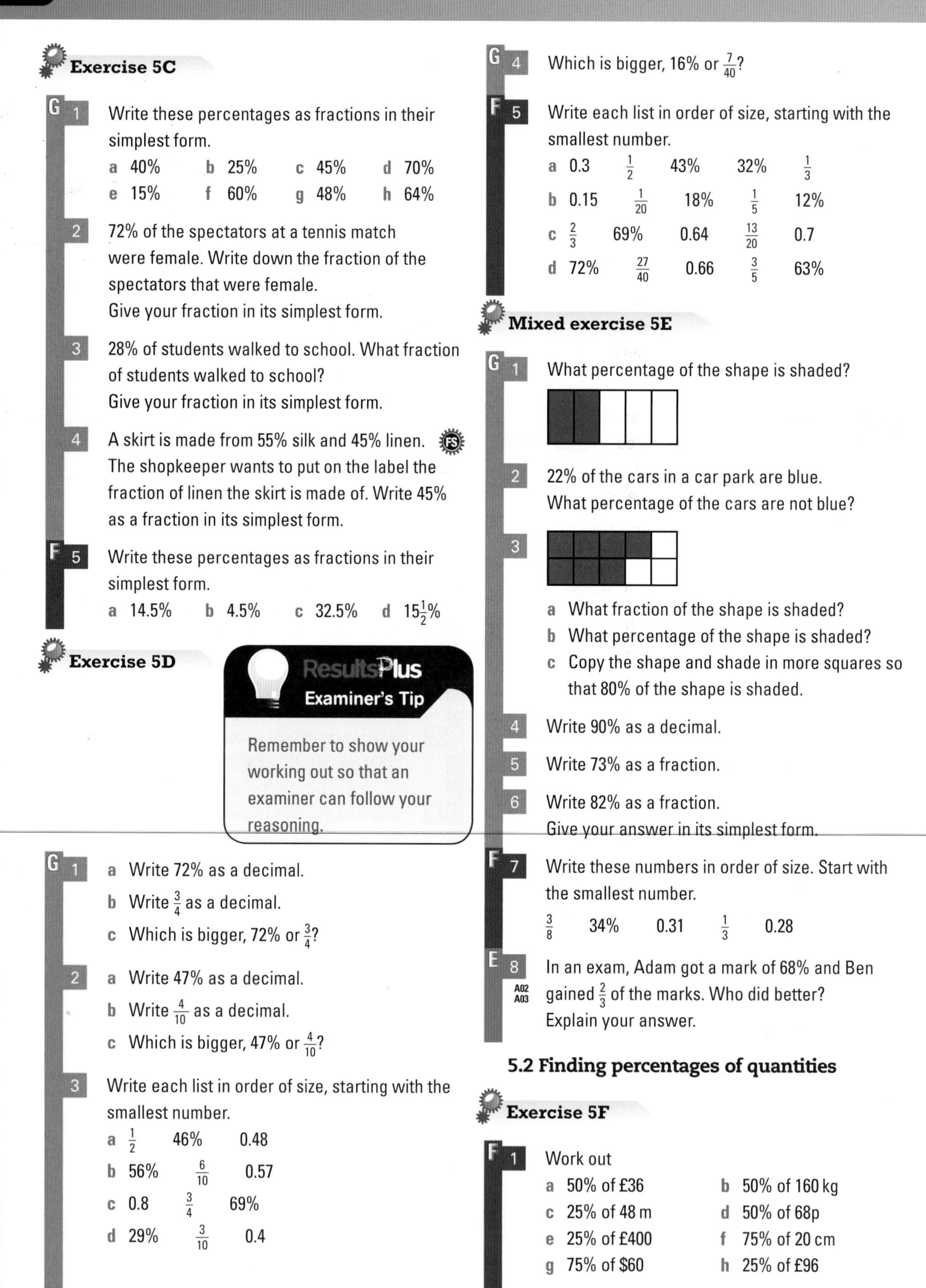

Exercise 5C

G

1. Write these percentages as fractions in their simplest form.
   a 40% b 25% c 45% d 70%
   e 15% f 60% g 48% h 64%

2. 72% of the spectators at a tennis match were female. Write down the fraction of the spectators that were female.
   Give your fraction in its simplest form.

3. 28% of students walked to school. What fraction of students walked to school?
   Give your fraction in its simplest form.

4. A skirt is made from 55% silk and 45% linen. The shopkeeper wants to put on the label the fraction of linen the skirt is made of. Write 45% as a fraction in its simplest form. FS

F

5. Write these percentages as fractions in their simplest form.
   a 14.5% b 4.5% c 32.5% d $15\frac{1}{2}$%

Exercise 5D

ResultsPlus Examiner's Tip

Remember to show your working out so that an examiner can follow your reasoning.

G

1. a Write 72% as a decimal.
   b Write $\frac{3}{4}$ as a decimal.
   c Which is bigger, 72% or $\frac{3}{4}$?

2. a Write 47% as a decimal.
   b Write $\frac{4}{10}$ as a decimal.
   c Which is bigger, 47% or $\frac{4}{10}$?

3. Write each list in order of size, starting with the smallest number.
   a $\frac{1}{2}$ 46% 0.48
   b 56% $\frac{6}{10}$ 0.57
   c 0.8 $\frac{3}{4}$ 69%
   d 29% $\frac{3}{10}$ 0.4

4. Which is bigger, 16% or $\frac{7}{40}$?

F

5. Write each list in order of size, starting with the smallest number.
   a 0.3 $\frac{1}{2}$ 43% 32% $\frac{1}{3}$
   b 0.15 $\frac{1}{20}$ 18% $\frac{1}{5}$ 12%
   c $\frac{2}{3}$ 69% 0.64 $\frac{13}{20}$ 0.7
   d 72% $\frac{27}{40}$ 0.66 $\frac{3}{5}$ 63%

Mixed exercise 5E

G

1. What percentage of the shape is shaded?

2. 22% of the cars in a car park are blue.
   What percentage of the cars are not blue?

3. a What fraction of the shape is shaded?
   b What percentage of the shape is shaded?
   c Copy the shape and shade in more squares so that 80% of the shape is shaded.

4. Write 90% as a decimal.

5. Write 73% as a fraction.

6. Write 82% as a fraction.
   Give your answer in its simplest form.

F

7. Write these numbers in order of size. Start with the smallest number.
   $\frac{3}{8}$ 34% 0.31 $\frac{1}{3}$ 0.28

E

8. In an exam, Adam got a mark of 68% and Ben gained $\frac{2}{3}$ of the marks. Who did better? Explain your answer. A02 A03

5.2 Finding percentages of quantities

Exercise 5F

F

1. Work out
   a 50% of £36 b 50% of 160 kg
   c 25% of 48 m d 50% of 68p
   e 25% of £400 f 75% of 20 cm
   g 75% of $60 h 25% of £96

F 2 Work out

a 10% of £70
b 10% of 90 km
c 20% of 80 km
d 30% of £180
e 20% of £75
f 15% of 30 kg
g 70% of 200 ml
h 35% of £60

E 3 Ajhar's salary last year was £37 400. FS
He saved 10% of his salary. Ajhar wants to buy a car costing £3750. Has he saved enough?

4 A packet of breakfast cereal contains 500 g of cereal plus '20% extra free'.
Work out how much extra cereal the packet contains.

D 5 A03 Jake earned £32 500 last year. FS
He was allowed £6400 tax free. He paid income tax of 20% on the remainder.
How much income tax did he pay?

6 Dina earns £7.20 per hour. She is given a pay rise of 5%. FS
Work out how much extra she gets per hour.

7 The price of a new dining table is £520. FS
Sophie pays a deposit of 15% of the price.
Work out the deposit she pays.

C 8 The normal cost of a suit is £120. FS
In a sale the cost of the suit is reduced by 25%.
Work out how much the cost of the suit is reduced by in the sale.

9 A02 A03 Gemma's income last year was £58 200. FS
After her tax-free allowance of £6400, she paid 20% income tax on the remaining amount.
Work out the income tax she paid last year.

Exercise 5G

D 1 Work out

a 14% of £40
b 68% of 45 kg
c 45% of £370
d 73% of 640 km
e 84% of 330 ml
f 32% of $90
g 6% of £170
h 29% of 1500 m

5.3 Finding the new amount after a percentage increase or decrease

Exercise 5G

D 1 Write down the single number you can multiply by to work out an increase of:

a 30% b 25% c 70% d 4%

2 a Increase 400 metres by 45%.
b Increase 20 litres by 60%.
c Increase £3000 by 15%.
d Increase £150 by 100%.

3 Write down the single number you can multiply by to work out a decrease of:

a 30% b 25% c 70% d 4%

4 a Decrease 90 kilometres by 40%.
b Decrease 2500 grams by 95%.
c Decrease £310 by 10%.
d Decrease 30 centimetres by 25%.

6 Ratio and proportion

Key Points

- **ratio:** a comparison of a part to a part, written in the form $a:b$.
- **unitary form:** a ratio written in form $1:n$.
- **equivalent ratios:** equal ratios like 15 : 20 and 3 : 4.
- **map scale:** a ratio in the form $1:n$ that tells us how the distance on a map relates to the real distance.
- **direct proportion:** when two quantities increase and decrease in the same ratio.
- **simplifying ratios:** divide all the quantities in the ratio by common factors (like simplifying fractions).
- **sharing in a given ratio:** work out the total number and divide it into the given ratio.
- **using the unitary method for problems involving direct proportion:** find the value of one item first, then multiply it by the value you want to find.

6.1 Introducing ratio

Exercise 6A

Questions in this chapter are targeted at the grades indicated.

E

1 For each pattern of tiles, write down the ratio of the number of white tiles to the number of blue tiles.

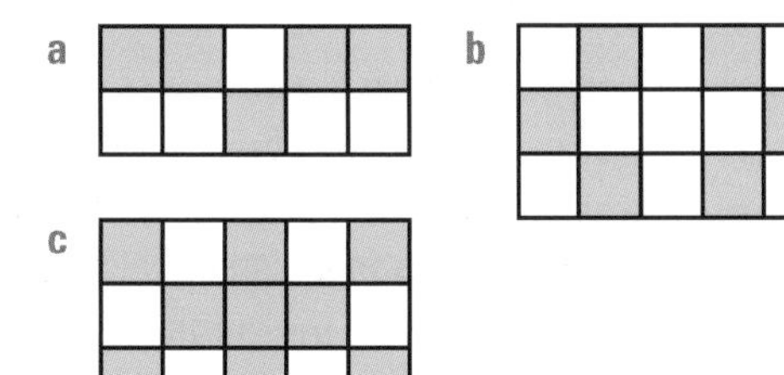

2 Craig is 15 years old. Saskia is 13 years old.

a What is the ratio of Craig's age to Saskia's age?

b What is the ratio of Saskia's age to Craig's age?

3 △ △ ○ ○ ○ □ □ □ □ □ □ □

a What is the ratio of the number of triangles to the number of circles?

b What is the ratio of the number of triangles to the number of squares?

c What is the ratio of the number of squares to the number of circles?

d What is the ratio of the number of triangles to the number of circles to the number of squares?

Exercise 6B

E

1 In an office, the ratio of the number of men to the number of women is 1 : 2.
What fraction of these people are women?

2 A box contains plain chocolates and milk chocolates in the ratio 3 : 2.

a What fraction of these chocolates are plain chocolates?

b What fraction of these chocolates are milk chocolates?

3 The ratio of children to adults visiting a theme park is 3 : 5.

a What fraction of the visitors are children?

b What fraction of the visitors are adults?

D

4 A box contains blue pens, red pens and black pens in the ratio 5 : 1 : 3.

a What fraction of these pens are red?

b What fraction of these pens are blue?

5 In a wildlife park, $\frac{2}{3}$ of the monkeys are female.
What is the ratio of female monkeys to male monkeys in this wildlife park?

6 In the wildlife park, $\frac{1}{2}$ of the meerkats are female.
What is the ratio of female meerkats to male meerkats in this wildlife park?

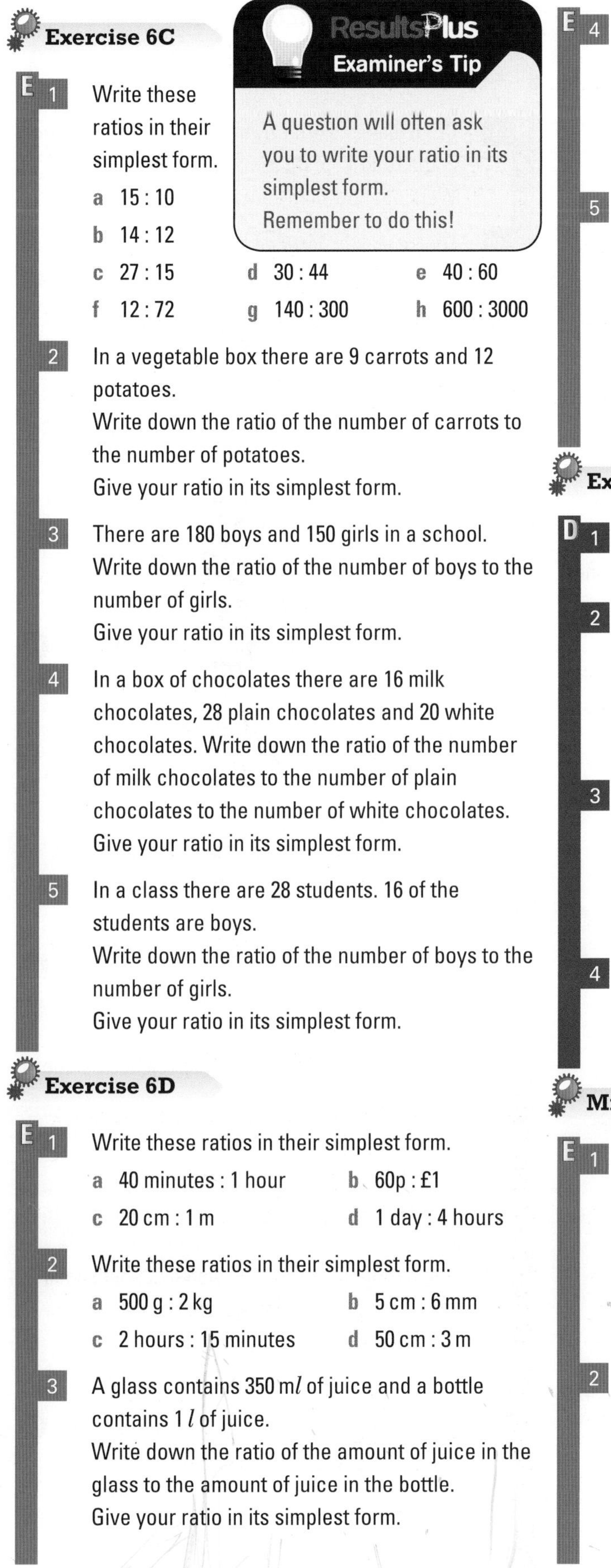

Exercise 6C

ResultsPlus
Examiner's Tip
A question will often ask you to write your ratio in its simplest form.
Remember to do this!

E

1 Write these ratios in their simplest form.

a 15 : 10
b 14 : 12
c 27 : 15
d 30 : 44
e 40 : 60
f 12 : 72
g 140 : 300
h 600 : 3000

2 In a vegetable box there are 9 carrots and 12 potatoes.
Write down the ratio of the number of carrots to the number of potatoes.
Give your ratio in its simplest form.

3 There are 180 boys and 150 girls in a school.
Write down the ratio of the number of boys to the number of girls.
Give your ratio in its simplest form.

4 In a box of chocolates there are 16 milk chocolates, 28 plain chocolates and 20 white chocolates. Write down the ratio of the number of milk chocolates to the number of plain chocolates to the number of white chocolates.
Give your ratio in its simplest form.

5 In a class there are 28 students. 16 of the students are boys.
Write down the ratio of the number of boys to the number of girls.
Give your ratio in its simplest form.

Exercise 6D

E

1 Write these ratios in their simplest form.

a 40 minutes : 1 hour
b 60p : £1
c 20 cm : 1 m
d 1 day : 4 hours

2 Write these ratios in their simplest form.

a 500 g : 2 kg
b 5 cm : 6 mm
c 2 hours : 15 minutes
d 50 cm : 3 m

3 A glass contains 350 ml of juice and a bottle contains 1 l of juice.
Write down the ratio of the amount of juice in the glass to the amount of juice in the bottle.
Give your ratio in its simplest form.

E

4 A small bag of flour weighs 325 g and a large bag of flour weighs 2 kg.
Write down the ratio of the weight of the small bag to the weight of the large bag.
Give your ratio in its simplest form.

5 A kitchen worktop has a length of 1.4 m and a width of 50 cm.

a Find, in its simplest form, the ratio of the length of the worktop to the width of the worktop.
b Find, in its simplest form, the ratio of the width of the worktop to the length of the worktop.

Exercise 6E

D

1 Write these ratios in the form 1 : n.

a 2 : 7
b 5 : 14
c 13 : 3
d 200 : 350

2 The length of a model car is 24 cm. The length of the real car is 360 cm.
Write down the ratio of the length of the model car to the length of the real car.
Give your answer in the form 1 : n.

3 In a school there are 120 teachers and 1640 students. The headteacher wants to display on his website the ratio of the number of teachers to the number of students in the form 1 : n.
Work out this ratio.

4 Write these ratios in the form 1 : n.

a 25p : £1
b 150 g : 1 kg
c 4 cm : 5 mm
d 20 minutes : 3 hours

Mixed exercise 6F

E

1 Write, in its simplest form, the ratio of the number of white squares to the number of blue squares.

2 A bracelet has brown beads and pink beads only.
The ratio of the number of brown beads to the number of pink beads is 7 : 2.

a What fraction of the beads are brown?
b What fraction of the beads are pink?

E

3

Ingredient	Weight in grams
Protein	8
Carbohydrate	20
Fat	4

Write, in its simplest form, the ratio of:

a the weight of carbohydrate to the weight of protein

b the weight of fat to the weight of carbohydrate

c the weight of protein to the weight of carbohydrate to the weight of fat.

4 Write these ratios in their simplest form.

a 2 hours : 20 min b 600 m : 2 km

5 A recipe uses 100 g of margarine and 1 kg of fruit. Write down the ratio of the weight of margarine to the weight of fruit.
Give your ratio in its simplest form.

D

6 Write the ratio 24 : 60

a in its simplest form b in the form 1 : n.

7 On a coach there are 12 adults and 33 children. Write the ratio of the number of adults to the number of children in the form 1 : n.

8 A03 A school is organising a museum trip. FS
The museum website recommends the following adult to pupil ratios.
Under 8s (Years 1–3): minimum ratio 1 : 6
Under 11s (Years 4–6): minimum ratio 1 : 10
Over 11 (Year 7 onwards): minimum ratio 1 : 15

The pupils going on the visit are shown in this table.

Year	Number of pupils
3	8
4	12
5	14
6	13

Work out the minimum number of adults required for the visit.

6.2 Solving ratio problems

Exercise 6G

D

1 Craig makes mortar by mixing cement and sand in the ratio 1 : 5. He uses 6 buckets of cement. Work out how many buckets of sand he uses.

2 Louise makes porridge by mixing oats and water in the ratio 1 : 2. Work out the number of cups of water she uses for:

a 4 cups of oats b 5 cups of oats
c 8 cups of oats.

3 An alloy contains iron and aluminium in the ratio 4 to 1 by weight.

a If there is 28 kg of iron, work out the weight of aluminium.

b If there is 12 kg of aluminium, work out the weight of iron.

4 In a recipe for making pastry the ratio of the weight of flour to the weight of margarine is 3 : 2. FS

a Work out the weight of margarine needed for:
i 80 g of flour ii 200 g of flour
iii 550 g of flour.

b Work out the weight of flour needed for:
i 40 g of margarine ii 150 g of margarine
iii 200 g of margarine.

5 Kimberley is making a fruit drink. FS
She mixes apple juice, mango juice and syrup in the ratio 3 : 4 : 1.

a If she uses 600 ml of apple juice, how much syrup will she need?

b If she uses 2 l of mango juice, how much apple juice will she need?

6 At an athletics meeting, the ratio of the number of bronze medals won by Great Britain to the number of bronze medals won by Canada was 5 : 4. Great Britain won 2 more bronze medals than Canada.
How many bronze medals did Great Britain win?

7 A03 Sue makes purple paint by mixing 3 parts of red paint with 5 parts of blue paint. She has 750 ml of red paint and 1.5 litres of blue paint. FS
What is the maximum amount of purple paint that Sue can make?

Exercise 6H

D

1 Simon uses a scale of 1 : 40 to draw a plan of his lounge. On the plan the length of the lounge is 8 cm. Work out the real length of the lounge.

2 Chloe makes a scale model of a house. She uses a scale of 1 : 15.
The height of the model house is 60 cm.
Work out the height of the real house.

3 A model of a building is made using a scale of 1 : 500. The length of the model building is 40 cm. Work out the length of the real building.

4 The length of a car is 3 metres.
Jay makes a model of the car. He uses a scale of 1 : 15. Work out the length, in centimetres, of the model car.

5 A company makes model cars using a scale of 1 : 16.
a Work out the length of the real car if the length of a model car is 20 cm.
b Work out the length of a model car if the length of the real car is 4.68 m.

6 The scale of a map is 1 : 200 000.
On the map the distance between two towns is 6 cm. Work out the real distance between the two towns.

7 The scale of a map is 1 : 40 000.
On the map the length of a road is 3.5 cm.
Work out the real length of the road.

8 The scale of a map is 1 : 100 000.
The real distance between two towns is 24 km.
Work out the distance between the towns on the map.

6.3 Sharing in a given ratio

Exercise 6I

D

1 Share £70 in the ratio 1 : 4.

2 Share £32 in the ratio 3 : 5.

3 Share £35 in the ratio 2 : 3.

Some students divide the total amount by the numbers in the ratio. Make sure you work out the number of shares first.

D

4 Alex and Ben share 60 sweets in the ratio 3 : 1.
Work out how many sweets each receives.

5 At school the technician is going to make some brass. Brass is made from copper and nickel in the ratio 17 : 3.
Work out how much copper and how much nickel he will need to make 600 g of brass. (FS)

6 The ratio of boys to girls in a group is 4 : 5.
There are 36 students in the group.
Work out the number of girls in the group.

C

7 Share £48 in the ratio 1 : 3 : 4.

8 Emily is making shortbread. She uses flour, sugar and butter in the ratio 3 : 1 : 2.
Work out how much of each ingredient she needs to make 960 g of shortbread. (FS)

9 Three boys shared £40 in the ratio 5 : 3 : 2.
Jake received the smallest amount.
Work out how much Jake received.

10 Raul made some compost. He mixed soil, manure and leaf mould in the ratio 3 : 2 : 1.
Raul made 180 litres of compost.
Work out how much manure he used.

Mixed exercise 6J

D

1 A model of a workshop is made using a scale of 1 : 64. The length of the model workshop is 20 cm. Work out the length, in metres, of the real workshop.

2 Guy is making cakes. In a recipe the ratio of the weight of margarine to the weight of caster sugar is 3 : 1.
Work out the weight of caster sugar Guy needs if the recipe uses 280 g of margarine. (FS)

3 The ratio of Rob's height to Alex's height is 8 : 7.
Rob's height is 160 cm. What is Alex's height?

4 In a school orchestra the ratio of the number of boys to the number of girls is 2 : 5. There is a total of 42 boys and girls in the orchestra.
Work out the number of girls in the orchestra.

5 The scale of a map is 1 : 300 000.
On the map the distance between two towns is 5.5 cm. Work out the real distance, in kilometres, between the towns.

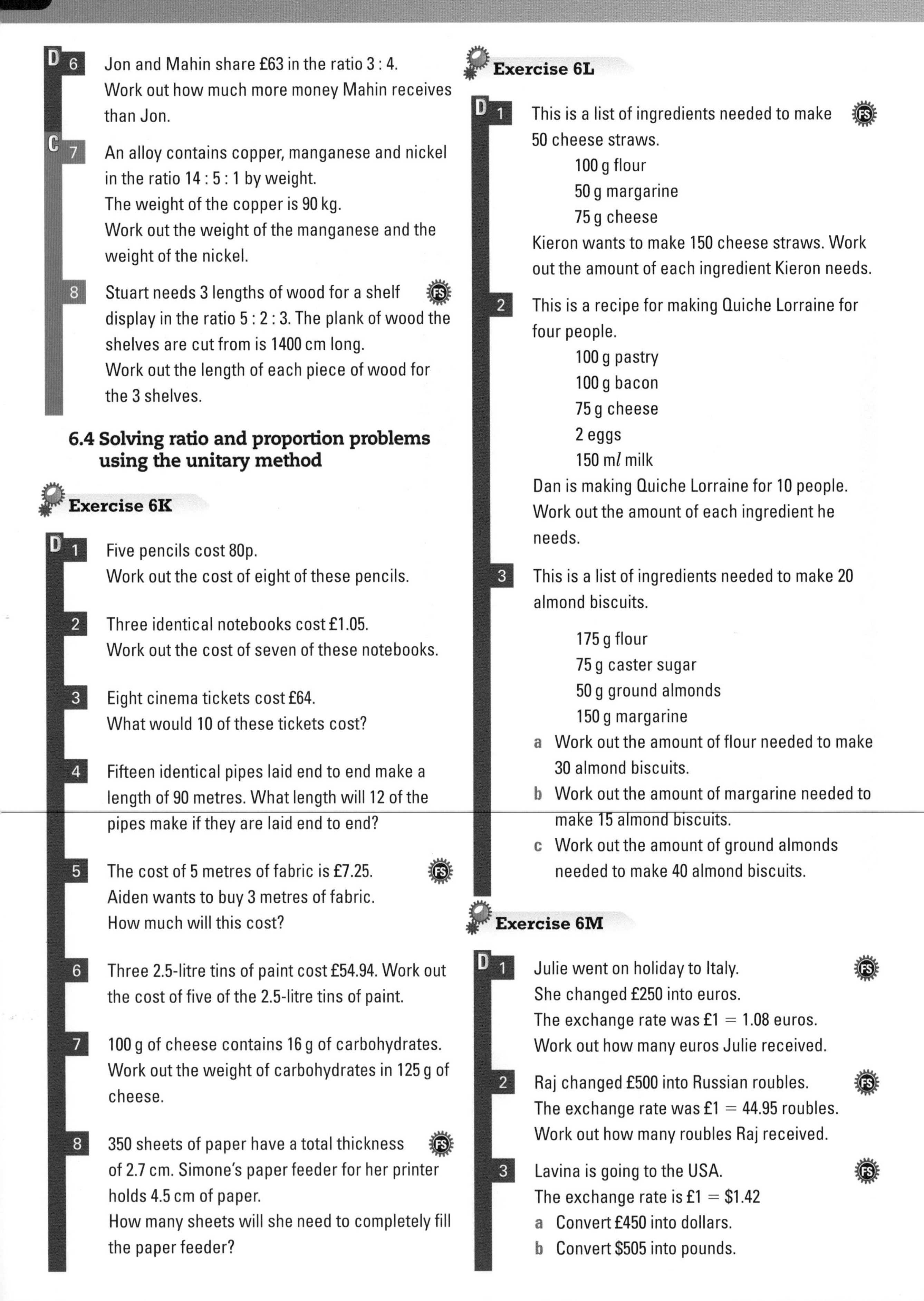

D 6 Jon and Mahin share £63 in the ratio 3 : 4.
Work out how much more money Mahin receives than Jon.

C 7 An alloy contains copper, manganese and nickel in the ratio 14 : 5 : 1 by weight.
The weight of the copper is 90 kg.
Work out the weight of the manganese and the weight of the nickel.

8 Stuart needs 3 lengths of wood for a shelf display in the ratio 5 : 2 : 3. The plank of wood the shelves are cut from is 1400 cm long.
Work out the length of each piece of wood for the 3 shelves.

6.4 Solving ratio and proportion problems using the unitary method

Exercise 6K

D 1 Five pencils cost 80p.
Work out the cost of eight of these pencils.

2 Three identical notebooks cost £1.05.
Work out the cost of seven of these notebooks.

3 Eight cinema tickets cost £64.
What would 10 of these tickets cost?

4 Fifteen identical pipes laid end to end make a length of 90 metres. What length will 12 of the pipes make if they are laid end to end?

5 The cost of 5 metres of fabric is £7.25.
Aiden wants to buy 3 metres of fabric.
How much will this cost?

6 Three 2.5-litre tins of paint cost £54.94. Work out the cost of five of the 2.5-litre tins of paint.

7 100 g of cheese contains 16 g of carbohydrates.
Work out the weight of carbohydrates in 125 g of cheese.

8 350 sheets of paper have a total thickness of 2.7 cm. Simone's paper feeder for her printer holds 4.5 cm of paper.
How many sheets will she need to completely fill the paper feeder?

Exercise 6L

D 1 This is a list of ingredients needed to make 50 cheese straws.

100 g flour
50 g margarine
75 g cheese

Kieron wants to make 150 cheese straws. Work out the amount of each ingredient Kieron needs.

2 This is a recipe for making Quiche Lorraine for four people.

100 g pastry
100 g bacon
75 g cheese
2 eggs
150 m*l* milk

Dan is making Quiche Lorraine for 10 people.
Work out the amount of each ingredient he needs.

3 This is a list of ingredients needed to make 20 almond biscuits.

175 g flour
75 g caster sugar
50 g ground almonds
150 g margarine

a Work out the amount of flour needed to make 30 almond biscuits.
b Work out the amount of margarine needed to make 15 almond biscuits.
c Work out the amount of ground almonds needed to make 40 almond biscuits.

Exercise 6M

D 1 Julie went on holiday to Italy.
She changed £250 into euros.
The exchange rate was £1 = 1.08 euros.
Work out how many euros Julie received.

2 Raj changed £500 into Russian roubles.
The exchange rate was £1 = 44.95 roubles.
Work out how many roubles Raj received.

3 Lavina is going to the USA.
The exchange rate is £1 = $1.42
a Convert £450 into dollars.
b Convert $505 into pounds.

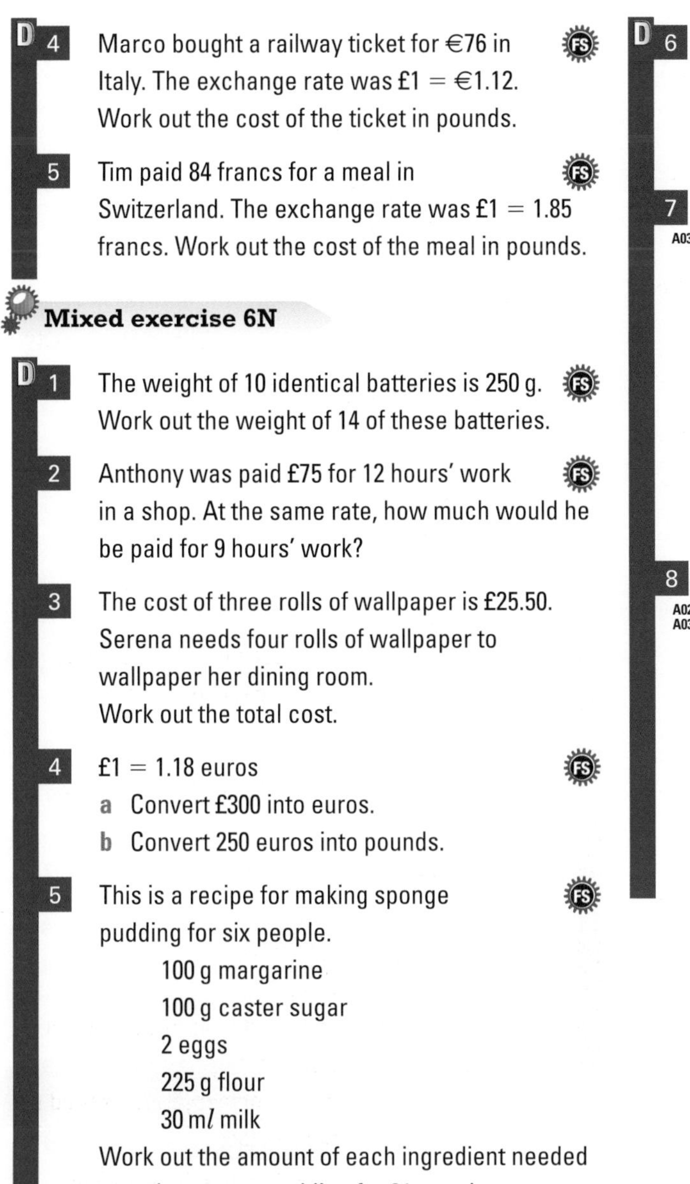

D

4 Marco bought a railway ticket for €76 in Italy. The exchange rate was £1 = €1.12.
Work out the cost of the ticket in pounds.

5 Tim paid 84 francs for a meal in Switzerland. The exchange rate was £1 = 1.85 francs. Work out the cost of the meal in pounds.

Mixed exercise 6N

D

1 The weight of 10 identical batteries is 250 g.
Work out the weight of 14 of these batteries.

2 Anthony was paid £75 for 12 hours' work in a shop. At the same rate, how much would he be paid for 9 hours' work?

3 The cost of three rolls of wallpaper is £25.50.
Serena needs four rolls of wallpaper to wallpaper her dining room.
Work out the total cost.

4 £1 = 1.18 euros

a Convert £300 into euros.

b Convert 250 euros into pounds.

5 This is a recipe for making sponge pudding for six people.

100 g margarine
100 g caster sugar
2 eggs
225 g flour
30 ml milk

Work out the amount of each ingredient needed to make sponge pudding for 21 people.

D

6 Kayla came back from a holiday in Australia.
She changed $184 into pounds.
The exchange rate was £1 = $2.05.
Work out how many pounds she received.

7 A03

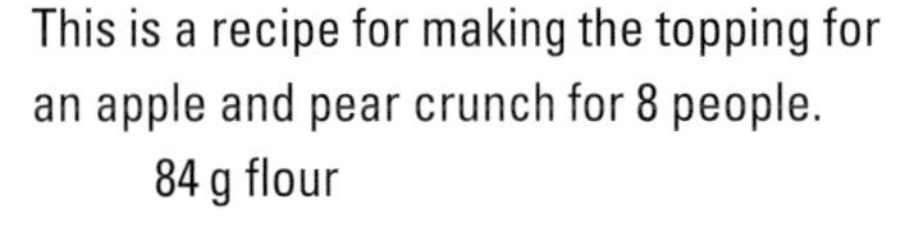

This is a recipe for making the topping for an apple and pear crunch for 8 people.

84 g flour
280 g porridge oats
110 g sugar
84 g margarine

Lena has only 147 g of flour and plenty of all the other ingredients.
How many people can she make the topping for?

8 A02 A03

Nikki is on a cruise. She can pay for a book in £, € or $. The book costs £7.99, €10 or $13. Which currency should she use? Explain your answer.

Exchange rates
£1 = 1.18 euros
£1 = $1.42

7 Algebra 1

Key Points

- **variable:** something that can be changed, and is written using a letter.
- **term:** a multiple of a variable. It can be a combination of variables and numbers such as x^3, ab, $3y^2$.
- **expression:** a collection of terms and variables.
- **like terms:** terms that use the same variable(s).
- **an equation:** has an equals sign and is used to find a numerical value for a variable.
- **formula:** where one variable is equal to an expression in a different variable(s).
- **simplifying expressions:** add, subtract and collect like terms, taking care with negative values.
- **multiplying algebraic terms**: combine them by writing them next to each other.
- **dividing algebraic expressions:** cancel the common factors using the same method as cancelling fractions.
- **expanding brackets:** multiply everything inside the bracket by everything outside the bracket.
- **factorising expressions:** take out the common factors and put in a bracket.

7.1 Using letters to represent numbers

Exercise 7A

Questions in this chapter are targeted at the grades indicated.

Find these.

F

1. $a + a + a =$
2. $b + b + b + b + b =$
3. $c + c + c =$
4. $5x - x =$
5. $7y - 3y =$

Exercise 7B

F

1. Use algebra to write:
   a 4 more than p
   b x plus 5
   c q take away 3
   d 2 less than g
   e 6 more than h
   f k minus 7
   g j with 3 taken away
   h a plus 2
   i y minus 5
   j m with 1 taken away
   k p with 8 added
   l 4 together with h.
2. Ian had c CDs. He buys 10 more.
   How many CDs does Ian have now?
3. Luma had a apples. She eats 2 of them.
   How many apples does she have now?
4. Adrian had d downloads on his MP3 player. He downloads 8 more.
   How many downloads has he altogether?
5. Aisha has g computer games. She sells 9 on the internet.
   How many computer games has she got now?
6. Amy and Todd go shopping. Todd buys x books and Amy buys y books.
   How many books do they have altogether?

Exercise 7C

ResultsPlus
Examiner's Tip

If a question asks for the answer in pence, you do not need to write p or pence and mess up your algebra.

E

1. Sausages are sold in packs of 6 and packs of 8.
   Carl buys f packs of 6 sausages and Ewan buys t packs of 8 sausages.
   How many sausages do they buy altogether?
2. Elisha buys candles in boxes of 6 and boxes of 12.
   One day she bought f boxes of 6 candles and t boxes of 12 candles.
   How many candles did she buy altogether?

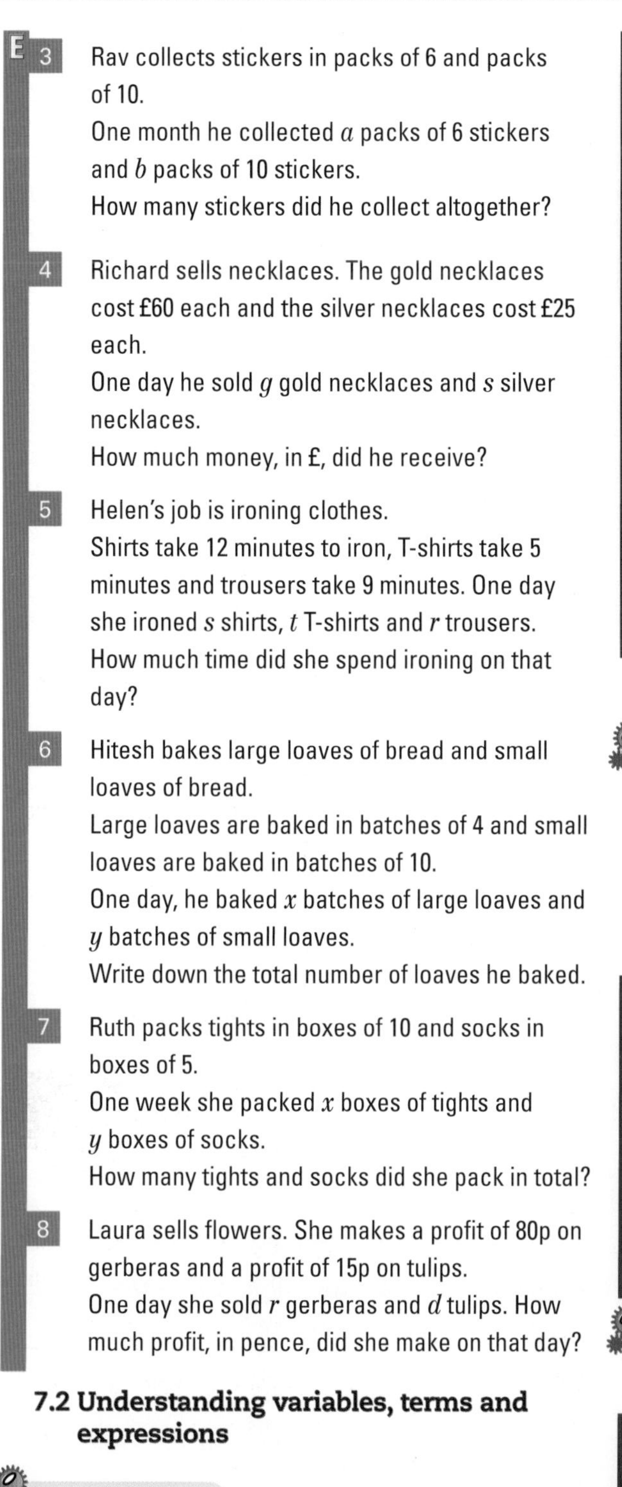

E

3 Rav collects stickers in packs of 6 and packs of 10.
One month he collected a packs of 6 stickers and b packs of 10 stickers.
How many stickers did he collect altogether?

4 Richard sells necklaces. The gold necklaces cost £60 each and the silver necklaces cost £25 each.
One day he sold g gold necklaces and s silver necklaces.
How much money, in £, did he receive?

5 Helen's job is ironing clothes.
Shirts take 12 minutes to iron, T-shirts take 5 minutes and trousers take 9 minutes. One day she ironed s shirts, t T-shirts and r trousers.
How much time did she spend ironing on that day?

6 Hitesh bakes large loaves of bread and small loaves of bread.
Large loaves are baked in batches of 4 and small loaves are baked in batches of 10.
One day, he baked x batches of large loaves and y batches of small loaves.
Write down the total number of loaves he baked.

7 Ruth packs tights in boxes of 10 and socks in boxes of 5.
One week she packed x boxes of tights and y boxes of socks.
How many tights and socks did she pack in total?

8 Laura sells flowers. She makes a profit of 80p on gerberas and a profit of 15p on tulips.
One day she sold r gerberas and d tulips. How much profit, in pence, did she make on that day?

7.2 Understanding variables, terms and expressions

Exercise 7D

F

1 Write down the letters that are the variables in these expressions.

a $a + 3b$ b $4x + y$ c $4a - 3t$
d $2x - y$ e $5t - 2d$ f $5a - 2s$
g $6b - 4$ h $6g + 9$ i $7t + 5$
j $5a + 2b$

2 Write down the terms in these expressions.

a $4a + 3b$ b $4x + y$ c $4a - 3t$
d $2x - y$ e $5t - 2d$ f $5a - 2s + 8$
g $6b - 4h$ h $6g + 9r + 4$ i $7t + 5s - 3$
j $5a + 2b$

3 Write down the variables in these terms and expressions.

a $2a$ b $3y$ c $5t$
d $3x$ e $6d$ f $5a - 2s$
g $6b - 4$ h $6g + 9$ i $7t + 5$
j $5a + 2b$

4 Use some of the terms in question 3 to make five new expressions.

5 Use some of the variables you identified in question 1 to make five new terms.

7.3 Collecting like terms

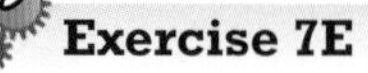

Exercise 7E

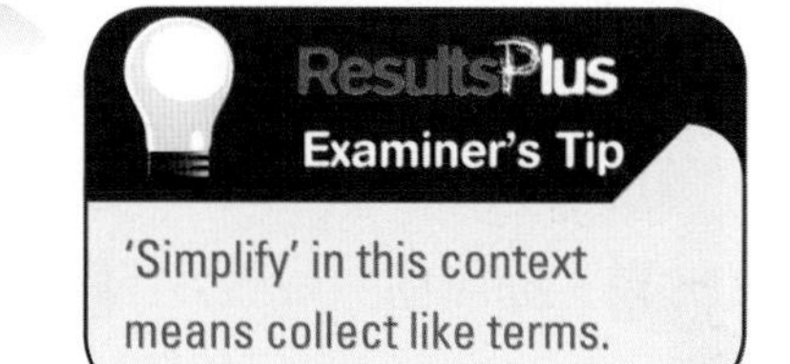

ResultsPlus
Examiner's Tip
'Simplify' in this context means collect like terms.

Simplify

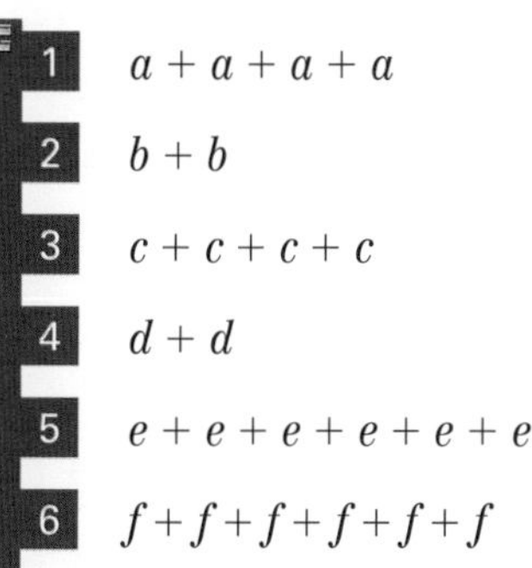

F

1 $a + a + a + a$

2 $b + b$

3 $c + c + c + c$

4 $d + d$

5 $e + e + e + e + e + e$

6 $f + f + f + f + f + f$

Exercise 7F

Simplify

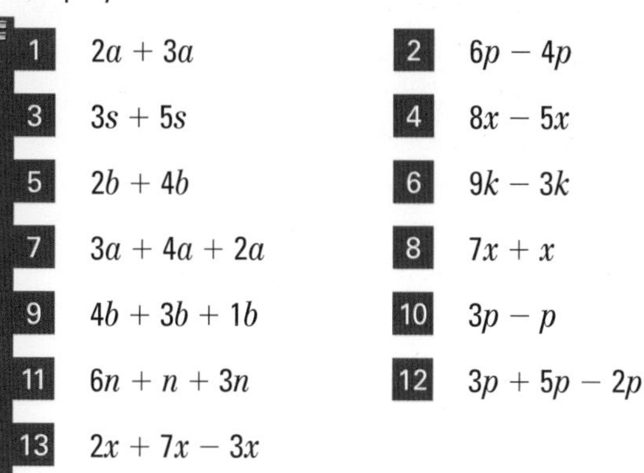

F

1 $2a + 3a$ 2 $6p - 4p$

3 $3s + 5s$ 4 $8x - 5x$

5 $2b + 4b$ 6 $9k - 3k$

7 $3a + 4a + 2a$ 8 $7x + x$

9 $4b + 3b + 1b$ 10 $3p - p$

11 $6n + n + 3n$ 12 $3p + 5p - 2p$

13 $2x + 7x - 3x$

Exercise 7G

ResultsPlus
Watch Out!
Do not try to combine the *a*s and the *b*s together when you are adding and taking away. They are different variables.

Simplify

E

1. $5a + 3b + 3a + 4b$
2. $8m + 6n + 2m + 3n$
3. $3s + 4t + 4s + 2t$
4. $6e + 3f - 4e$
5. $5g + 4h - 2g + 3h$
6. $5p - 4p + 6r - 2r$
7. $4j + 3k - 2j - k$
8. $8m + 7n - m$
9. $4a + 5b + 3a - 3b$
10. $4r + 5s - r - 3s$
11. $8v + 6w - 2v - 3w$
12. $p + 4q + 6p - 2q$
13. $6e + 3f - 2e - f$
14. $8g + 6h - 4g - 5h$
15. $7p - 5r + 6r - 5p$
16. $4j + 6k + 3j - 2k$
17. $2m + 5n - m$
18. $9c + 7d - 5c - 2d$
19. $7p - 3j - 4p + 4j$
20. $7p + 3t - 7p - t$
21. $5x + 2x - 4x - x$
22. $3x + 2m - 3x - 2m$
23. $2k + 3g - 2k - g$
24. $5m - 4m + 2m + m$

Exercise 7H

Simplify

E

1. $2s + 5 + 3 + 5s$
2. $4m + 3 + 2m + 6$
3. $4 + 3p + 7 + 2p$
4. $8e + 3 - 5e$
5. $4 + 5h - 6 + 2h$
6. $7g + 9 - 4 - 6g$
7. $5j + 6 - 2j - 5$
8. $6m + 8 - m - 6$

D

9. $5a + 6b - 7 + 3a - 3b + 6$
10. $3t + 5 - t - 2 + 4r$
11. $5m + 6n - 2 + 10$
12. $4p + 7q - 6 + p - 2q + 8$
13. $4e + 3f - e - f + 3$
14. $2p + 5 - 3p + 4r - 2r - 3$
15. $5p - 7r - 6 + 3r - 2p + 4$

Exercise 7I

Simplify

E

1. $x^2 + 2x^2$
2. $2p^2 + 2p^2$
3. $7a^2 - 3a^2$

D

4. $4a^2 + 5b^3 + 3a^2 + 4b^3$
5. $3m^2 + 2n + 6m^2 + 5n$
6. $5p^3 + 4pq + 2p^3 - 3pq$
7. $3lm + 5lm - 6lm$
8. $5g^2 + 3h^3 - 4g^2 + 4h^3$
9. $7pq - 6pq + 5r^3 - 2r^2$
10. $5jk + 6jk - 2jk - 3jk$
11. $4m^3 + 5n - m^3$
12. $6s^2 + 4t^2 + 3s^2 - 3t^2$
13. $8a^3 + 5b^2 - 3a^3 - 2b^2$
14. $6m + 6n^2 - 3m - 2n^2$
15. $3pq + 4pq + 5p^3 - 6q^2$
16. $4e^3 + 3f^2 - e^3 - f^2$
17. $7gh + 7h^3 - 5gh - 6h^3$
18. $7lmn - 6lmn + 5lmn - 2lmn$

Exercise 7J

Simplify

D

1. $4a - 7a$
2. $5m + 2n - 8m + 3n$
3. $p + q - 4p - 5q$
4. $3e - 7e + 3 - 5$
5. $8c + 3d - 3c - 6d$
6. $4k^2 - k^2 + 2l^3 - 6l^3$
7. $2j + 3k - 2j - 8k$
8. $2m^3 - 7n - 8m^3$
9. $3a - 5b - 6a + 2b$
10. $3ab + 2ab - 9ab$
11. $6m + 4 - 9m - 7$
12. $3y + 6 - 5y - 11$
13. $8d + e - 6d - 3e$
14. $4g^2 + 3 - 8g^2 - 8$
15. $9p - 5r + 3 + 8r - 12p$
16. $5j + 6k - 8j - 2k$
17. $2m - 6n - 2m - 7$
18. $9a - 3b - 5a - 6b$
19. $7p - 2j - 9p - 8j$
20. $8h - 6 - 8h - 3$
21. $x + 5 - 3 - 7x$
22. $6y^3 + 7 - 6y^3 - 7$
23. $2k^2 + 6g^3 - 2k^2 - 9g^3$
24. $4af - 7f^2 - 11af + 8f^2$

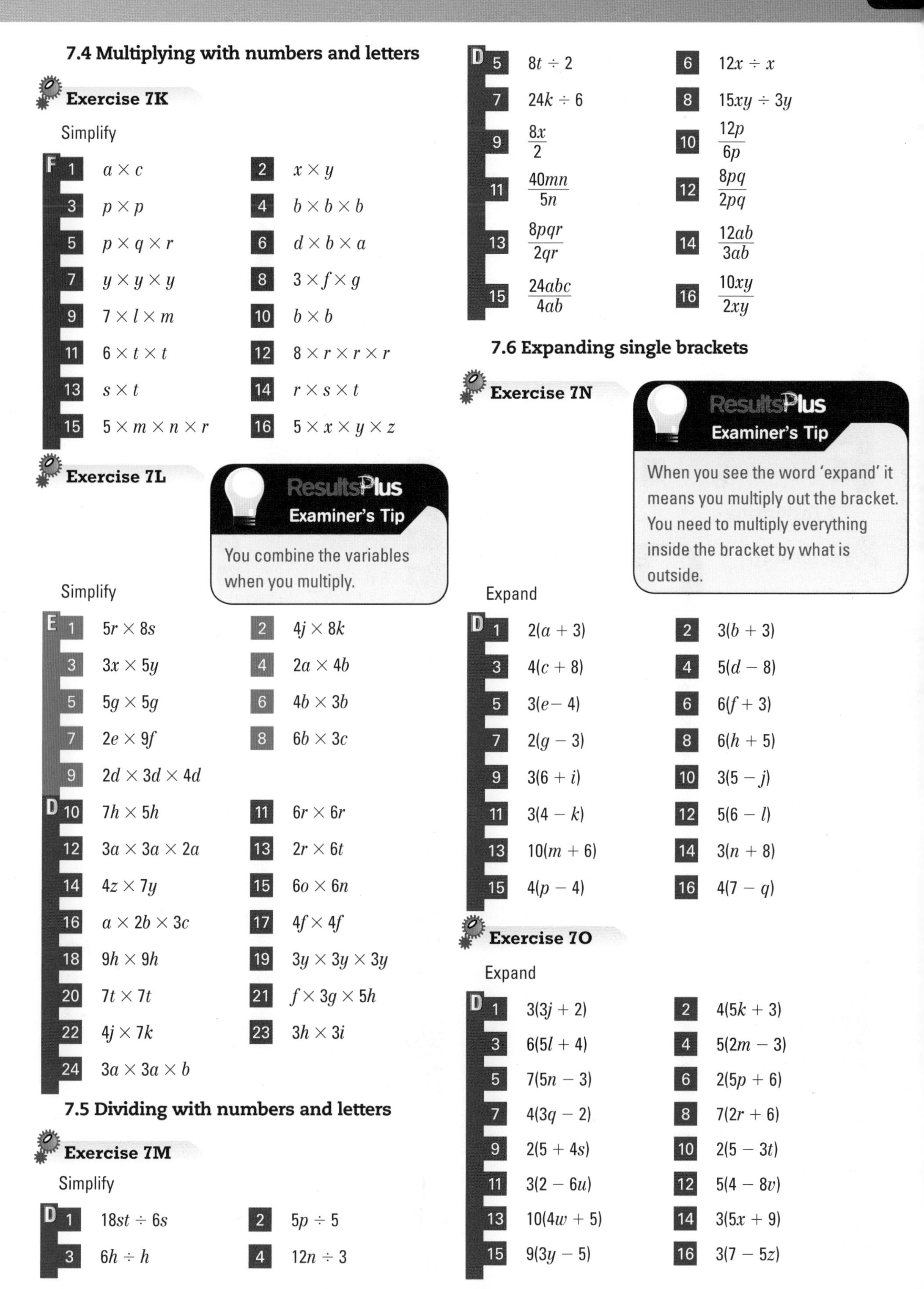

7.4 Multiplying with numbers and letters

Exercise 7K

Simplify

F

1 $a \times c$ 2 $x \times y$

3 $p \times p$ 4 $b \times b \times b$

5 $p \times q \times r$ 6 $d \times b \times a$

7 $y \times y \times y$ 8 $3 \times f \times g$

9 $7 \times l \times m$ 10 $b \times b$

11 $6 \times t \times t$ 12 $8 \times r \times r \times r$

13 $s \times t$ 14 $r \times s \times t$

15 $5 \times m \times n \times r$ 16 $5 \times x \times y \times z$

Exercise 7L

ResultsPlus Examiner's Tip

You combine the variables when you multiply.

Simplify

E

1 $5r \times 8s$ 2 $4j \times 8k$

3 $3x \times 5y$ 4 $2a \times 4b$

5 $5g \times 5g$ 6 $4b \times 3b$

7 $2e \times 9f$ 8 $6b \times 3c$

9 $2d \times 3d \times 4d$

D

10 $7h \times 5h$ 11 $6r \times 6r$

12 $3a \times 3a \times 2a$ 13 $2r \times 6t$

14 $4z \times 7y$ 15 $6o \times 6n$

16 $a \times 2b \times 3c$ 17 $4f \times 4f$

18 $9h \times 9h$ 19 $3y \times 3y \times 3y$

20 $7t \times 7t$ 21 $f \times 3g \times 5h$

22 $4j \times 7k$ 23 $3h \times 3i$

24 $3a \times 3a \times b$

7.5 Dividing with numbers and letters

Exercise 7M

Simplify

D

1 $18st \div 6s$ 2 $5p \div 5$

3 $6h \div h$ 4 $12n \div 3$

5 $8t \div 2$ 6 $12x \div x$

7 $24k \div 6$ 8 $15xy \div 3y$

9 $\frac{8x}{2}$ 10 $\frac{12p}{6p}$

11 $\frac{40mn}{5n}$ 12 $\frac{8pq}{2pq}$

13 $\frac{8pqr}{2qr}$ 14 $\frac{12ab}{3ab}$

15 $\frac{24abc}{4ab}$ 16 $\frac{10xy}{2xy}$

7.6 Expanding single brackets

Exercise 7N

ResultsPlus Examiner's Tip

When you see the word 'expand' it means you multiply out the bracket. You need to multiply everything inside the bracket by what is outside.

Expand

D

1 $2(a + 3)$ 2 $3(b + 3)$

3 $4(c + 8)$ 4 $5(d - 8)$

5 $3(e - 4)$ 6 $6(f + 3)$

7 $2(g - 3)$ 8 $6(h + 5)$

9 $3(6 + i)$ 10 $3(5 - j)$

11 $3(4 - k)$ 12 $5(6 - l)$

13 $10(m + 6)$ 14 $3(n + 8)$

15 $4(p - 4)$ 16 $4(7 - q)$

Exercise 7O

Expand

D

1 $3(3j + 2)$ 2 $4(5k + 3)$

3 $6(5l + 4)$ 4 $5(2m - 3)$

5 $7(5n - 3)$ 6 $2(5p + 6)$

7 $4(3q - 2)$ 8 $7(2r + 6)$

9 $2(5 + 4s)$ 10 $2(5 - 3t)$

11 $3(2 - 6u)$ 12 $5(4 - 8v)$

13 $10(4w + 5)$ 14 $3(5x + 9)$

15 $9(3y - 5)$ 16 $3(7 - 5z)$

Exercise 7P

Expand

D

1 $a(a + 3)$
2 $b(b + 4)$
3 $c(d + 8)$
4 $e(2e - 3)$
5 $f(f - 6)$
6 $g(g + 7)$
7 $h(2h - 4)$
8 $i(3i + 5)$
9 $j(4 + j)$
10 $k(6 - 2k)$
11 $l(7 - 3l)$
12 $m(4 - 6m)$
13 $3n(n + p)$
14 $2q(q + 9)$
15 $5r(r + s)$
16 $3t(9 - t)$
17 $5u(2u + 5)$
18 $5v(3v - w)$
19 $2y(x + 8y)$
20 $5z(4 - 3z)$

7.7 Factorising

Exercise 7Q

ResultsPlus Examiner's Tip

When you see the word 'factorise' it means you take out the factors and put the bracket back in.

Factorise

D

1 $5p - 15$
2 $2n + 10$
3 $3s + 3$
4 $3k + 15$
5 $3f - 12$
6 $2a + 8$
7 $5r + 25$
8 $3x - 18$
9 $8p - 48$
10 $3m - 18$
11 $4q + 16$
12 $2a - 14$
13 $5a - 30$
14 $6x + 36$
15 $7w + 21$
16 $6y - 6$

Exercise 7R

Factorise

D

1 $x^3 - 6x$
2 $a^2 + 9a$
3 $y^3 + 7y$
4 $p^3 + 5p$
5 $s^2 - 10s$
6 $a^2 + 3a$
7 $p^2 + 7p$
8 $a^2 - 3a$
9 $j^3 - 8j$
10 $m^3 - 3m^2$
11 $c^3 + 9c$
12 $p^3 - 2p$
13 $x^3 - x^2$
14 $x^2 + 9x$
15 $a + a^2$
16 $y^2 - y$

Exercise 7S

ResultsPlus Examiner's Tip

When you see 'factorise completely' it means there is more than one factor to take outside the bracket.

Factorise completely

C

1 $12p^2 + 6p$
2 $3a^2 + 6a$
3 $4y^3 + 8y$
4 $16c^3 + 8c$
5 $3s^2 - 6s$
6 $10x^3 - 2x$
7 $6a^2 + 3a$
8 $8a^2 - 2a$
9 $2p^3 + 10p$
10 $6x^2 + 30x$
11 $9j^3 - 3j$
12 $18a + 6a^2$
13 $6x^3 - 3x^2$
14 $6m^3 - 2m^2$
15 $8p^3 - 24p$
16 $30y^2 - 5y$

7.8 Understanding expressions, equations and formulae

Exercise 7T

State whether each of the following is an equation, expression or formula.

D

1 $8x = 56$
2 $V = \frac{4}{3}\pi r^3$
3 $2x + 4 = 2(x + 2)$
4 $v = u + at$
5 $5p^2 - 5p$
6 $y^2 + 3xy + x$
7 $2a^2 + ab + b^2$
8 $2a + 3b$
9 $A = \dfrac{bh}{2}$
10 $pqr + p^2$

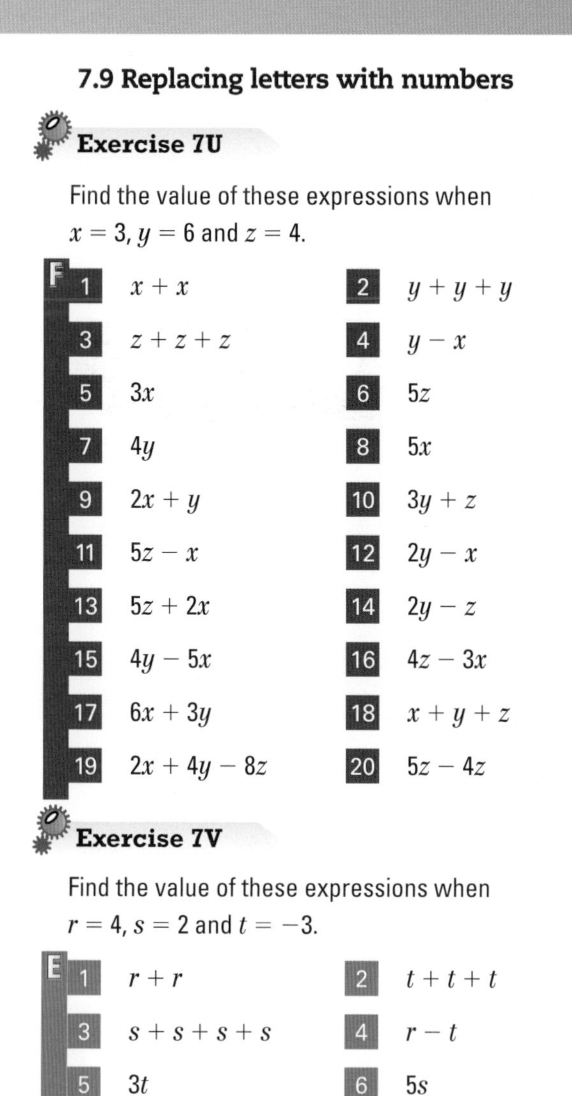

7.9 Replacing letters with numbers

Exercise 7U

Find the value of these expressions when $x = 3$, $y = 6$ and $z = 4$.

F

1. $x + x$
2. $y + y + y$
3. $z + z + z$
4. $y - x$
5. $3x$
6. $5z$
7. $4y$
8. $5x$
9. $2x + y$
10. $3y + z$
11. $5z - x$
12. $2y - x$
13. $5z + 2x$
14. $2y - z$
15. $4y - 5x$
16. $4z - 3x$
17. $6x + 3y$
18. $x + y + z$
19. $2x + 4y - 8z$
20. $5z - 4z$

Exercise 7V

Find the value of these expressions when $r = 4$, $s = 2$ and $t = -3$.

E

1. $r + r$
2. $t + t + t$
3. $s + s + s + s$
4. $r - t$
5. $3t$
6. $5s$
7. $4r$
8. $5t + r$

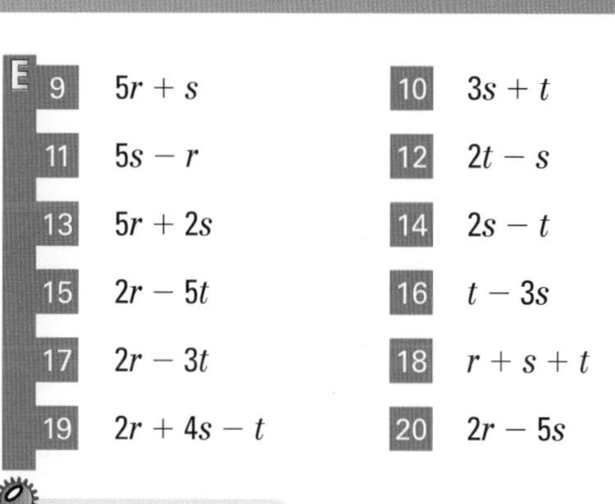

9. $5r + s$
10. $3s + t$
11. $5s - r$
12. $2t - s$
13. $5r + 2s$
14. $2s - t$
15. $2r - 5t$
16. $t - 3s$
17. $2r - 3t$
18. $r + s + t$
19. $2r + 4s - t$
20. $2r - 5s$

Exercise 7W

Find the value of these expressions when $e = 5, f = 8$ and $g = 6$.

D

1. $2(e + f)$
2. $3(f + g)$
3. $4(e + g)$
4. $5(f - e)$
5. $3(e + 2f)$
6. $5(e + 2g)$
7. $4(g - f)$
8. $5(2e - f)$
9. $2(e + 2f)$
10. $3(4f + 2g)$
11. $5(g - e)$
12. $2(3f - 2e)$
13. $5(g + 2e)$
14. $2(2f - 3g)$
15. $4(5f - 3e)$
16. $4(2f - 3g)$
17. $6(e + 3f)$
18. $2(e + f + g)$
19. $2(e - 4f)$
20. $5(g - e - f)$

8 Algebra 2

Key Points

- **power or index:** the 3 in 4^3. Gives the number of times 4 must be multiplied by itself.
- **indices:** the plural of index.
- **base:** the 4 in 4^3.
- **BIDMAS:** the order of number operations. **B**rackets, **I**ndices, **D**ivide, **M**ultiply, **A**dd, **S**ubtract.
- **solving an equation like $5^x = 125$:** work out how many times the base has to be multiplied by itself to give the answer.
- **using index laws:**
  - to multiply powers of the same number, add indices: $x^m \times x^n = x^{m+n}$
  - to divide powers of the same number, subtract indices: $\frac{x^m}{x^n} = x^{m-n}$
  - any variable raised to the power of 1 equals the number itself: $x^1 = x$
  - any variable raised to the power of 0 equals 1: $x^0 = 1$
  - to raise the power of a number to a further power, multiply indices: $(x^m)^n = x^{m \times n}$
- **expanding brackets:** multiply each term inside the bracket by the term outside the bracket.
- **simplifying brackets:** expand the brackets and collect like terms.
- **factorising:** find a common factor and write outside the bracket.

8.1 Calculating with powers

Exercise 8A

Questions in this chapter are targeted at the grades indicated.

E **1** Find the value of
a 2^4
b 4 to the power 3
c 1^7
d 10 to the power 5
e 5^3
f 6 to the power 5

2 Write these using index notation.
a $2 \times 2 \times 2 \times 2 \times 2$
b $4 \times 4 \times 4 \times 4$
c $1 \times 1 \times 1 \times 1 \times 1$
d $8 \times 8 \times 8 \times 8 \times 8$
e $3 \times 3 \times 8 \times 8 \times 8 \times 8 \times 8$
f $4 \times 4 \times 4 \times 4 \times 2 \times 2 \times 2 \times 2 \times 2$

3 Work out the value of
a 2^5 b 3^4 c 6^4
d 5^4 e 8^2 f $2^3 \times 9^4$
g $2^5 \times 4^6$ h $5^4 \times 3^3$ i $2^5 \times 3^6$
j 4×4^3

E **4** Copy and complete the table for powers of 10.

Power of 10	Index	Value	Value in words
10^3			One thousand
	2	100	
		1 000 000	One million
10	1		
10^5		100 000	

5 Work out the value of
a $4^3 \times 10^3$ b 6×10^2
c 7×10^3 d $10^3 \div 5^2$
e $10^3 \div 2^4$ f $4^3 \div 2^3$

D **6** Find x when
a $5^x = 25$ b $3^x = 243$
c $2^x = 128$ d $10^x = 100\,000$
e $9^x = 729$ f $3^x = 81$
g $2^x = 32$ h $7^x = 343$

8.2 Writing expressions as a single power of the same number

Exercise 8B

Simplify these expressions by writing as a single power of the number.

C

1 a $3^8 \times 3^3$ b $7^3 \times 7^5$ c $4^4 \times 4^2$

2 a $5^3 \div 5^2$ b $2^6 \div 2^3$ c $8^5 \div 8$

3 a $3^2 \times 3^3$ b $6^3 \div 6$ c $7^9 \div 7^8$

4 a $3^6 \times 3^4 \times 3^3$ b $5^3 \times 5^7 \times 5$

5 a $10^2 \times 10^2 \times 10^3$ b $7^4 \div 7^4$

6 a $4^3 \times 4^7 \times 4$ b $6^2 \times 6^2 \times 6^2$

7 a $2^5 \times 2 \times 2^2$ b $8^7 \times \frac{8^5}{8^6}$

8 a $\frac{5^8}{5^2} \times 5^3$ b $3^8 \times \frac{3^4}{3^7}$ c $\frac{6^9}{6^2} \times 6^5$

9 a $(4^2)^3$ b $(9^4)^2$

8.3 Using powers in algebra to simplify expressions

Exercise 8C

Simplify the following.

C

1 a $a^7 \times a^4$ b $b^2 \times b^6$ c $c^8 \times c^6$

2 a $x^3 \times x^9$ b $y^4 \times y^4$ c $z^6 \times z^5$

3 a $m^4 \div m^2$ b $n^{11} \div n^4$ c $p^6 \div p^4$

4 a $p^7 \div p^2$ b $q^4 \div q^3$ c $r^{17} \div r^{13}$

5 a $t^3 \times t^3 \times t^2$ b $u^4 \times u^4 \times u^3$
c $v^5 \times v^4 \times v^3$

6 a $4p^2 \times 2p^3$ b $6q^9 \times 3q^{20}$ c $6r^8 \times 5r^2$

7 a $8x^8 \div 4x^3$ b $25y^5 \div 5y^3$ c $6z^5 \div 2z^2$

8 a $(a^2)^4$ b $(b^4)^2$ c $(c^3)^3$
d $(d^6)^9$

9 a $(e^7)^4$ b $(f^3)^2$ c $(g^3)^0$
d $(h^0)^{22}$

10 a $(3j^2)^7$ b $(4k)^3$ c $(3l^{151})^0$

11 a $x^4 \times \frac{x^5}{x^9}$ b $y^7 \times \frac{y}{y^4}$
c $z^3 \times \frac{z^4}{z^2} \times z^5$

12 a $4e^9 \times 2e$ b $8f^8 \div 4f^4$ c $(4g^2)^2$

8.4 Understanding order of operations

Exercise 8D

E

1 Use BIDMAS to help you find the value of these expressions.

a $6 + (3 + 1)$ b $6 - (3 + 1)$
c $6 \times (2 + 3)$ d $6 \times 2 + 3$
e $5 \times (4 + 3)$ f $6 \times 4 + 3$
g $20 \div 5 + 1$ h $24 \div (5 + 1)$
i $5 + 4 \div 2$ j $(5 + 4) \div 3$
k $24 \div (5 - 2)$ l $25 \div 5 - 2$
m $8 - (4 + 2)$ n $8 - 4 + 2$
o $((15 - 5) \times 3) \div ((2 + 3) \times 3)$

2 Make these expressions correct by replacing the • with + or − or × or ÷ and using brackets if you need to. The first one is done for you.

a $4 \bullet 5 = 20$ becomes $4 \times 5 = 20$
b $4 \bullet 5 = 9$ c $2 \bullet 3 \bullet 4 = 1$
d $3 \bullet 2 \bullet 7 = 7$ e $5 \bullet 2 \bullet 2 = 6$
f $4 \bullet 2 \bullet 7 = 9$ g $5 \bullet 4 \bullet 5 \bullet 2 = 6$
h $5 \bullet 4 \bullet 5 \bullet 2 = 3$

3 Work out

a $(3 + 4)^3$ b $3^3 + 4^3$
c $3 \times (4 + 5)^3$ d $3 \times 4^3 + 3 \times 5^3$
e $2 \times (4 + 2)^3$ f $2^2 + 3^3$
g $3 \times (3^2 + 2)$ h $\frac{(4 + 2)^3}{3^2 + 3}$
i $\frac{5^2 + 2^3}{3}$ j $4^3 - 2^2$
k $3^5 - 5^2$ l $4^4 - 8^2$

8.5 Multiplying out brackets in algebra

Exercise 8E

Expand and simplify.

C

1 $3(a + 2) + 2(a + 4)$

2 $4(2b - 1) + 3(4b + 7)$

3 $5(3c + 2) + 4(2c + 1)$

4 $7(3 - 2d) + 3(2d - 3)$

5 $6(4 - 2e) - 3(5 + 3e)$

6 $4(3 - 2f) + 3(1 - 5f)$

7 $3(3x - 5y) + 2(2x - 4y)$

8 $6(6y + 2x) - 5(3x + 2y)$

C

9 $4(2x - 3y) - 3(5x + 6y)$

10 $5(2x + 3y) - 3(x + y)$

11 $5(3y - 2) - 4(y - 2)$

12 $3(3x + 6) - 2(2x - 5)$

13 $6(3 - 2x) - 5(5 - 3x)$

14 $3(3 - y - 2x) - 2(4x - 3y)$

15 $5(2x - 3y) + 3(3x - 2y)$

16 $4(3y - 5x) - 3(x - 3y)$

17 $(4x - 3y) + 3(3x - 2y)$

18 $(3x - 5y) - 4(x - 3y)$

19 $2(2y + 1) + x(3y + 1)$

20 $x(3y + 1) + 2y(2x + 1)$

21 $3y(3x - 2) + 2x(2 - 3y)$

22 $2x(2y - 5x) + 4y(x - y)$

8.6 Factorising expressions

Exercise 8F

Factorise each of the expressions in questions 1–6.

D

1
a $2x + 8$ b $6y + 4$ c $15b - 10$
d $4r - 6$ e $5x + 3xy$ f $12x + 4y$
g $12x - 20$ h $9 - 6x$ i $12 + 15g$

C

2
a $4x^2 + 3x$ b $5y^2 - 2y$ c $3a^2 + a$
d $5b^2 - 3b$ e $7c - 2c^2$ f $d^2 + 4d$
g $7m^2 - m$ h $xy + 4x$ i $n^3 - 6n^2$

3
a $8x^2 + 2x$ b $9p^2 + 3p$ c $6x^2 - 2x$
d $15b^2 - 9b$ e $12a + 6a^2$ f $15c - 5c^2$
g $21x^4 + 7x^3$ h $20y^3 - 12y^2$ i $6d^4 - 2d^2$

4
a $ab^2 + ab$ b $qr^2 - qr$ c $xb^2 - xb$
d $pr^2 + p^2$ e $b^2x + bx^2$ f $x^2y - xy^2$
g $12a^3 - 9a^2$ h $8x^3 - 2x^4$ i $18x^3 + 6x^5$

5
a $12a^2b + 24ab^2$ b $8x^2y - 2xy^2$
c $24a^2b + 8ab^2 + 12ab$ d $4x^2y + 6xy^2 - 2xy$
e $12ax^2 + 6a^2x - 3ax$ f $a^2bx + ab^2x + abx^2$

6
a $5x + 25$ b $12y - 18$ c $5x^2 + 3x$
d $5y - 3y^2$ e $8a + 2a^2$ f $12b^2 - 6b$
g $dy^2 + dy$ h $3dx^2 - 9dx$ i $12c^2d + 15cd^2$

9 Sequences

Key Points

- **sequence:** a pattern of numbers or shapes that follows a rule.
- **terms of a sequence:** the numbers in a number sequence.
- **term-to-term rule:** the rule of a number pattern that says how to find a term from the one before it.
- **term number:** the position of each term in the pattern.
- **number machine:** used to produce a number pattern with a term number to term rule.
- ***n*th term:** a term that can be used to find any number in a sequence, given the term number.
- **completing a sequence table of value:** use a number machine to find the sequence term from the term number.
- **identifying whether a number is in a pattern:** use the *n*th term to see if the number belongs to the sequence.

9.1 Sequences

Exercise 9A

Questions in this chapter are targeted at the grades indicated.

F

1 Find the two missing numbers in these number patterns. For each pattern, write down the term to term rule.

a 6, 9, 12, __, __, 21, 24
b 7, 11, 15, __, __, 27, 31
c 10, 15, 20, 25, __, __, 40, 45
d 7, 12, 17, 22, __, __, 37, 42
e 4, 7, 10, 13, __, __, 22, 25
f 7, 9, 11, 13, __, __, 19, 21
g 8, 13, 18, 23, __, __, 38, 43
h 7, 10, 13, 16, __, __, 25, 28
i 6, 10, 14, 18, __, __, 30, 34
j 20, 30, 40, __, __, 70, 80

2 a Write down the next two numbers in these sequences.

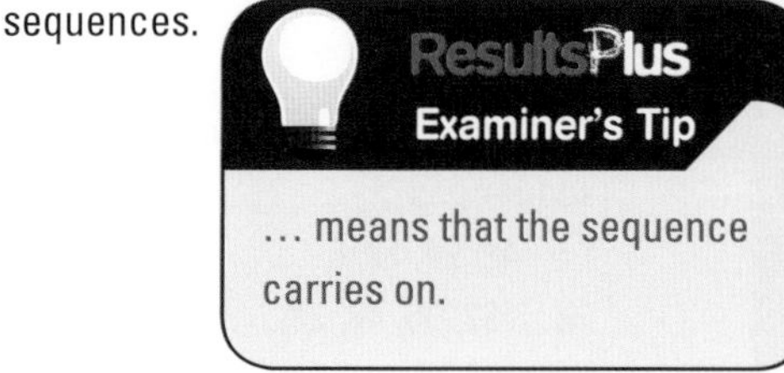

i 5, 9, 13, 17, 21, ...
ii 5, 8, 11, 14, 17, ...
iii 7, 11, 15, 19, 23, ...
iv 8, 12, 16, 20, 24, ...
v 8, 11, 14, 17, 20, ...
vi 11, 17, 23, 29, ...
vii 6, 10, 14, 18, 22, ...
viii 7, 13, 19, 25, 31, ...
ix 11, 19, 27, 35, 43, ...
x 9, 13, 17, 21, 25, ...

b Write down the rule you used to find the missing numbers in each sequence.

E

3 (A03) Leona saves £10 each month.
Here is the pattern of how her money grows.

Month	Savings (£)
Jan	10
Feb	20
Mar	30
Apr	
May	
Jun	
Jul	

a Copy and complete the table.
b Leona needs to save £140 for a school holiday. In which month will she have saved enough money?

Exercise 9B

F

1 Find the two missing numbers in these number patterns. Write down the rule for each number pattern.

a 36, 34, 32, 30, __, __, 24
b 25, 23, 21, 19, __, __, 13
c 60, 55, 50, 45, __, __, 30
d 49, 44, 39, 34, __, __, 19
e 27, 24, 21, 18, __, __, 9
f 43, 41, 39, 37, __, __, 31
g 47, 40, 33, 26, __, __, 5
h 32, 29, 26, 23, __, __, 14
i 50, 46, 42, 38, __, __, 26
j 90, 80, 70, __, __, 40

F 2 a Write down the next two numbers in these sequences.

i 35, 31, 27, 23, …
ii 21, 18, 15, 12, …
iii 43, 39, 35, 31, …
iv 28, 25, 22, 19, …
v 47, 44, 41, 38, …
vi 54, 48, 42, 36, …
vii 32, 30, 28, 26, …
viii 57, 52, 47, 42, …
ix 49, 42, 35, 28, …
x 7, 5, 3, 1, −1, −3, …

b Write down the rule you used to find the missing numbers in each sequence.

3 Find the 10th number of each of the number patterns in questions 1 and 2.

E 4 A03 Asif's mother gives him £12 each week to buy his lunch. His lunch costs him £2 each day.
Here is the pattern of how he spends his money.

Day	Money left at end of day (£)
Mon	10
Tue	8
Wed	
Thur	
Fri	

How much money will Asif have left at the end of the week?

Exercise 9C

F 1 Find the missing numbers in these number patterns.
For each pattern, write down the rule.

a 1, 2, 4, 8, 16, __, __, 128
b 2, 8, 32, 128, __, 2048
c 2, 10, __, 250, __, 6250
d 1, 10, 100, 1000, __, __
e 4, 8, 16, 32, __, __, 256
f 3, 9, 27, __, __, 729
g 3, 12, 48, __, __, 3072
h 4, 40, 400, 4000, __, __, 4 000 000
i 2, 10, 50, 250, __, 6250
j 4, 20, 100, __, 2500

F 2 a Write down the next two numbers in these sequences.

i 3, 6, 12, 24, …
ii 2, 6, 18, 54, …
iii 4, 16, 64, 256, …
iv 3, 15, 75, 375, …
v 8, 16, 32, 64, …
vi 4, 12, 36, 108, …
vii 9, 27, 81, 243, …
viii 6, 60, 600, 6000, …
ix 7, 14, 28, 56, …
x 7, 42, 252, 1512, …

b Write down the rule you used to find the missing numbers in each sequence.

E 3 Find the 10th number of each of the number patterns in questions 1 and 2.

4 The number of stray cats at a rescue centre doubled every month for 6 months.
The table shows the beginning of the pattern.

Month	Jan	Feb	Mar	Apr
Number of stray cats	1	2		

a Copy and complete the table.

A02 b How many cats were in the rescue centre in month 6?

Exercise 9D

F 1 Find the missing numbers in these number patterns.
Write down the rule for each number pattern.

a 64, 32, 16, __, __, 2, 1
b 1024, 256, __, 16, 4
c 3125, 625, __, __, 5, 1
d 200 000, 20 000, 2000, __, __, 2
e 448, 224, 112, 56, __, __, 7
f 1152, 576, 288, 144, __, __, 18
g 3645, 1215, 405, 135, __, 15
h 400 000, 40 000, 4000, __, __, 4
i 9375, 1875, 375, __, __, 3
j 6000, 600, 60, __, __, 0.06

2 a Write down the next two numbers in these sequences.

i 48, 24, 12, 6, …
ii 729, 243, 81, 27, …
iii 192, 96, 48, 24, …
iv 3125, 625, 125, 25, …
v 160, 80, 40, 20, …
vi 2916, 972, 324, 108, …
vii 1620, 540, 180, 60, …
viii 70 000, 7000, 700, 70, …
ix 144, 72, 36, 18, …
x 7776, 1296, 216, 36, …

b Write down the rule you used to find the missing number in each sequence.

E 3 Find the 8th number of each of the number patterns in questions 1 and 2.

4 Vita conducted an experiment to find out how long a saucer containing water would take to evaporate in the sun. She measured the volume each hour.

A03

a Copy and complete the table.

Time	1000	1100	1200	1300	1400
Volume of water (m*l*)	500	250	125		

b What was the volume of water left at 1500?

Exercise 9E

E 1 For these patterns:

i draw the next two patterns

ii write down the rule in words to find the next pattern

iii use your rule to find the 10th term.

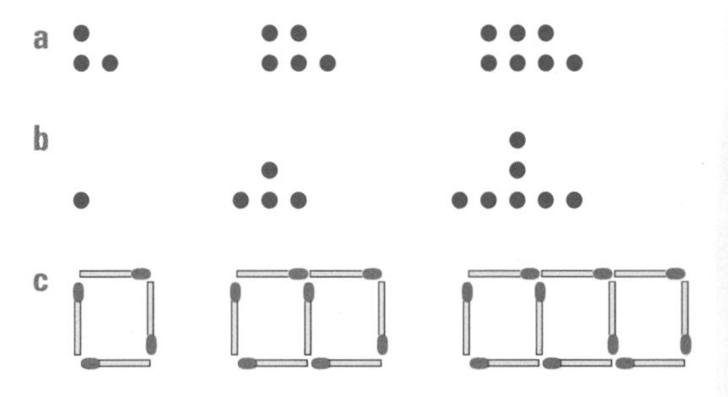

2 a Write down the number of matches in each of these patterns.

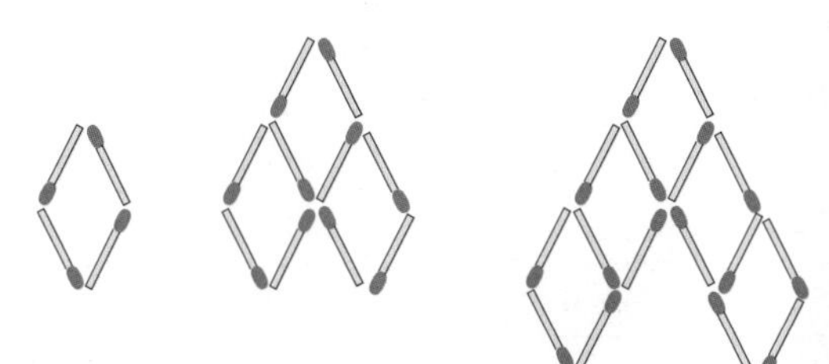

b Draw the next two patterns.

c Write down the rule in words to continue the pattern.

d Use your rule to find the number of matches needed for pattern number 10.

9.2 Using input and output machines to investigate number patterns

Exercise 9F

For each of these questions:

a copy and complete the table of values for the number machine

b write down the rule for finding the term from the term number

c write down the rule for finding the next term from the term before it.

F 1

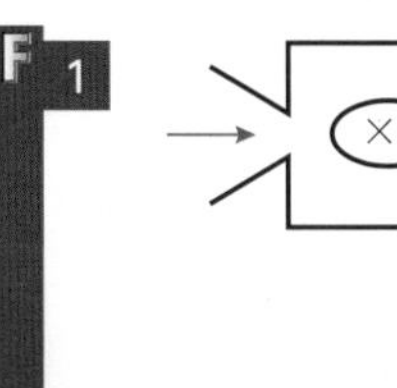

Term number	Term
1	1
2	2
3	
4	

2

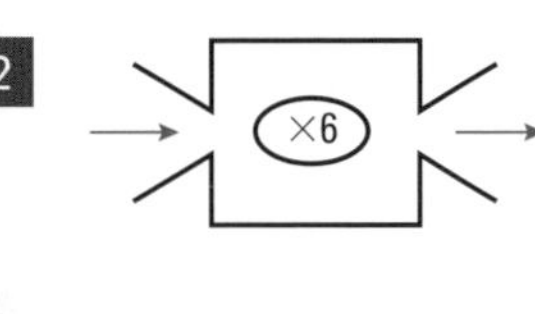

Term number	Term
1	6
2	12
3	
4	

3

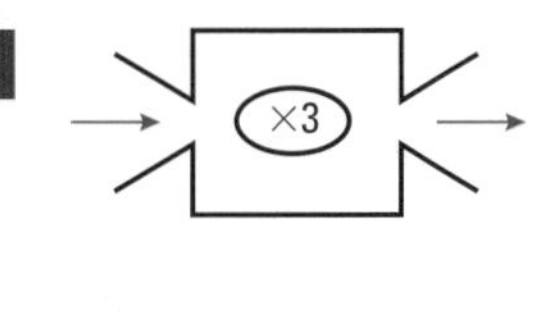

Term number	Term
1	3
2	6
3	
4	

4

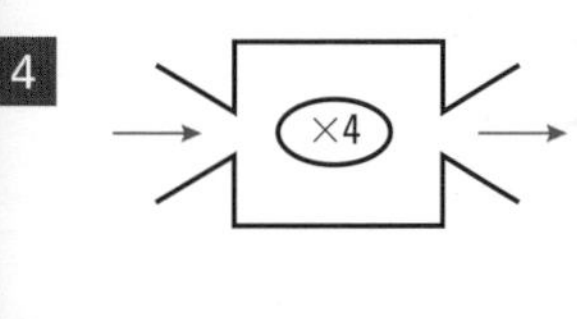

Term number	Term
1	4
2	8
3	
4	

5

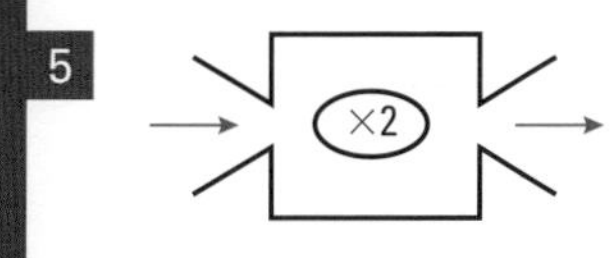

Term number	Term
1	2
2	4
3	
4	

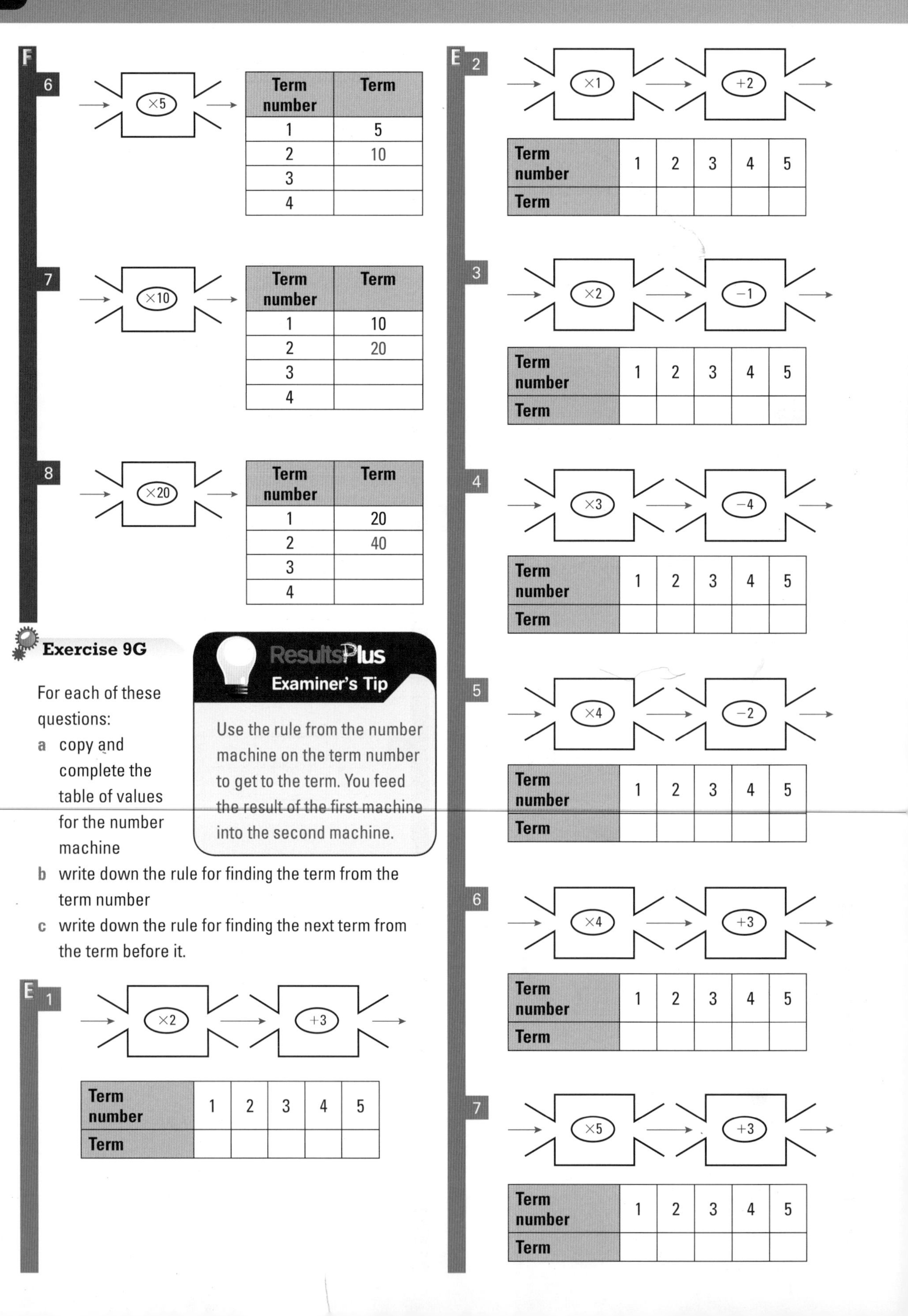

F

6 → ×5 →

Term number	Term
1	5
2	10
3	
4	

7 → ×10 →

Term number	Term
1	10
2	20
3	
4	

8 → ×20 →

Term number	Term
1	20
2	40
3	
4	

Exercise 9G

ResultsPlus
Examiner's Tip

Use the rule from the number machine on the term number to get to the term. You feed the result of the first machine into the second machine.

For each of these questions:

a copy and complete the table of values for the number machine

b write down the rule for finding the term from the term number

c write down the rule for finding the next term from the term before it.

E

1 → ×2 → +3 →

Term number	1	2	3	4	5
Term					

2 → ×1 → +2 →

Term number	1	2	3	4	5
Term					

3 → ×2 → −1 →

Term number	1	2	3	4	5
Term					

4 → ×3 → −4 →

Term number	1	2	3	4	5
Term					

5 → ×4 → −2 →

Term number	1	2	3	4	5
Term					

6 → ×4 → +3 →

Term number	1	2	3	4	5
Term					

7 → ×5 → +3 →

Term number	1	2	3	4	5
Term					

E **8**

→ ×5 → −2 →

Term number	1	2	3	4	5
Term					

9

→ ×4 → +5 →

Term number	1	2	3	4	5
Term					

10

→ ×5 → −3 →

Term number	1	2	3	4	5
Term					

Exercise 9H

Copy and complete these tables of values.

ResultsPlus
Examiner's Tip

Don't forget Bidmas: you do the × before the −. You met Bidmas in Chapter 9.

E **1**

×2 → +2	
Term number	**Term**
1	4
2	
3	
4	
5	
↓	↓
10	
↓	↓
	34

E **2**

×3 → −1	
Term number	**Term**
1	2
2	
3	
4	
5	
↓	↓
10	
↓	↓
	35

3

×3 → +5	
Term number	**Term**
1	8
2	
3	
4	
5	
↓	↓
10	
↓	↓
	50

4

×4 → −2	
Term number	**Term**
1	2
2	
3	
4	
5	
↓	↓
10	
↓	↓
	78

E

5

×10 → +2	
Term number	Term
1	12
2	
3	
4	
5	
↓	↓
10	
↓	↓
	182

6

×4 → −3	
Term number	Term
1	1
2	
3	
4	
5	
↓	↓
10	
↓	↓
	53

7 a Find the 8th number in this number pattern.
2, 5, 8, 11, …
b What is the term number for the term that is 47?

8 a Find the 8th number in this number pattern.
4, 8, 12, 16, …
b What is the term number for the term that is 64?

9 a Find the 8th number in this number pattern.
6, 10, 14, 18, …
b What is the term number for the term that is 50?

9.3 Finding the nth term of a number pattern

Exercise 9I

C

1 For questions 1, 2 and 3 in Exercise 13H, find the nth term of each of the number patterns.

2 Write each pattern in a table and use the table to find the nth term of these number patterns.
Use your nth term to find the 10th term in each of these number patterns.

a 3, 5, 7, 9, 11, …
b 7, 9, 11, 13, …
c 5, 8, 11, 14, 17, …
d 11, 14, 17, 20, …
e 5, 9, 13, 17, 21, …
f 2, 6, 10, 14, 18, …
g 7, 12, 17, 22, 27, …
h 4, 9, 14, 19, 24, …
i 13, 18, 23, 28, …
j 5, 7, 9, 11, …
k 40, 35, 30, 25, …
l 38, 36, 34, 32, …
m 35, 32, 29, 26, …
n 20, 18, 16, 14, …
o 19, 17, 15, 13, …
p 190, 180, 160, …

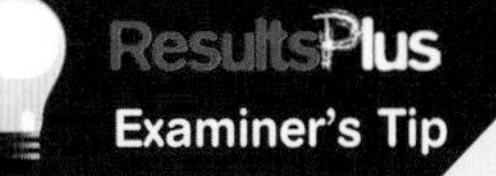

To find the nth term of a sequence that gets smaller you subtract a multiple of n from a fixed number.
e.g. $15 - 2n$ is the nth term of 13, 11, 9, 7, …

9.4 Deciding whether or not a number is in a number pattern

Exercise 9J

Are the numbers in brackets part of the sequence in each question?

C

1 2, 4, 6, 8, 10, … (17, 32)
2 4, 6, 8, 10, 12, … (72, 83)
3 1, 4, 7, 10, 13, … (54, 61)
4 4, 7, 10, 13, 16, … (52, 71)
5 3, 7, 11, 15, 19, … (92, 95)
6 4, 8, 12, 16, 20, … (111, 120)
7 1, 6, 11, 16, 21, … (67, 71)
8 2, 8, 14, 20, 26, … (32, 37)
9 60, 55, 50, 45, 40, … (10, 7)
10 44, 42, 40, 38, 36, … (24, 21
11 4, 9, 14, 19, 24, … (43, 34)
12 3, 6, 9, 12, 15, … (97, 99)

10 Graphs 1

Key Points

- **coordinate:** a point on a grid (x, y).
- **x-coordinate:** the number of units horizontally, given first.
- **y-coordinate:** the number of units vertically, given second.
- **origin:** point O with coordinates (0, 0).
- **x-axis:** the horizontal axis.
- **y-axis:** the vertical axis.
- **quadrants:** the four regions on a coordinate grid divided by the x- and y-axes.
- **finding the midpoint of a line:** add the x-coordinates and divide by 2, and add the y-coordinates and divide by 2.

10.1 Coordinates of points in the first quadrant

Exercise 10A

Questions in this chapter are targeted at the grades indicated.

G **1** Here is the plan of part of an adventure park. It is drawn on a coordinate grid.

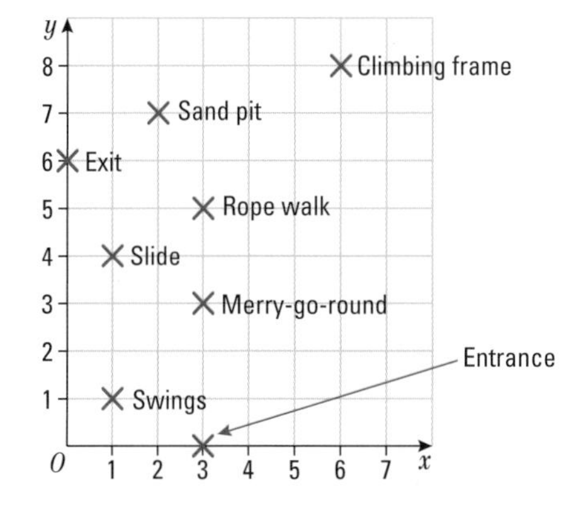

a Write down the names of the equipment at the following coordinates.

i (1, 1) ii (3, 3) iii (6, 8)
iv (0, 6) v (2, 7)

b Write down the coordinates where you will find the:

i entrance ii merry-go-round iii slide
iv swings v rope walk.

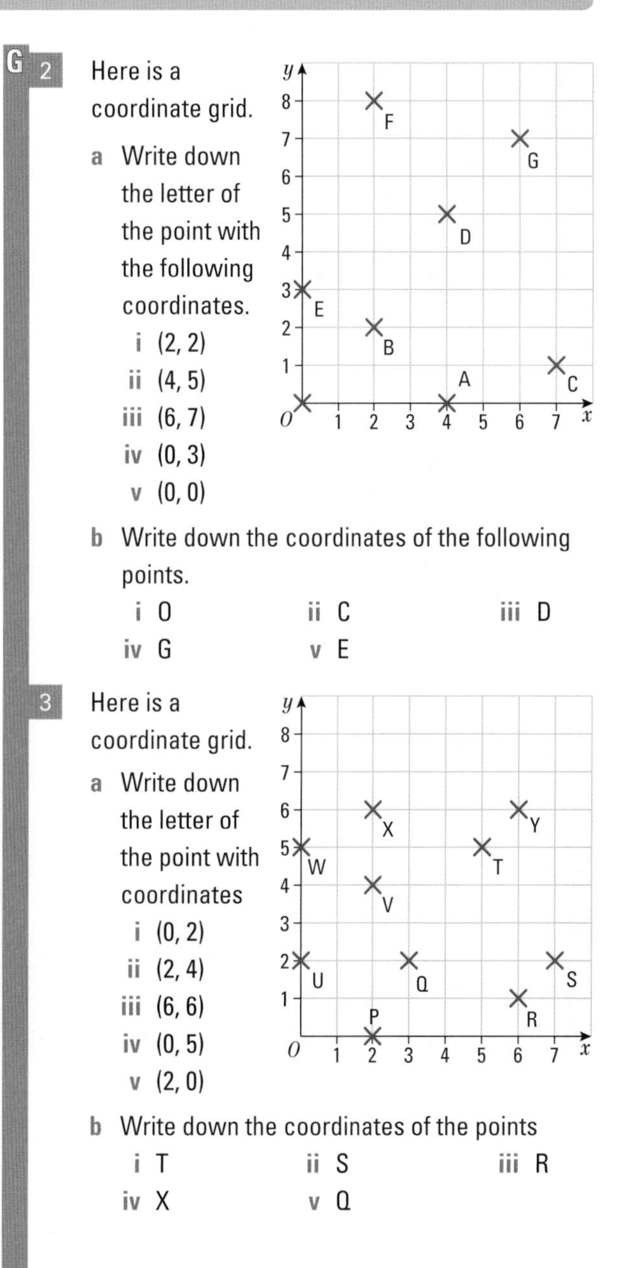

G **2** Here is a coordinate grid.

a Write down the letter of the point with the following coordinates.

i (2, 2)
ii (4, 5)
iii (6, 7)
iv (0, 3)
v (0, 0)

b Write down the coordinates of the following points.

i O ii C iii D
iv G v E

3 Here is a coordinate grid.

a Write down the letter of the point with coordinates

i (0, 2)
ii (2, 4)
iii (6, 6)
iv (0, 5)
v (2, 0)

b Write down the coordinates of the points

i T ii S iii R
iv X v Q

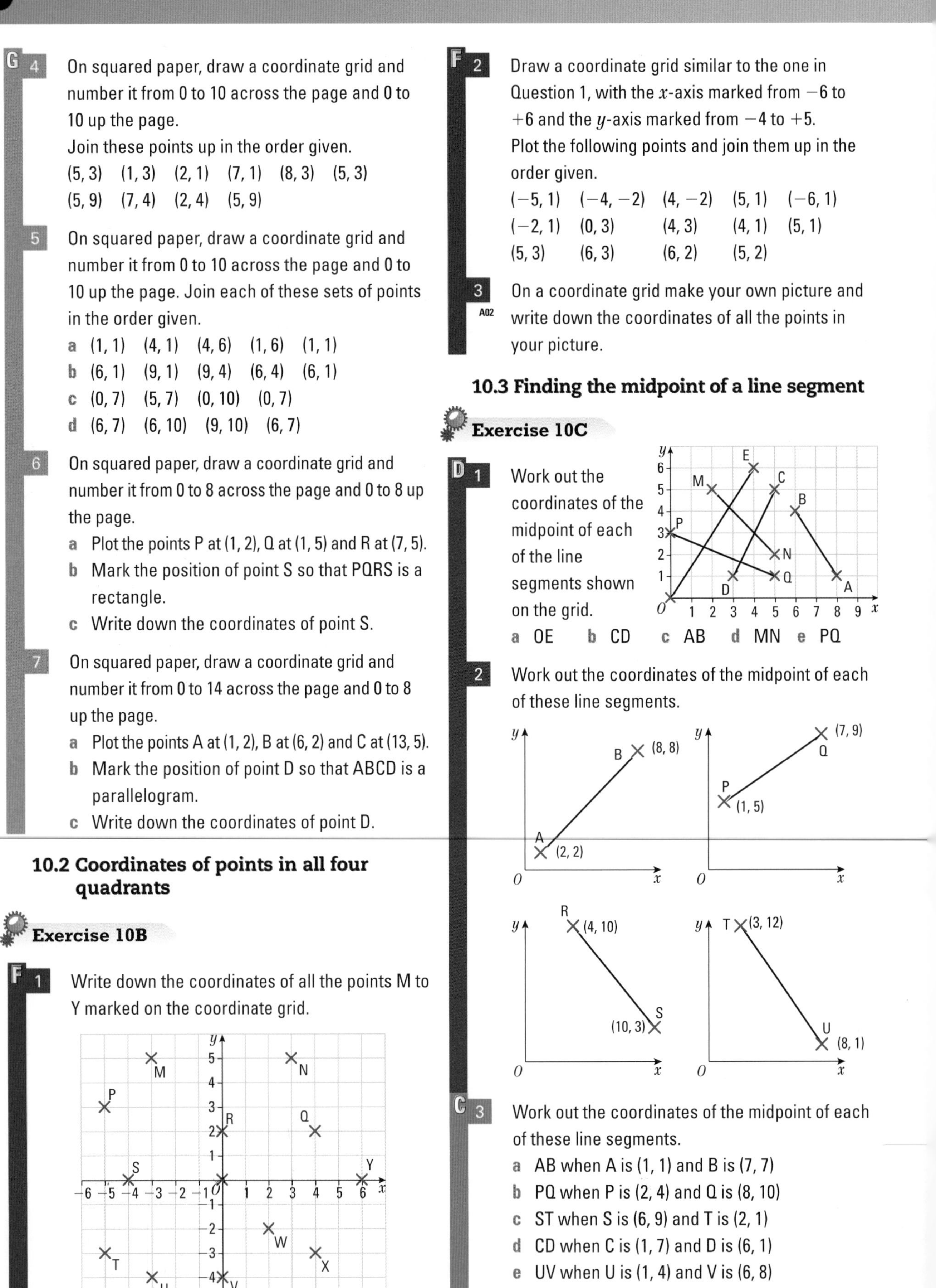

G

4 On squared paper, draw a coordinate grid and number it from 0 to 10 across the page and 0 to 10 up the page.
Join these points up in the order given.
(5, 3) (1, 3) (2, 1) (7, 1) (8, 3) (5, 3)
(5, 9) (7, 4) (2, 4) (5, 9)

5 On squared paper, draw a coordinate grid and number it from 0 to 10 across the page and 0 to 10 up the page. Join each of these sets of points in the order given.
a (1, 1) (4, 1) (4, 6) (1, 6) (1, 1)
b (6, 1) (9, 1) (9, 4) (6, 4) (6, 1)
c (0, 7) (5, 7) (0, 10) (0, 7)
d (6, 7) (6, 10) (9, 10) (6, 7)

6 On squared paper, draw a coordinate grid and number it from 0 to 8 across the page and 0 to 8 up the page.
a Plot the points P at (1, 2), Q at (1, 5) and R at (7, 5).
b Mark the position of point S so that PQRS is a rectangle.
c Write down the coordinates of point S.

7 On squared paper, draw a coordinate grid and number it from 0 to 14 across the page and 0 to 8 up the page.
a Plot the points A at (1, 2), B at (6, 2) and C at (13, 5).
b Mark the position of point D so that ABCD is a parallelogram.
c Write down the coordinates of point D.

10.2 Coordinates of points in all four quadrants

Exercise 10B

F

1 Write down the coordinates of all the points M to Y marked on the coordinate grid.

2 Draw a coordinate grid similar to the one in Question 1, with the x-axis marked from -6 to $+6$ and the y-axis marked from -4 to $+5$.
Plot the following points and join them up in the order given.
(−5, 1) (−4, −2) (4, −2) (5, 1) (−6, 1)
(−2, 1) (0, 3) (4, 3) (4, 1) (5, 1)
(5, 3) (6, 3) (6, 2) (5, 2)

3 On a coordinate grid make your own picture and write down the coordinates of all the points in your picture.
A02

10.3 Finding the midpoint of a line segment

Exercise 10C

D

1 Work out the coordinates of the midpoint of each of the line segments shown on the grid.
a OE **b** CD **c** AB **d** MN **e** PQ

2 Work out the coordinates of the midpoint of each of these line segments.

C

3 Work out the coordinates of the midpoint of each of these line segments.
a AB when A is (1, 1) and B is (7, 7)
b PQ when P is (2, 4) and Q is (8, 10)
c ST when S is (6, 9) and T is (2, 1)
d CD when C is (1, 7) and D is (6, 1)
e UV when U is (1, 4) and V is (6, 8)
f GH when G is (1, 5) and H is (7, 3)

Exercise 10D

D **1** Work out the coordinates of the midpoint of each of the line segments shown on the grid.

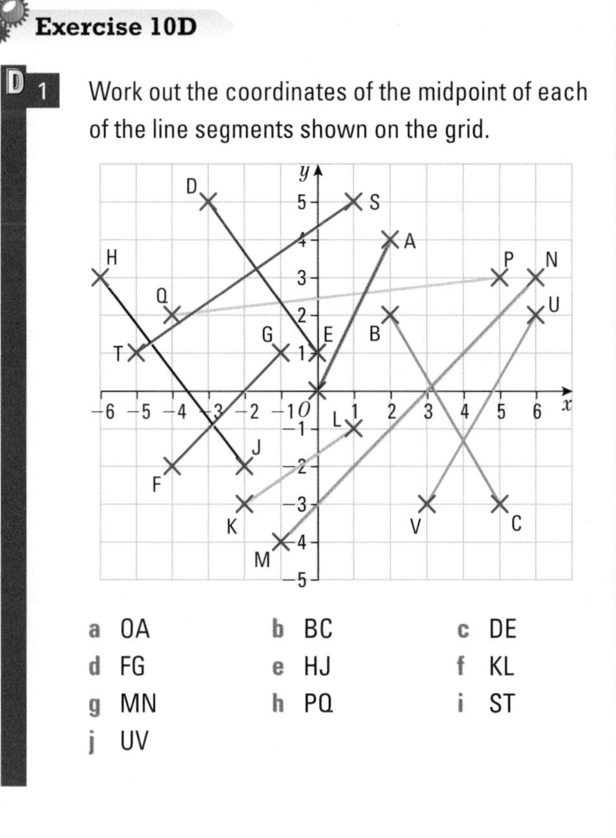

a OA **b** BC **c** DE
d FG **e** HJ **f** KL
g MN **h** PQ **i** ST
j UV

D **2** Work out the coordinates of the midpoint of each of these line segments.

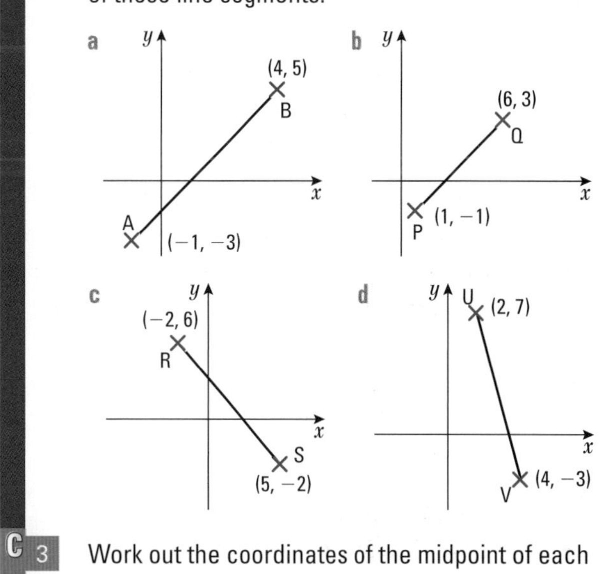

C **3** Work out the coordinates of the midpoint of each of these line segments.

a AB when A is (−1, −1) and B is (7, 7)

b PQ when P is (2, −5) and Q is (−6, 8)

c ST when S is (5, −8) and T is (−3, 4)

d CD when C is (2, 7) and D is (−7, 3)

e UV when U is (−3, 5) and V is (6, −8)

f GH when G is (−3, −5) and H is (7, 3)

11 Graphs 2

Key Points

- **horizontal line:** a line with the form $y = n$ where n is the y-coordinate.
- **vertical line:** a line with the form $x = n$ where n is the x-coordinate.
- **gradient:** the slant of a line.
  - **positive gradient:** a line that slants upwards
  - **negative gradient:** a line that slants downwards
- **drawing a straight-line graph from an equation:** make a table of values for some values of x, substitute x into the equation and plot the points on a grid.
- **drawing a straight-line graph without a table:** use the equation of the straight line $y = mx + c$, where m is the gradient and c is where the line crosses the y-axis.
- **finding the equation of a straight line:** find the gradient of the line and the point it crosses the y-axis. Put these into the form $y = mx + c$, where m is the gradient and c is where the line crosses the y-axis.

11.1 Drawing and naming horizontal and vertical lines

Exercise 11A

Questions in this chapter are targeted at the grades indicated.

ResultsPlus Examiner's Tip

The x-axis has equation $y = 0$.
The y-axis has equation $x = 0$.

E

1 Write down the equations of the lines marked **a** to **d** in this diagram.

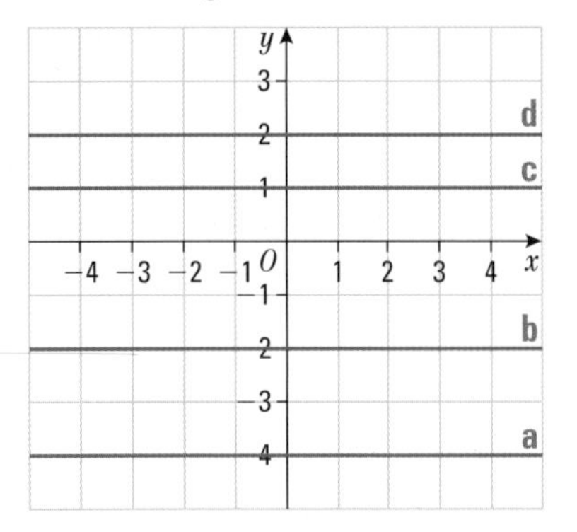

E

2 Write down the equations of the lines labelled **a** to **d** in this diagram.

3 Draw a coordinate grid with x- and y-axes labelled from -5 to 5.
On the grid draw and label the graphs of
a $x = 3$ **b** $x = -3$ **c** $x = -2$
d $x = 2$

4 Draw a coordinate grid with axes labelled from -5 to 5. On the grid draw and label the graphs of
a $y = 5$ **b** $y = -3$ **c** $y = -2$
d $y = 3$

5 Draw a coordinate grid with axes labelled from -5 to 5. On the grid draw and label the graphs of
a $y = 1$ **b** $x = -4$
c Write down the coordinates of the point where the two lines cross.

11.2 Drawing slanting lines

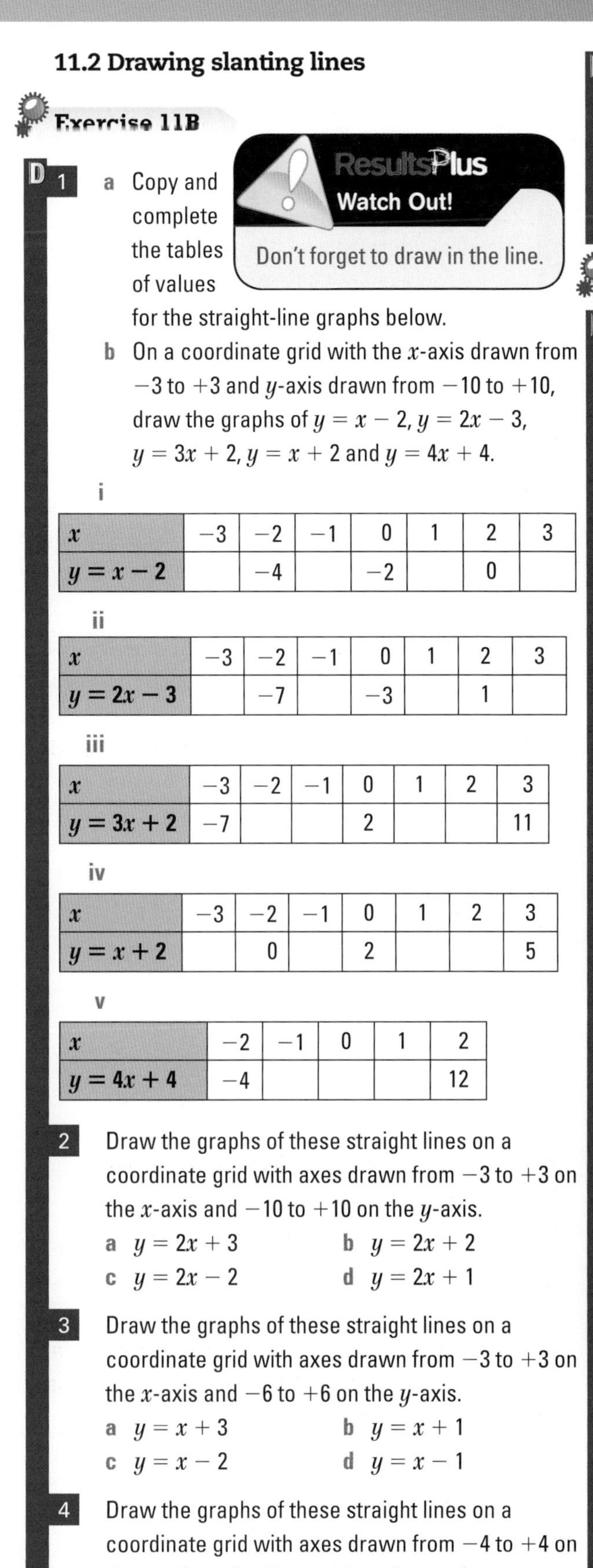

Exercise 11B

> **ResultsPlus Watch Out!**
> Don't forget to draw in the line.

D **1** **a** Copy and complete the tables of values for the straight-line graphs below.

b On a coordinate grid with the x-axis drawn from -3 to $+3$ and y-axis drawn from -10 to $+10$, draw the graphs of $y = x - 2$, $y = 2x - 3$, $y = 3x + 2$, $y = x + 2$ and $y = 4x + 4$.

i

x	-3	-2	-1	0	1	2	3
$y = x - 2$		-4		-2		0	

ii

x	-3	-2	-1	0	1	2	3
$y = 2x - 3$		-7		-3		1	

iii

x	-3	-2	-1	0	1	2	3
$y = 3x + 2$	-7			2			11

iv

x	-3	-2	-1	0	1	2	3
$y = x + 2$		0		2			5

v

x	-2	-1	0	1	2
$y = 4x + 4$	-4				12

2 Draw the graphs of these straight lines on a coordinate grid with axes drawn from -3 to $+3$ on the x-axis and -10 to $+10$ on the y-axis.

a $y = 2x + 3$ **b** $y = 2x + 2$
c $y = 2x - 2$ **d** $y = 2x + 1$

3 Draw the graphs of these straight lines on a coordinate grid with axes drawn from -3 to $+3$ on the x-axis and -6 to $+6$ on the y-axis.

a $y = x + 3$ **b** $y = x + 1$
c $y = x - 2$ **d** $y = x - 1$

4 Draw the graphs of these straight lines on a coordinate grid with axes drawn from -4 to $+4$ on the x-axis and -10 to $+10$ on the y-axis.

a $y = 2x - 2$ **b** $y = 3x + 1$
c $y = 2x - 1$ **d** $y = 3x - 1$

D **5** Draw the graphs of these straight lines on a coordinate grid with axes drawn from -4 to $+4$ on the x-axis and -6 to $+6$ on the y-axis.

a $y = \frac{1}{2}x + 2$ **b** $y = \frac{1}{2}x + 1$
c $y = \frac{1}{2}x - 2$ **d** $y = \frac{1}{2}x - 1$

Exercise 11C

D **1** **a** Copy and complete the tables of values for the straight-line graphs below.

b Draw the graphs of these straight lines on a coordinate grid with the x-axis drawn from -3 to $+3$ and y-axis drawn from -10 to $+10$.

i

x	-3	-2	-1	0	1	2	3
$y = -x - 2$		0		-2		-4	

ii

x	-3	-2	-1	0	1	2	3
$y = -2x - 3$		1		-3		-7	

iii

x	-3	-2	-1	0	1	2	3
$y = -3x + 2$	11			2			-7

iv

x	-3	-2	-1	0	1	2	3
$y = -x + 3$		5		3			0

v

x	-2	-1	0	1	2
$y = -4x + 2$	10				-6

2 Draw the graphs of these straight lines on a coordinate grid with axes drawn from -3 to $+3$ on the x-axis and -10 to $+10$ on the y-axis.

a $y = -2x + 3$ **b** $y = -2x + 2$
c $y = -2x - 2$ **d** $y = -2x + 1$

3 Draw the graphs of these straight lines on a coordinate grid with axes drawn from -3 to $+3$ on the x-axis and -6 to $+6$ on the y-axis.

a $y = -x - 1$ **b** $y = -x + 2$
c $y = -x - 2$ **d** $y = -x + 1$

4 Draw the graphs of these straight lines on a coordinate grid with axes drawn from -4 to $+4$ on the x-axis and -10 to $+10$ on the y-axis.

a $y = -2x - 2$ **b** $y = -3x + 2$
c $y = -2x - 1$ **d** $y = -3x - 1$

D **5** Draw the graphs of these straight lines on a coordinate grid with axes drawn from -4 to $+4$ on the x-axis and -6 to $+6$ on the y-axis.

a $y = -\frac{1}{2}x + 2$ b $y = -\frac{1}{2}x + 1$
c $y = -\frac{1}{2}x - 2$ d $y = -\frac{1}{2}x - 1$

11.3 Drawing straight-line graphs without a table of values

Exercise 11D

For each question, draw the graphs of all the straight lines on the same coordinate grid with the x-axis drawn from -3 to $+3$ and y-axis drawn from -10 to $+10$. What do you notice about the set of graphs for each question?

D **1** a $y = x + 2$ b $y = x + 1$
c $y = x + 4$ d $y = x - 2$
e $y = x - 1$

2 a $y = 2x + 2$ b $y = 2x + 1$
c $y = 2x + 4$ d $y = 2x - 2$
e $y = 2x - 1$

3 a $y = 3x + 2$ b $y = 3x + 1$
c $y = 3x + 4$ d $y = 3x - 2$
e $y = 3x - 1$

4 a $y = 4x + 2$ b $y = 4x + 1$
c $y = 4x + 4$ d $y = 4x - 2$
e $y = 4x - 1$

5 a $y = \frac{1}{2}x + 2$ b $y = \frac{1}{2}x + 1$
c $y = \frac{1}{2}x + 4$ d $y = \frac{1}{2}x - 2$
e $y = \frac{1}{2}x - 1$

Exercise 11E

For each question, draw the graphs of all the straight lines on the same coordinate grid with the x-axis drawn from -3 to $+3$ and y-axis drawn from -10 to $+10$. What do you notice about the set of graphs for each question?

D **1** a $y = -x + 2$ b $y = -x + 1$
c $y = -x + 4$ d $y = -x - 2$
e $y = -x - 1$

2 a $y = -2x + 2$ b $y = -2x + 1$
c $y = -2x + 4$ d $y = -2x - 2$
e $y = -2x - 1$

D **3** a $y = -3x + 2$ b $y = -3x + 1$
c $y = -3x + 4$ d $y = -3x - 2$
e $y = -3x - 1$

4 a $y = -4x + 2$ b $y = -4x + 1$
c $y = -4x + 4$ d $y = -4x - 2$
e $y = -4x - 1$

5 a $y = -\frac{1}{2}x + 2$ b $y = -\frac{1}{2}x + 1$
c $y = -\frac{1}{2}x + 4$ d $y = -\frac{1}{2}x - 2$
e $y = -\frac{1}{2}x - 1$

Exercise 11F

For each question, draw the graphs of all the straight lines on the same coordinate grid with the x-axis drawn from -3 to $+3$ and y-axis drawn from -8 to $+8$.

D **1** a $x + y = 1$ b $x + y = 2$ c $x + y = 3$

2 a $x + y = -4$ b $x + y = -5$ c $x + y = 4$

3 a $x + y = 0$ b $x + y = 1.5$ c $x + y = 1$

11.4 Naming straight-line graphs

Exercise 11G

C **1** Write down the equations of these straight lines.

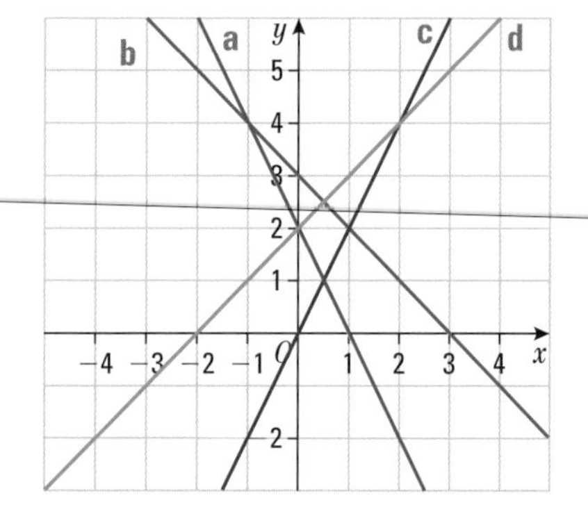

2 Write down the equations of these straight lines.

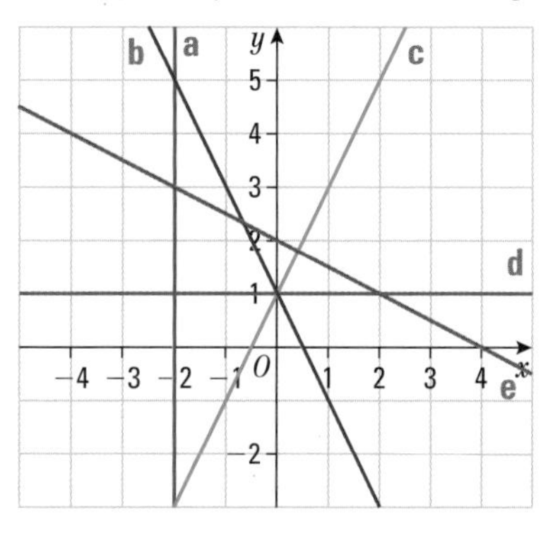

C 3 Write down the equations of these straight lines.

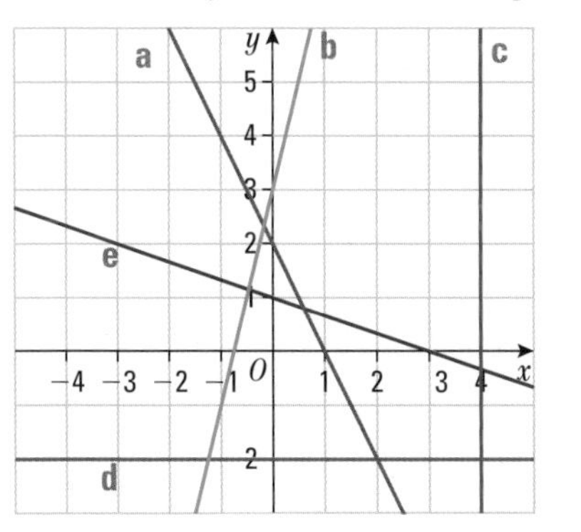

C 4 Write down the equations of these straight lines.

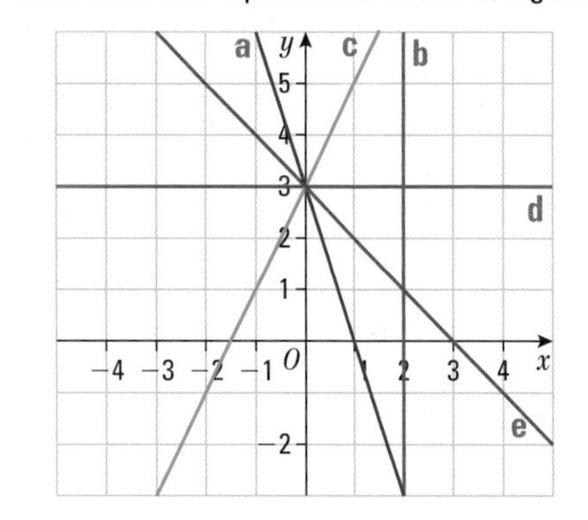

12 Graphs 3

Key Points

- **conversion graph:** a graph used to change measurements in one unit to a different unit.
- **distance–time graph:** a graph showing movement over time.
  - time is on the horizontal axis
  - distance is on the vertical axis
  - a slanting line shows movement
  - a horizontal line shows no movement
  - the steeper the slope the quicker the speed
- **calculating speed:** use a distance–time graph to find the distance travelled and the time taken, and use
  - **speed** $= \dfrac{\textbf{distance travelled}}{\textbf{time taken}}$

12.1 Interpreting and drawing the graphs you meet in everyday life

Exercise 12A

Questions in this chapter are targeted at the grades indicated.

F

1 The table shows the cost of apples per kg.

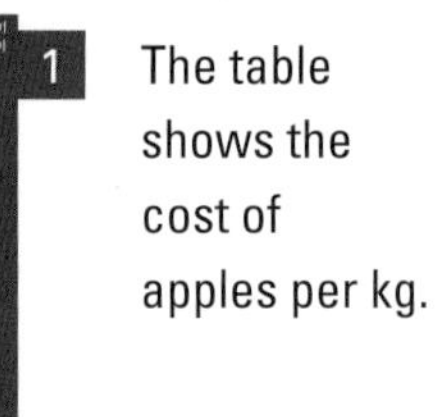

ResultsPlus
Examiner's Tip
Make sure you understand the scale of your graph before you draw it or read values from it.

Weight in kg	1	2	3	4	5
Cost in pence	40	70	100	130	160

a Draw a graph for this table.
b Work out how much 2.5 kg of apples would cost.
c Extend the graph to work out the cost of 6 kg of apples.

2 The table shows the cost of notepads.

Number of notepads	Cost in pence
1	30
2	60
3	90
4	120
5	150

a Draw a graph for the cost of notepads from the table.
b Extend the graph and then use it to work out the cost of
i 8 notepads ii 6 notepads.

F

3 The table shows the number of litres of petrol left in a car's petrol tank on a journey.

Travelling time in hours	Number of litres left
1	50
2	45
3	40
4	35
5	30
6	25
7	20
8	15

a Draw a graph from the information given in the table.
b How many litres were in the tank at the start of the journey (after 0 hours)?
c How many litres were in the tank after $5\frac{1}{2}$ hours?

4 A car uses 2 litres of petrol for every 10 km it travels.

a Copy and complete the table showing how much petrol the car uses.

Distance travelled in km	Petrol used in litres
0	0
10	2
20	4
30	
40	
50	

b Draw a graph from the information in your table.

F

c Work out how much petrol is used to travel 8 km.

d Work out how many kilometres have been travelled by the time 15 litres of petrol have been used.

5 The water in a reservoir is 130 m deep. During a dry period the water level falls by 3 m each week.

a Copy and complete this table showing the expected depth of water in the reservoir.

Week	Expected depth of water in m
0	130
1	127
2	
3	
4	
5	
6	
7	
8	

b Draw a graph from the information in your table.

c How deep would you expect the reservoir to be after 10 weeks?

If the water level falls to 96 m, the water company will divert water from another reservoir.

d After how long will the water company divert water?

6 Nisha has a pay-as-you-go mobile phone. She pays 30p for each minute she uses her phone.

a Copy and complete this table of values for the cost of using Nisha's phone.

Minutes used	Cost in pounds
0	0
5	
10	3
15	
20	
25	
30	
35	
40	
45	
50	15

F

b Plot the points in the table on a coordinate grid and draw a graph to show the cost of using Nisha's phone.

c Use your graph to find the cost of using her phone for 32 minutes.

d One month Nisha paid £8.40 to use her phone. For how many minutes did Nisha use her phone that month?

7 Will has a contract phone. He pays £10 each month and then 5p for each minute he uses his phone.

a Copy and complete this table of values for the cost of using Will's phone.

Minutes used	Cost in pounds
0	10
5	
10	
15	
20	11
25	
30	
35	
40	12
45	
50	

b Plot the points in the table on a coordinate grid and draw a graph to show the cost of using Will's phone.

c Use your graph to find the cost of Will using his phone for 32 minutes.

d One month Will paid £17 to use his phone. For how many minutes did Will use his phone that month?

8 Samir buys his gas from a company that charges 40p for each unit of gas he uses.

a Copy and complete the table (overleaf) of values for the cost of gas used by Samir.

F

Units used	Cost in pounds
0	0
10	
20	
30	
40	
50	20
60	
70	
80	
90	
100	40

b Plot the points in the table on a coordinate grid and draw a graph to show the cost of using gas.

c Use your graph to find the cost of using 32 units of gas.

d One month Samir paid £45 for gas. How many units of gas did Samir use that month?

E **9** Jack buys his electricity from a company that charges £15 each month and then 20p for each unit of electricity he uses.

a Copy and complete this table of values for the cost of using electricity for Jack.

Units used	Cost in pounds
0	15
10	
20	19
30	
40	
50	
60	
70	
80	
90	
100	35

b Plot the points in the table on a coordinate grid and draw a graph to show the cost of using electricity.

c Use your graph to find the cost of using 32 units of electricity.

d One month Jack paid £38 for electricity. How many units of electricity did Jack use that month?

C **10** The graph shows the cost of using internet broadband for one month on three different tariffs.

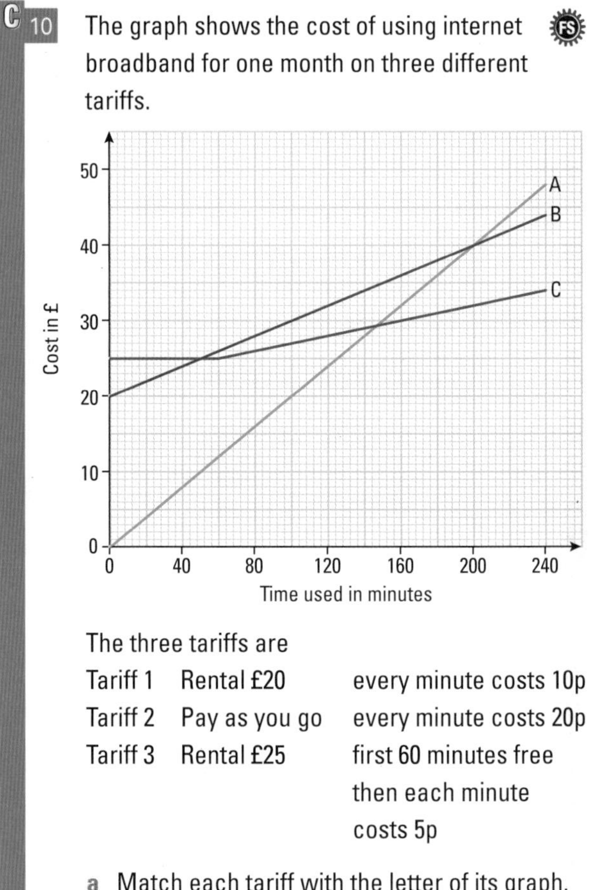

The three tariffs are

Tariff 1	Rental £20	every minute costs 10p
Tariff 2	Pay as you go	every minute costs 20p
Tariff 3	Rental £25	first 60 minutes free then each minute costs 5p

a Match each tariff with the letter of its graph.

Gemma uses the internet for more than 2 hours each month.

* A03 **b** Explain which tariff would be the cheapest for her to use.
You must give the reasons for your answer.

11 * A02 A03 Kizzy wants a smartphone. These are the tariffs offered.

Tariff	Cost/mth	Call cost/min
1	£10	30p
2	£12	20p
3	£15	10p
4	£20	5p

If Kizzy wants to use her phone for at least 60 mins per month, which tariff should she choose?

12.2 Drawing and interpreting conversion graphs

Exercise 12B

F 1 This graph can be used to convert between pounds (£) and Hong Kong dollars.

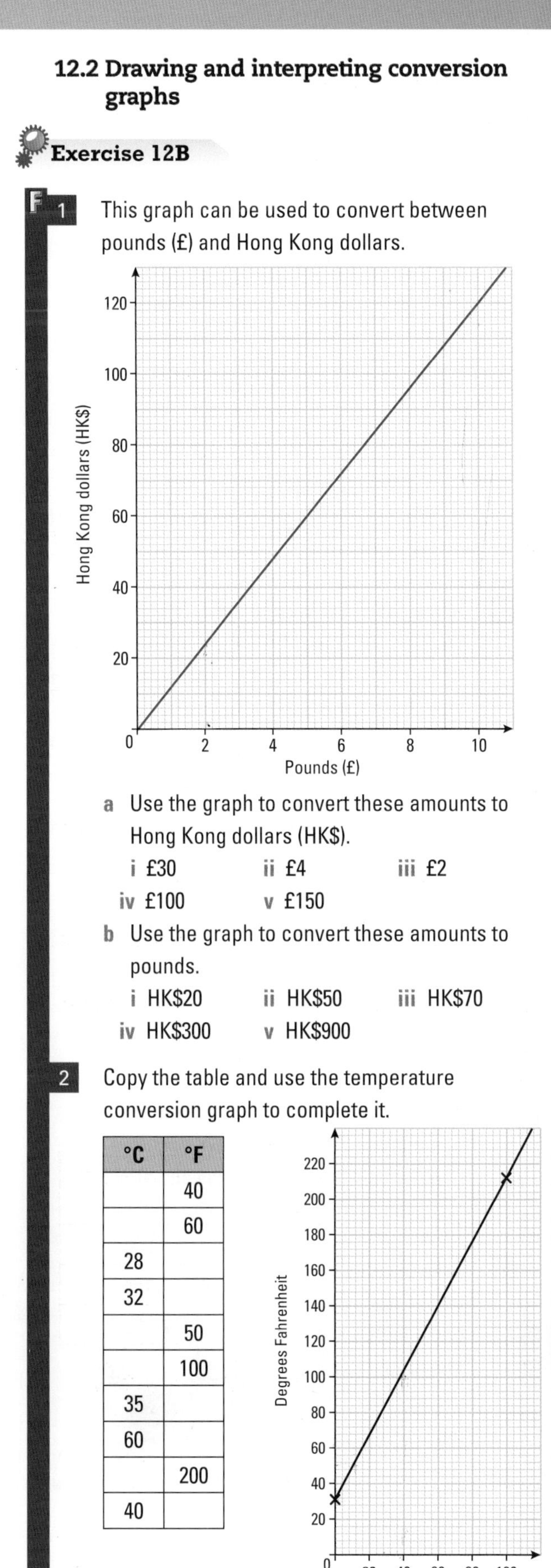

a Use the graph to convert these amounts to Hong Kong dollars (HK$).
i £30 ii £4 iii £2
iv £100 v £150

b Use the graph to convert these amounts to pounds.
i HK$20 ii HK$50 iii HK$70
iv HK$300 v HK$900

2 Copy the table and use the temperature conversion graph to complete it.

°C	°F
	40
	60
28	
32	
	50
	100
35	
60	
	200
40	

F 3 a Draw a conversion graph from pounds to kilograms. Use the fact that 0 pounds is 0 kilograms and 50 kg is approximately 110 pounds.
On your graph draw axes for kilograms and pounds using scales of 1 cm = 10 pounds and 1 cm = 10 kg. Plot the points (0, 0) and (50, 110) and join them with a straight line.

b Copy and complete this table using your conversion chart to help you.

Kilograms	Pounds
0	0
	10
9	
	99
30	
	33
23	
	14
	77
50	110

4 Copy this table and then use the information in the table to draw a conversion graph from inches into centimetres.
Use your graph to help you fill in the missing values.

Inches	Centimetres
0	0
1	
	5
	10
6	
8	
9	
	20
10	
12	30

5 Copy this table and then use the information in the table to draw a conversion graph from miles into kilometres.
Use your graph to help you fill in the missing values.

Miles	Kilometres
0	0
	8
10	
	64
	36
30	
45	
	20
24	
50	80

F 6 Copy this table and then use the information in the table to draw a conversion graph from acres into hectares.
Use your graph to help you fill in the missing values.

Hectares	Acres
0	0
8	
	12
	30
15	
17	
	24
18	
3	
20	50

12.3 Drawing and interpreting distance–time graphs

Exercise 12C

C 1 Jane walked to the library, chose some books then walked home again.

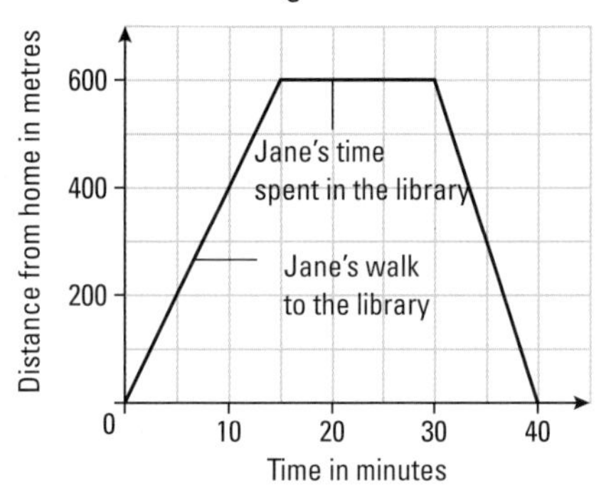

a How many minutes did it take Jane to walk to the library?

b How far away was the library?

c How many minutes did Jane spend in the library?

d How many minutes did it take Jane to walk home?

e Work out the speed at which Jane walked to the library.
First give your answer in metres per minute, then change it to km per hour.

f Work out the speed at which Jane walked back from the library.
First give your answer in metres per minute, then change it to km per hour.

C 2 A02 This graph shows Adam's journey by bike to deliver a present to his friend.

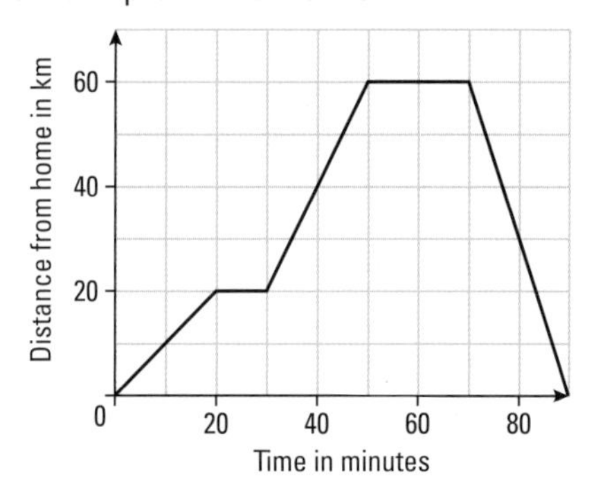

a Write a story of the journey. Explain what is happening in each part.

b What is Adam's speed in km/h for each part of his journey?

3 A02 Philip and Fiona need to travel to business meetings. This graph shows their journeys. Fiona catches a train from town A at 08:00 for a meeting in town B at 10:00. She has to change trains to get there. Philip drives from town B for a meeting in town A.

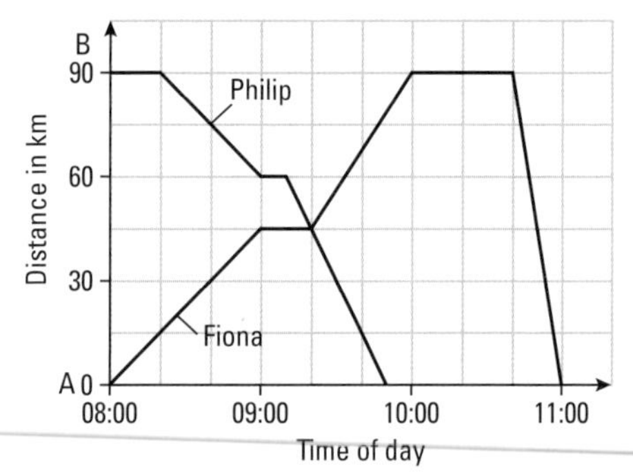

a Describe Fiona's journey in detail.

b Describe Philip's journey in detail.

c When did their paths cross and at what distance from A?

4* A02 Jake decides to give his dog a bath. The graph shows the depth of the water.
Explain what is happening in each part of the graph.

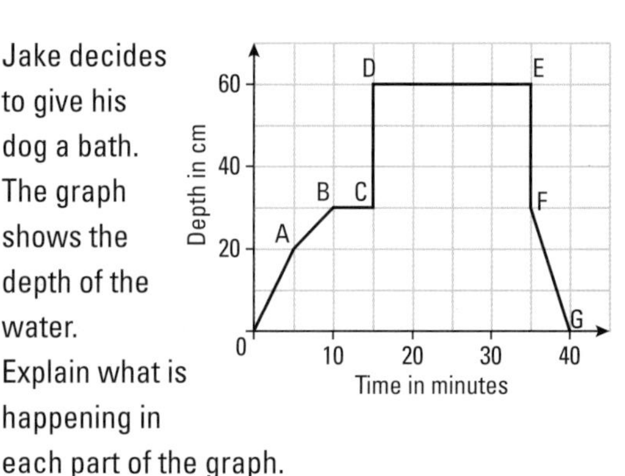

C **5*** A02 Iqra went to put a school wheelie bin out for collection at the end of the drive. It took her 5 minutes to take it the 200 metres and 2 minutes to position it. She walked back in 3 minutes.
Use graph paper to draw a distance–time graph for her journey.

6* A02 Esther was taken up in a glider. It rose 400 metres in the air in 30 minutes and stayed at this height for one and a quarter hours. It took 15 minutes to come back to the ground.
Use graph paper to draw a distance–time graph for her flight.

7 A02 Ben is going to college, 1.25 km from his home. He walks 250 metres to the bus stop. This takes him 4 minutes. A bus arrives after 5 minutes and the journey takes 6 minutes.

a Draw a distance–time graph for his journey.

b Work out the speed of the bus in

i m/min **ii** km/h.

8* A02 A03 Gary went to visit his Dad who lives 12 miles away. It took him 30 minutes to get there. He stayed 45 minutes and left. He had driven 4 miles in 10 minutes when he stopped to buy a newspaper, which took 5 minutes. He then drove straight home taking another 15 minutes.
Draw a distance–time graph for Gary's journey.

9 A03 A stone is thrown in the air. This graph illustrates its flight. Use the graph to find

a the height of the stone after

i 1.5 seconds **ii** 7.5 seconds

b the time taken for the stone to reach

i 80 metres on the way up

ii 80 metres on the way down.

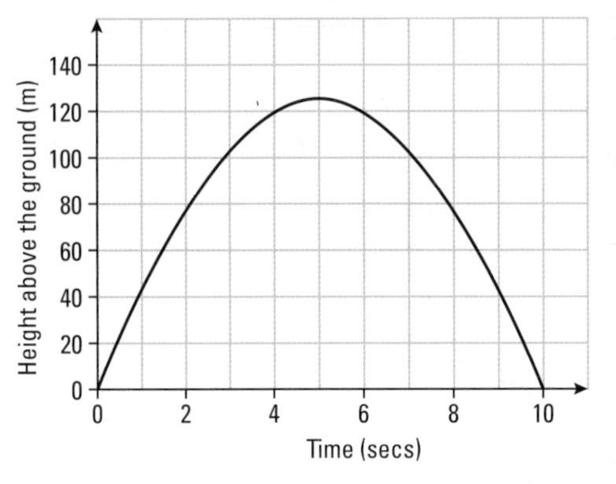

C **10**

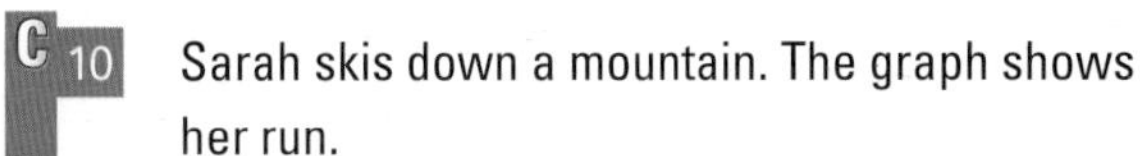

Sarah skis down a mountain. The graph shows her run.

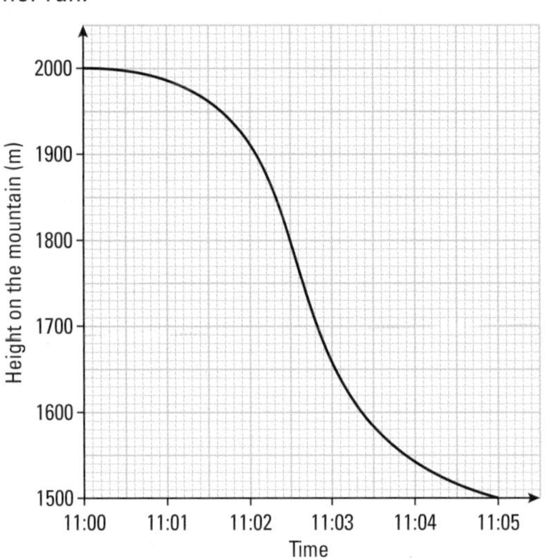

a From the graph, write down the height Sarah was at after

i 1 minute 30 seconds

ii 3 minutes 30 seconds

iii 4 minutes.

b Use the graph to write down the time at which Sarah was at the following heights.

i 1850 m

ii 1725 m

iii 1650 m.

13 Formulae

Key Points

- **word formula:** where words represent a relationship between quantities.
- **algebraic formula:** where one variable is equal to an expression in different variables, showing a relationship between quantities.
- **subject of the formula:** the variable that appears on its own on one side of the = sign. E.g. A is the subject of the formula $A = lw$.
- **substituting numbers into expressions:** replace the variables in the expression with the numeric values given.

13.1 Using word formulae

Exercise 13A

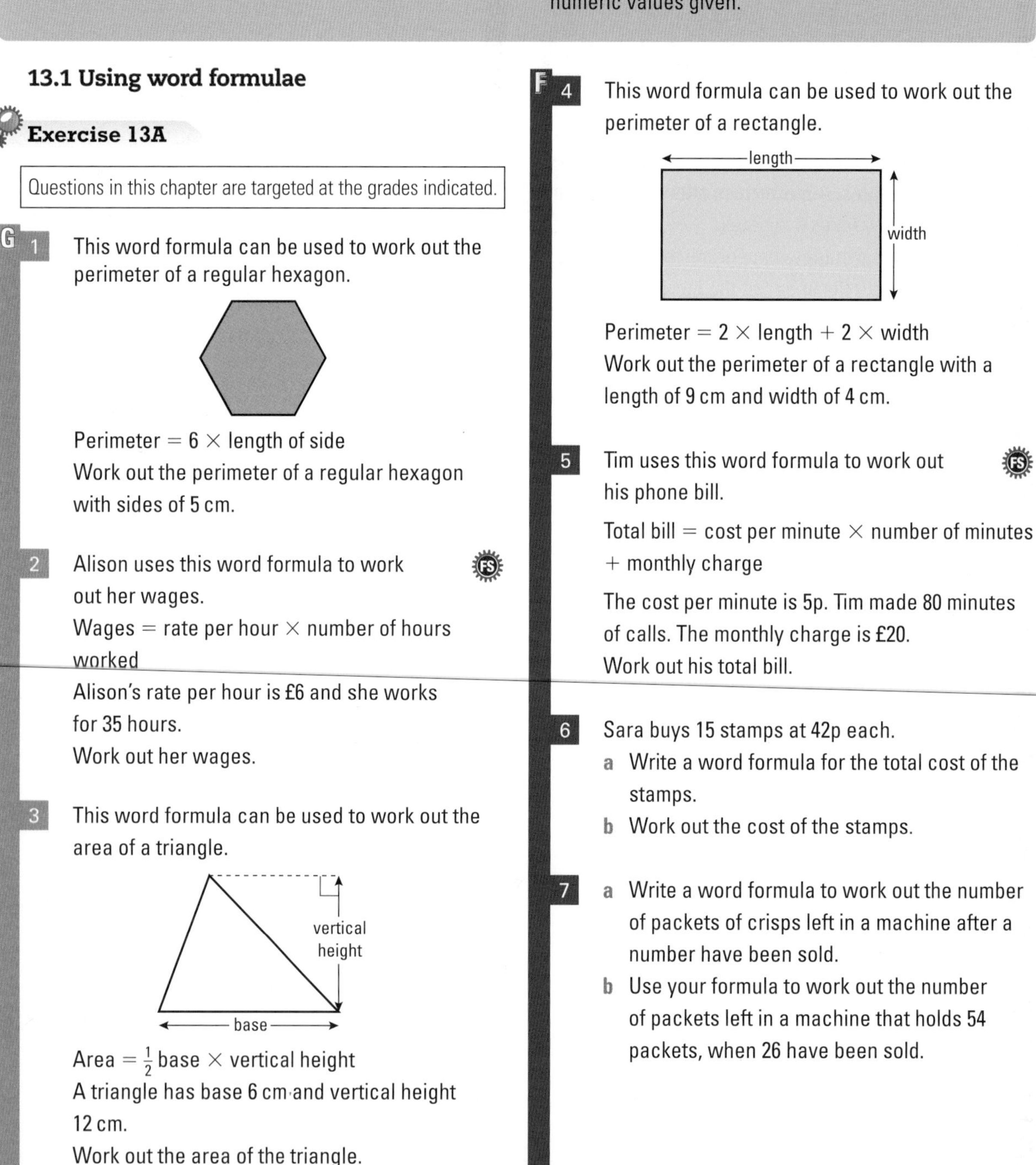

Questions in this chapter are targeted at the grades indicated.

G

1 This word formula can be used to work out the perimeter of a regular hexagon.

Perimeter = 6 × length of side

Work out the perimeter of a regular hexagon with sides of 5 cm.

2 Alison uses this word formula to work out her wages.

Wages = rate per hour × number of hours worked

Alison's rate per hour is £6 and she works for 35 hours.

Work out her wages.

3 This word formula can be used to work out the area of a triangle.

Area = $\frac{1}{2}$ base × vertical height

A triangle has base 6 cm and vertical height 12 cm.

Work out the area of the triangle.

F

4 This word formula can be used to work out the perimeter of a rectangle.

Perimeter = 2 × length + 2 × width

Work out the perimeter of a rectangle with a length of 9 cm and width of 4 cm.

5 Tim uses this word formula to work out his phone bill.

Total bill = cost per minute × number of minutes + monthly charge

The cost per minute is 5p. Tim made 80 minutes of calls. The monthly charge is £20.

Work out his total bill.

6 Sara buys 15 stamps at 42p each.

a Write a word formula for the total cost of the stamps.

b Work out the cost of the stamps.

7 a Write a word formula to work out the number of packets of crisps left in a machine after a number have been sold.

b Use your formula to work out the number of packets left in a machine that holds 54 packets, when 26 have been sold.

F **8** Petra uses this word formula to work out her take-home pay. (FS)

Take-home pay = rate per hour × number of hours worked − deductions

Petra's rate per hour is £6. She worked for 38 hours and her deductions were £94.

Work out her take-home pay.

9 For each part write a word formula, then use it to calculate the answers.

a Judy shared a bag of sweets equally between herself and her three brothers. There were 96 sweets in the bag. How many did each person have?

b At a buffet lunch there were 33 slices of pizza. Every person at the lunch had 3 slices. All the slices were eaten. How many people were at the lunch?

10 This word formula can be used to work out the average speed for a journey.

$$\text{Average speed} = \frac{\text{total distance travelled}}{\text{time taken}}$$

Jamal travels 225 miles in 5 hours. Work out his average speed in miles per hour (mph).

11 This word formula can be used to work out the angle sum, in degrees, of a polygon.

Angle sum = (number of sides − 2) × 180

Work out the angle sum of a polygon with 8 sides.

12 This word formula can be used to work out the size, in degrees, of each exterior angle of a regular polygon.

$$\text{Exterior angle} = \frac{360}{\text{number of sides}}$$

Work out the size of each exterior angle of a regular polygon with 5 sides.

13 This word formula can be used to work out the area inside a circle.

Area = π × radius × radius

Work out the area of a circle with a radius of 5 cm.

Give your answer to the nearest whole number.

13.2 Substituting numbers into expressions

Exercise 13B

D **1** $p = 4, q = 3, r = 2$ and $s = 0$

Work out the value of the following expressions.

a rs b $5p - 2r$ c $p + r$
d $pr - pq$ e pqr f $4(p + 7)$
g $p(r - q)$ h $6pq + 3qr$ i $p(q + 4)$
j $2p^3$ k $r^2 + 1$ l $(q + 1)^2$
m $(p - q)^3$ n $(p + r)^2$

2 $p = \frac{1}{4}, q = \frac{3}{4}, r = 1$ and $s = 2$

Work out the value of the following expressions.

a $4q$ b $4p - 6q$ c $7qr$
d $qr - pq$ e qrs f $4qr - 5pq$
g $r(r - 2)$ h $6(r - 2)$ i p^2
j $3r^3 - 2$ k $6q^3$ l $(r - 4)^2$
m $(p + q)^3$

3 $p = 0.2, q = 3, r = 2$ and $s = 0.25$

Work out the value of the following expressions.

a pq b $7p + 8s$ c $6p + 4q$
d $pr + 4q$ e $qr + rs$ f $6pr - 7qs$
g $q(p + r)$ h $5(p + q)$ i $5p^2$
j $5q^2 + 7$ k $p^2 + r^2$ l r^3
m $p^3 - q^3$

Exercise 13C

In this exercise $a = -6, b = 5, c = -3, d = \frac{1}{4}$ and $e = 1$.

Work out the value of the following expressions.

E
1 $a + b$ **2** $a - b$ **3** $b - a$
4 $a - c$ **5** $b - c$ **6** $a + b + c$
7 $3a + 7$ **8** $4a + 3b$ **9** $2b + 5c$
10 $2a - 5c$ **11** $3b - 2a$ **12** ab

D
13 acd **14** $3bc$ **15** $bd - 1$
16 $ab - bc$ **17** $2ab + 3ac$ **18** $3ac - 2bc$
19 $abcd$ **20** $3d(a + 1)$ **21** $b(c - a)$
22 $c(a + b)$

C
23 $5(c - 1)$ **24** $a(b + c)$ **25** a^2
26 $3c^2d$ **27** $4a^2 - 3$ **28** $5c^2 + 3c$

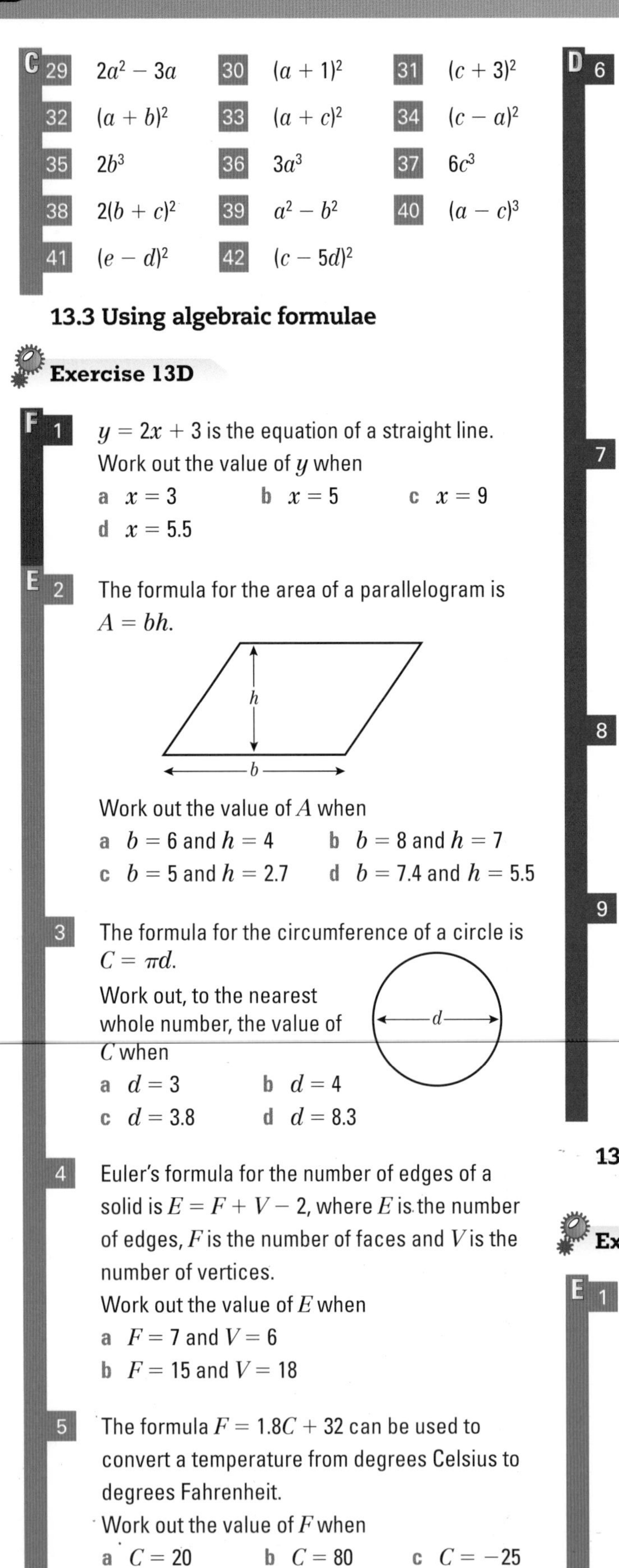

C

29 $2a^2 - 3a$

30 $(a + 1)^2$

31 $(c + 3)^2$

32 $(a + b)^2$

33 $(a + c)^2$

34 $(c - a)^2$

35 $2b^3$

36 $3a^3$

37 $6c^3$

38 $2(b + c)^2$

39 $a^2 - b^2$

40 $(a - c)^3$

41 $(e - d)^2$

42 $(c - 5d)^2$

13.3 Using algebraic formulae

Exercise 13D

F

1 $y = 2x + 3$ is the equation of a straight line.
Work out the value of y when

a $x = 3$ b $x = 5$ c $x = 9$
d $x = 5.5$

E

2 The formula for the area of a parallelogram is $A = bh$.

Work out the value of A when

a $b = 6$ and $h = 4$ b $b = 8$ and $h = 7$
c $b = 5$ and $h = 2.7$ d $b = 7.4$ and $h = 5.5$

3 The formula for the circumference of a circle is $C = \pi d$.
Work out, to the nearest whole number, the value of C when

a $d = 3$ b $d = 4$
c $d = 3.8$ d $d = 8.3$

4 Euler's formula for the number of edges of a solid is $E = F + V - 2$, where E is the number of edges, F is the number of faces and V is the number of vertices.
Work out the value of E when

a $F = 7$ and $V = 6$
b $F = 15$ and $V = 18$

5 The formula $F = 1.8C + 32$ can be used to convert a temperature from degrees Celsius to degrees Fahrenheit.
Work out the value of F when

a $C = 20$ b $C = 80$ c $C = -25$
d $C = 0$

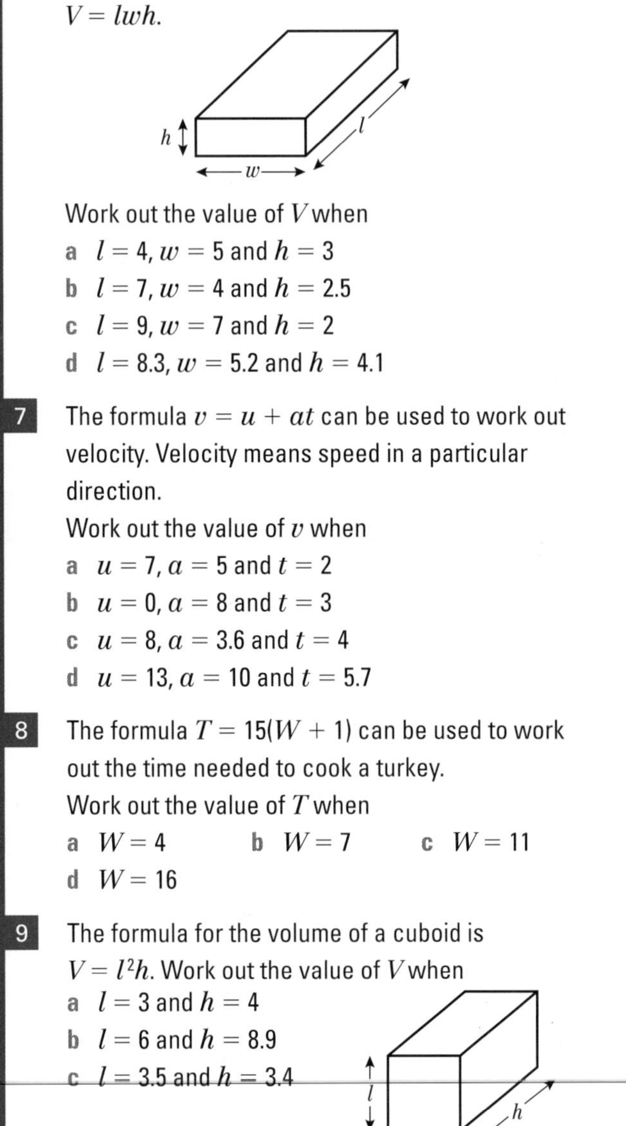

D

6 The formula for the volume of a cuboid is $V = lwh$.

Work out the value of V when

a $l = 4, w = 5$ and $h = 3$
b $l = 7, w = 4$ and $h = 2.5$
c $l = 9, w = 7$ and $h = 2$
d $l = 8.3, w = 5.2$ and $h = 4.1$

7 The formula $v = u + at$ can be used to work out velocity. Velocity means speed in a particular direction.
Work out the value of v when

a $u = 7, a = 5$ and $t = 2$
b $u = 0, a = 8$ and $t = 3$
c $u = 8, a = 3.6$ and $t = 4$
d $u = 13, a = 10$ and $t = 5.7$

8 The formula $T = 15(W + 1)$ can be used to work out the time needed to cook a turkey.
Work out the value of T when

a $W = 4$ b $W = 7$ c $W = 11$
d $W = 16$

9 The formula for the volume of a cuboid is $V = l^2h$. Work out the value of V when

a $l = 3$ and $h = 4$
b $l = 6$ and $h = 8.9$
c $l = 3.5$ and $h = 3.4$

13.4 Writing an algebraic formula to represent a problem

Exercise 13E

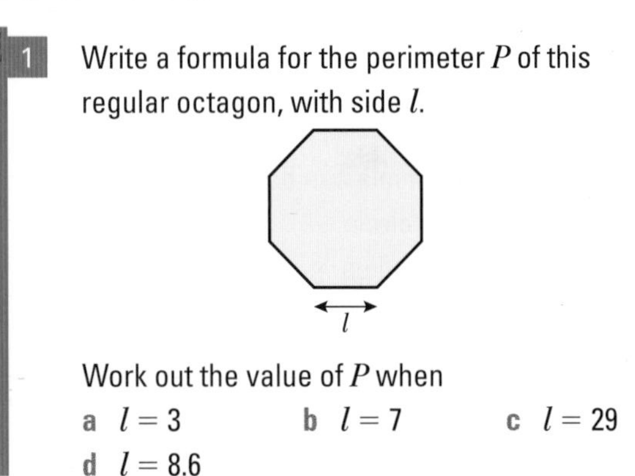

E

1 Write a formula for the perimeter P of this regular octagon, with side l.

Work out the value of P when

a $l = 3$ b $l = 7$ c $l = 29$
d $l = 8.6$

E **2** Write a formula for the perimeter of this isosceles triangle.

Work out the perimeter when

a $a = 6$ and $b = 3$

b $a = 12$ and $b = 8$

c $a = 5.3$ and $b = 2.4$

d $a = 3.7$ and $b = 8.5$

3 a Write an algebraic formula for the price of a number of pens that cost 60p each.

b Use your formula to work out the cost of:

i 3 pens ii 5 pens iii 10 pens.

4 Write a formula for the volume of this cube.

Work out the volume when

a $s = 3$ cm

b $s = 5.5$ cm

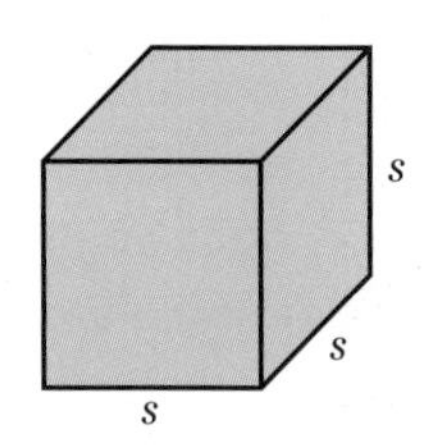

E **5** Write a formula for the surface area of the cube in question 4.

Work out the surface area when

a $s = 3$ cm b $s = 5.5$ cm

D **6** Write a formula for the length of the side of a square given the area.

Work out the length of the side when the area is

a $9\,\text{cm}^2$ b $1.32\,\text{cm}^2$

14 Angles 1

Key Points

- **an angle:** a measure of turn that is formed when two lines meet.
  - **full turn:** 360°
  - **half turn:** 180°
  - **quarter turn:** 90°, also known as a right angle.
- **types of angle:**

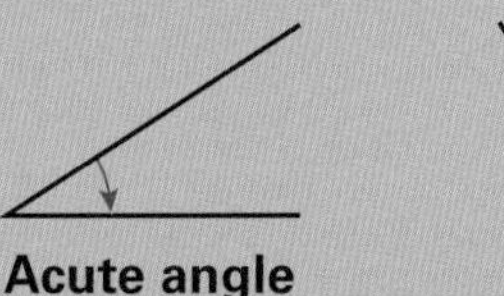

Acute angle
Less than $\frac{1}{4}$ turn

Obtuse angle
More than $\frac{1}{4}$ turn

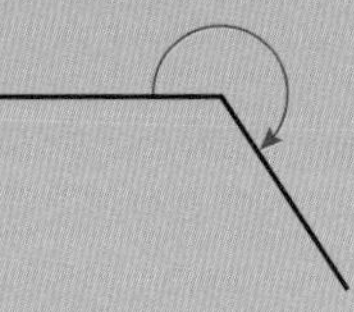

Reflex angle
More than $\frac{1}{2}$ turn

Right angle
90° or $\frac{1}{4}$ turn

- **line AB:** a line that starts at point A and ends at point B.
- **angle ABC (∠ABC):** the angle at B, between line AB and line BC.
- **protractor:** an instrument used to measure angles.
- **angle facts:**
  - **interior angles of a triangle** add up to 180°
  - **angles on a straight line** add up to 180°
  - **angles around a point** add up to 360°
  - when two lines cross, **vertically opposite angles** are equal
- **types of triangle:**
  - **isosceles:** 2 equal sides, 2 equal angles
  - **equilateral:** 3 equal sides, 3 equal angles
  - **right-angled:** 1 right angle
- **naming sides and angles of a shape:** label the corners of the shape with letters.
- **estimating angle measurements:** estimate the size of an angle before measuring it. This is useful for checking that an answer is sensible.
- **drawing with a protractor and ruler:** angles must be accurate to within 2° and lines to within 2 mm.

14.1 Fractions of a turn and degrees

Exercise 14A

Questions in this chapter are targeted at the grades indicated.

G 1 Write down the size of the following angles in degrees.

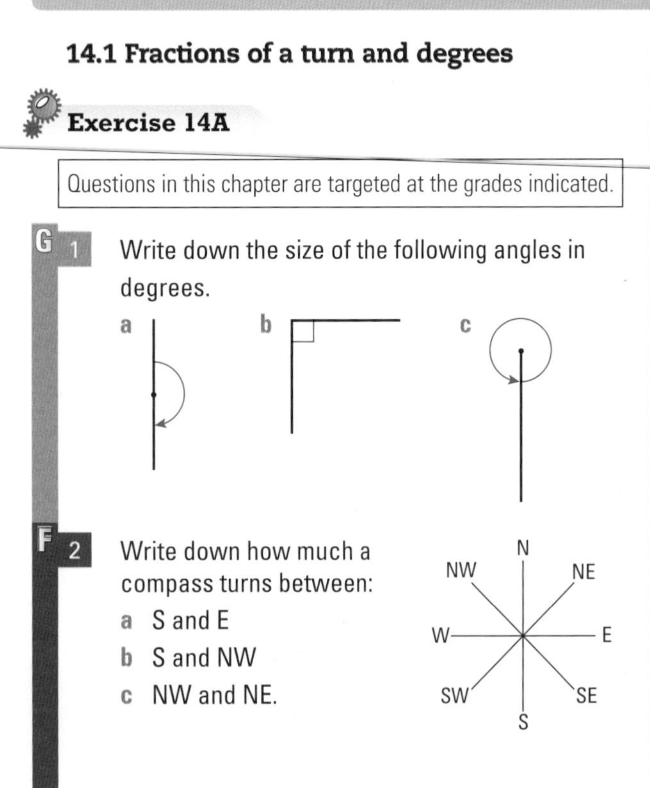

F 2 Write down how much a compass turns between:

a S and E
b S and NW
c NW and NE.

F 3 On a clock, how many degrees does the hour hand turn between:

a 4 pm and 10 pm
b 4 am and 7 am
c 7 pm and 2 am

C 4 (A02 A03) A lifeboat has to go 5 km due East and then 5 km due North to reach a boat in distress. It then goes back to the harbour by the shortest distance. What compass bearing does it use to travel back to the harbour?

5 (A02 A03) A doctor, Esther, has to make a house call at 8 am. It takes her 10 mins to reach her patient. Esther is with the patient for 35 minutes. It then takes her 5 minutes to get to the health centre. On a clock, how many degrees has the minute hand turned, from the time Esther drives to the patient to the time she reaches the health centre?

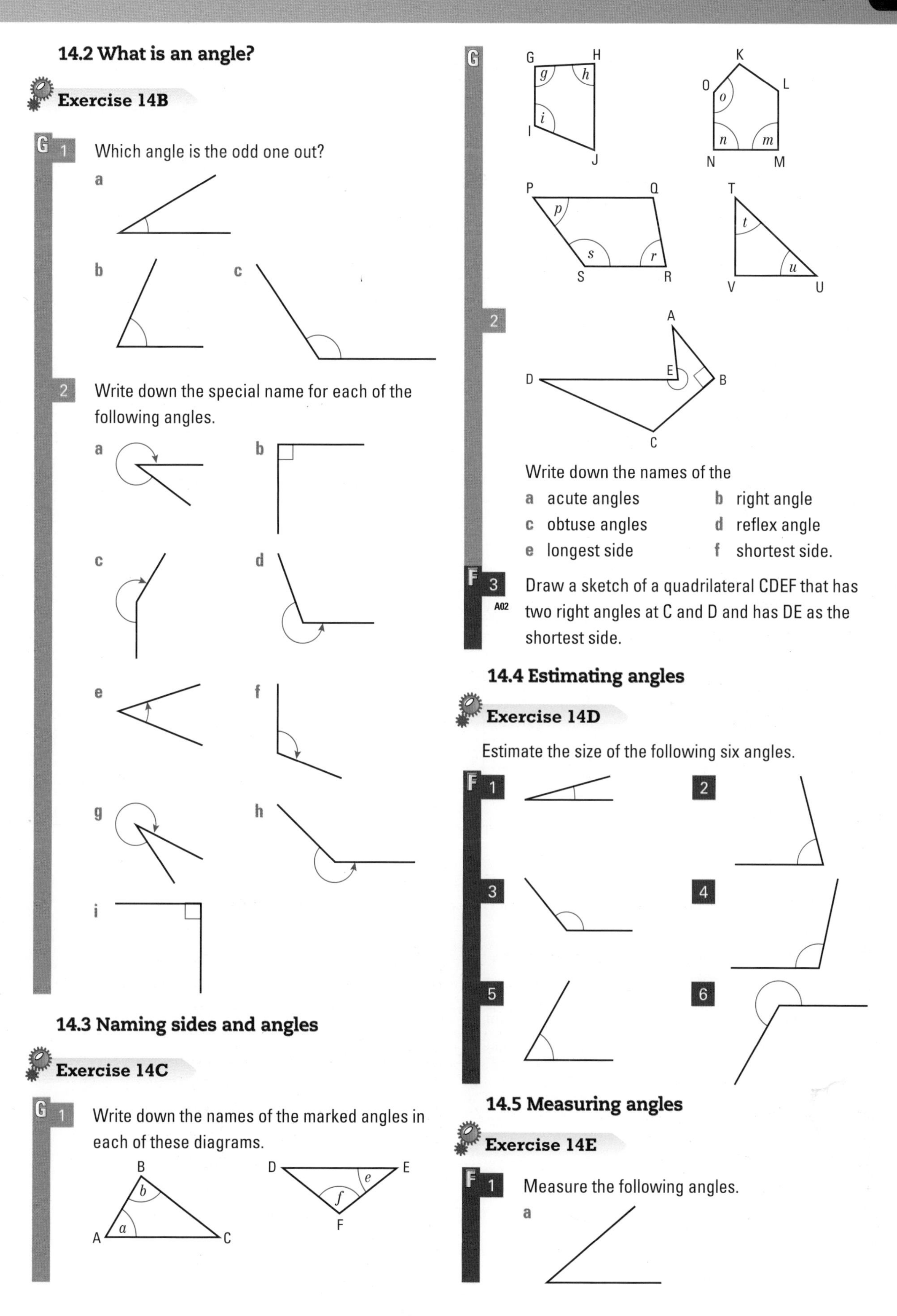

14.2 What is an angle?

Exercise 14B

G

1 Which angle is the odd one out?

a

b c

2 Write down the special name for each of the following angles.

a b c d e f g h i

14.3 Naming sides and angles

Exercise 14C

G

1 Write down the names of the marked angles in each of these diagrams.

2

Write down the names of the

a acute angles

b right angle

c obtuse angles

d reflex angle

e longest side

f shortest side.

F

3 (A02) Draw a sketch of a quadrilateral CDEF that has two right angles at C and D and has DE as the shortest side.

14.4 Estimating angles

Exercise 14D

Estimate the size of the following six angles.

F

1 2 3 4 5 6

14.5 Measuring angles

Exercise 14E

F

1 Measure the following angles.

a

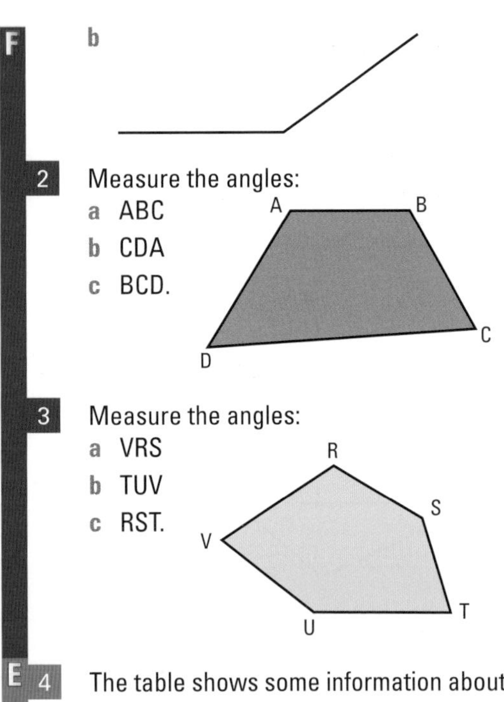

b

2 Measure the angles:

a ABC

b CDA

c BCD.

3 Measure the angles:

a VRS

b TUV

c RST.

4 The table shows some information about the angles used in cutting different materials with a chisel.

Material	Cutting angle	Angle of inclination
Aluminium	30°	22°
Medium steel	65°	39.5°
Mild steel	55°	34.5°
Brass	50°	32°
Copper	45°	29.5°
Cast iron	60°	37°

The diagram shows the angle of inclination for the chisel.

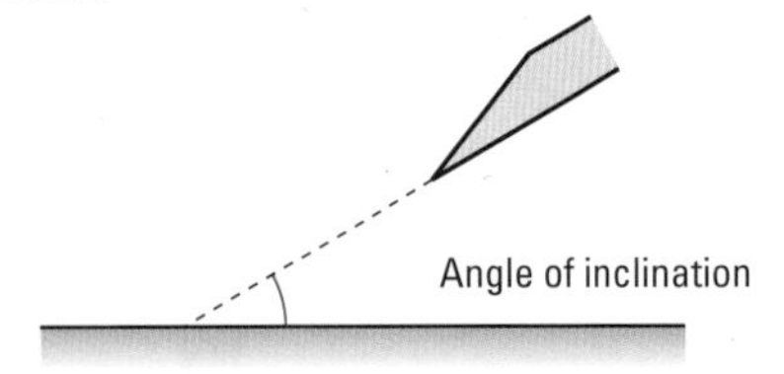

By measuring the angle shown in the diagram, write down the material being cut by the chisel.

14.6 Drawing angles

Exercise 14F

1 Use a protractor to draw the following angles.

a 60° b 150° c 65°

d 125° e 53° f 77°

g 127° h 172° i 18°

j 86°

2 Draw and label the following angles.

a ABC = 40° b DEF = 110°

c GHK = 75° d LMN = 58°

e PQR = 172° f STU = 95°

14.7 Special triangles

Exercise 14G

1 Work out the sizes of the missing angles in the triangles below.

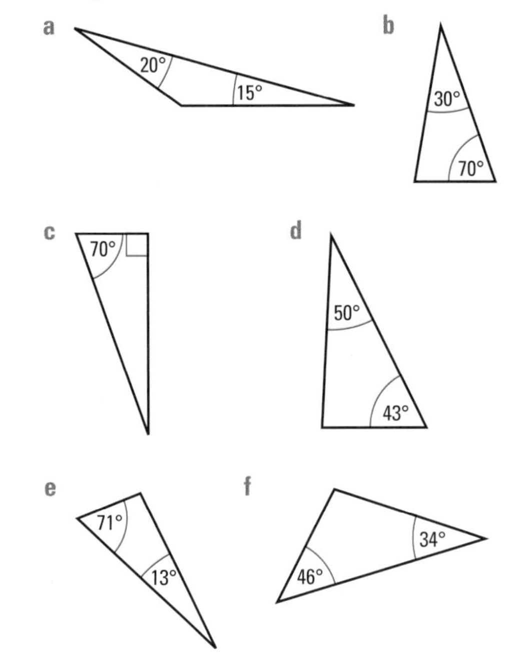

2 Work out the sizes of the missing angles in the following triangles.

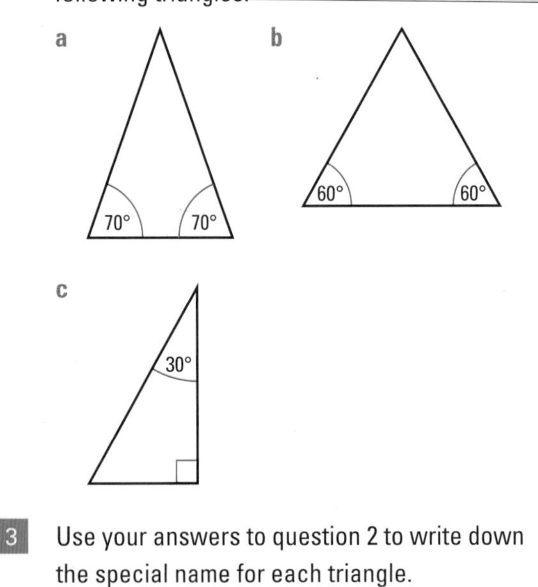

3 Use your answers to question 2 to write down the special name for each triangle.

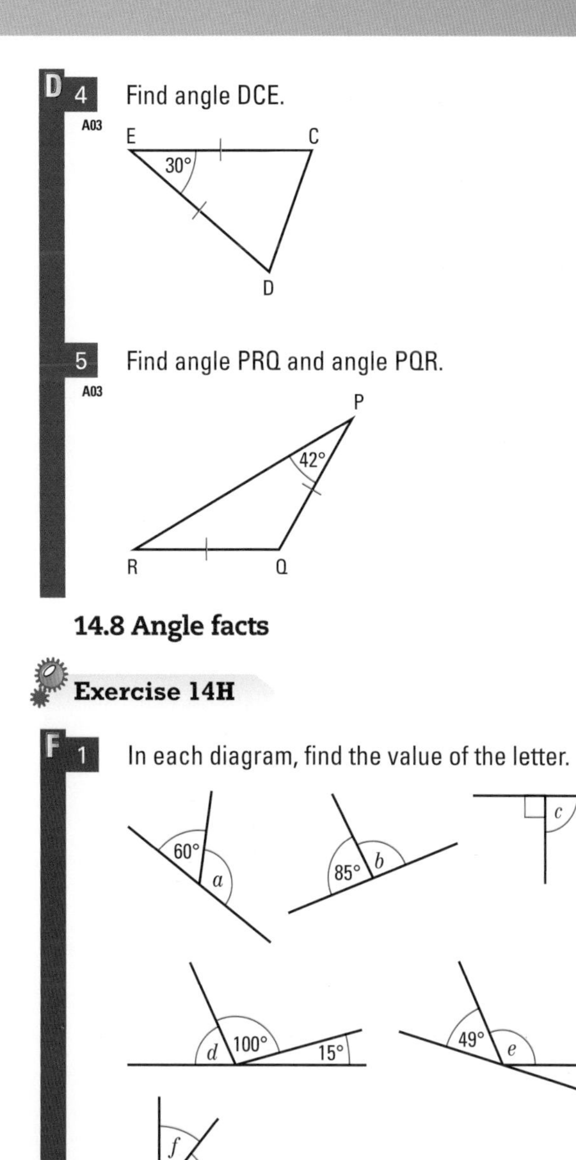

D

4 A03 Find angle DCE.

5 A03 Find angle PRQ and angle PQR.

14.8 Angle facts

Exercise 14H

F

1 In each diagram, find the value of the letter.

F

2 Find the value of the letter in each of the following diagrams.

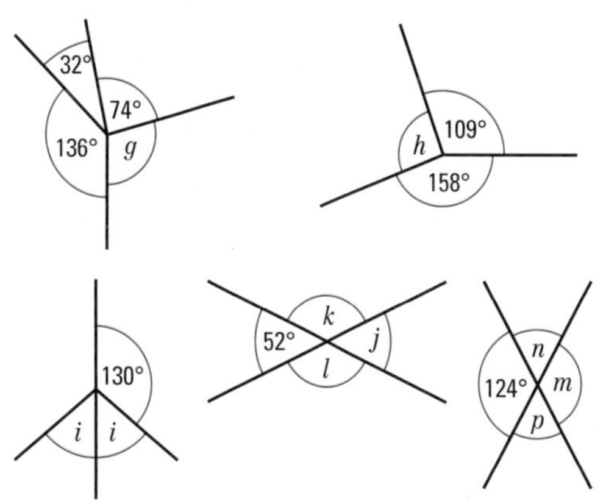

3 Find the value of the letters in the diagrams below.
Give reasons for your answers.

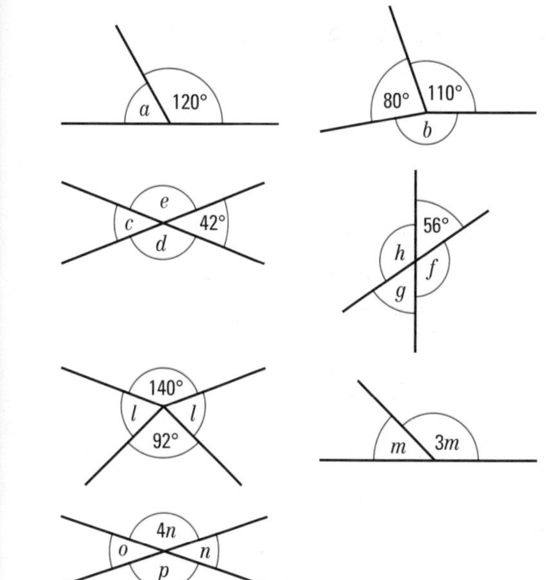

15 Two-dimensional shapes

Key Points

- **types of triangle:**
  - **isosceles:** 2 equal angles, 2 equal sides
  - **equilateral:** 3 equal angles, 3 equal sides
  - **right-angled:** 1 right angle
  - **scalene:** 3 different angles, 3 different sides
  - **acute-angled:** 3 acute angles
  - **obtuse-angled:** 1 obtuse angle
- **quadrilateral:** a four-sided shape with diagonals joining opposite corners.
- **types of quadrilateral:**
  - **square:** 4 equal sides, 4 right angles
  - **rectangle:** 2 pairs of equal and parallel sides, 4 right angles
  - **parallelogram:** 2 pairs of equal and parallel sides
  - **trapezium:** 1 pair of parallel sides
  - **kite:** 2 pairs of equal adjacent sides
  - **rhombus:** 4 equal sides, 2 pairs of parallel sides

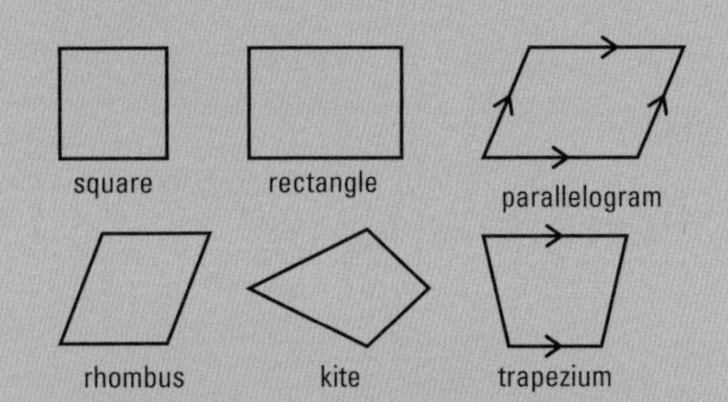

- **parts of a circle:** circumference, diameter, radius, tangent, chord, arc, segment, sector.
- **similar shapes:** when one shape is an enlargement of another shape.
- **symmetrical shape:** a shape that can be folded in half with one half a mirror image of the other half. The dividing line is the **line of symmetry** or **mirror line**.
- **rotational symmetry:** occurs if a shape looks the same as is did at 0° during a 360° rotation. The number of times during the rotation that it looks the same is the **order of rotational symmetry**.

15.1 Triangles

Exercise 15A

Questions in this chapter are targeted at the grades indicated.

G **1** Match each triangle (A to F) with its mathematical names (1 to 6).
Each shape can be matched with two names.

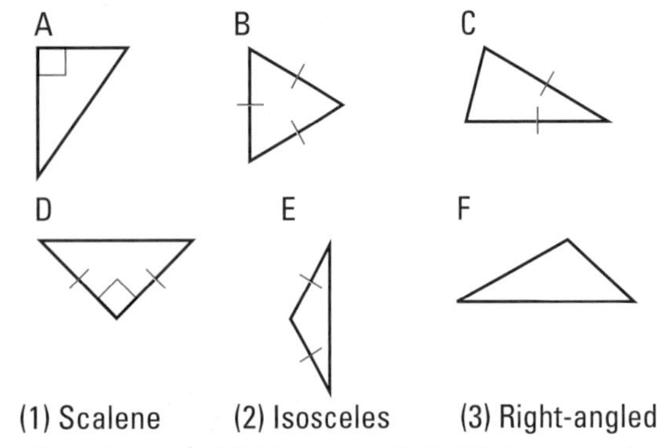

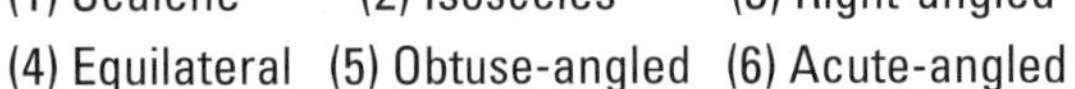
(1) Scalene (2) Isosceles (3) Right-angled
(4) Equilateral (5) Obtuse-angled (6) Acute-angled

G **2**

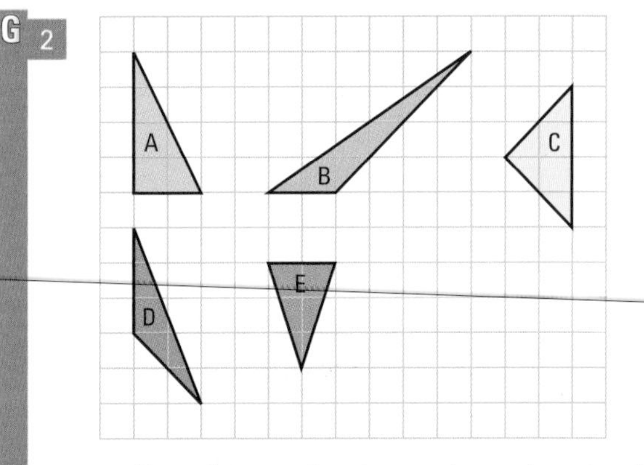

a Two of these triangles are isosceles triangles. Which two?

b Which of these triangles are right-angled triangles?

c Two of these triangles are obtuse-angled triangles. Which two?

d Which is a scalene triangle?

3 Draw a sketch of a right-angled triangle that is also isosceles.

F 4 (AO3) Adam says this triangle is an acute-angled triangle.

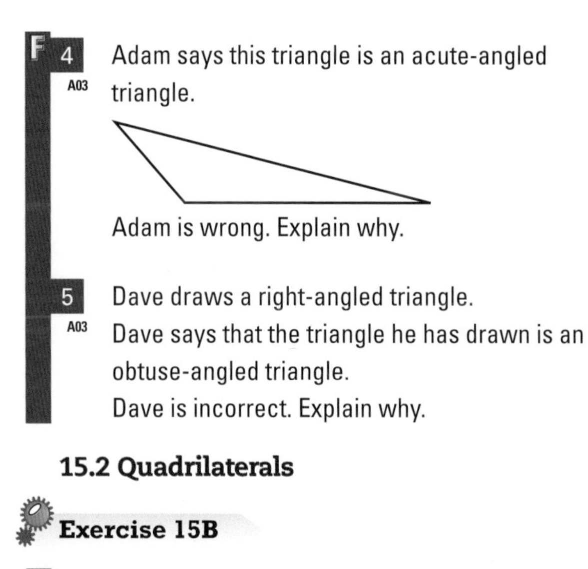

Adam is wrong. Explain why.

5 (AO3) Dave draws a right-angled triangle.
Dave says that the triangle he has drawn is an obtuse-angled triangle.
Dave is incorrect. Explain why.

15.2 Quadrilaterals

Exercise 15B

F 1

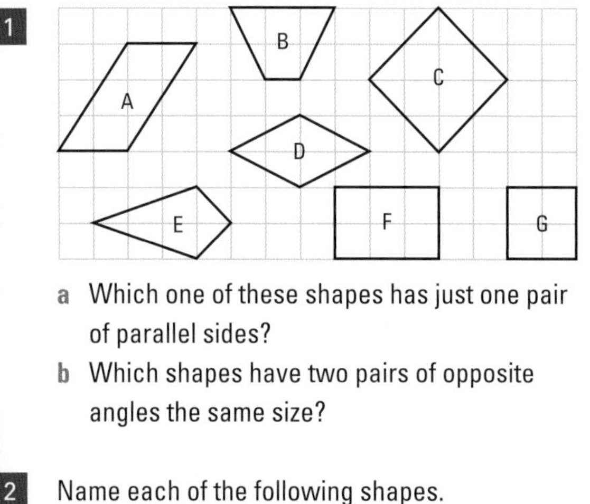

a Which one of these shapes has just one pair of parallel sides?
b Which shapes have two pairs of opposite angles the same size?

2 Name each of the following shapes.
a This shape has four equal sides and its diagonals bisect each other at right angles.
b This shape has twice the number of sides as a triangle.
c This shape has two pairs of parallel sides and its angles are all right angles.
d This shape has half the number of sides as a hexagon and has three equal angles.

3 ABCD is a rhombus.
Which of the following statements are true and which are false?

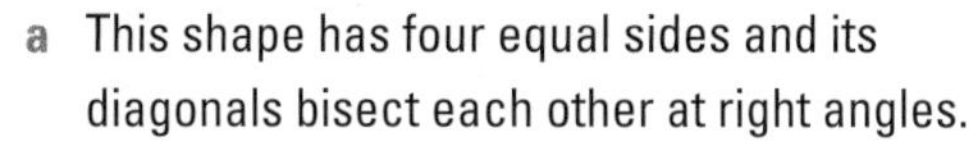

a AD = BC
b angle BCD = angle CBA
c AB is perpendicular to BC
d AD is parallel to BC

F 4 EFGH is a kite.
FH and EG meet at J.

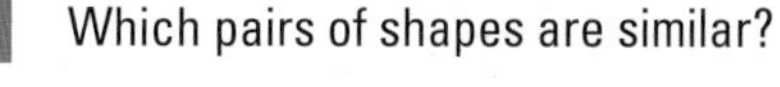

a Name a side equal in length to FG.
b Is angle EJF the same size as angle GJF?
c Name an angle equal to angle EHG.
d Is triangle EHG a scalene triangle?
e Write down the mathematical name for triangle FEH.

5 (AO3) Which one of the following pairs of lines could be the diagonals of a kite?

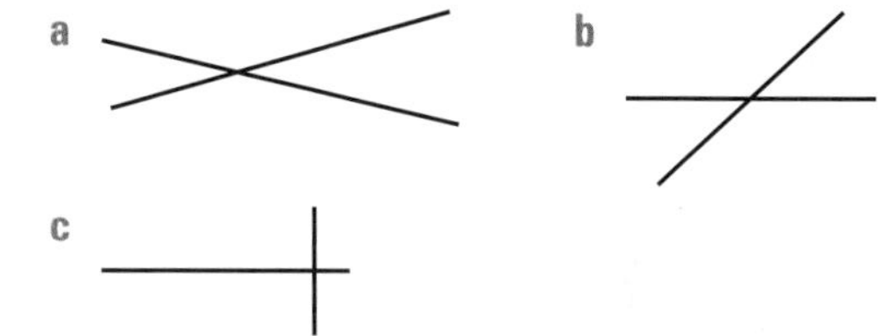

E 6 (AO2 AO3) A factory manufactures belt buckles. They ask Ben to design a new logo looking like a buckle. Draw a diagram made up of two different quadrilaterals that could be the new logo. Write down the names of the quadrilaterals used.

15.3 Similar shapes

Exercise 15C

E 1 Write down the letters of the pair of shapes that are similar.

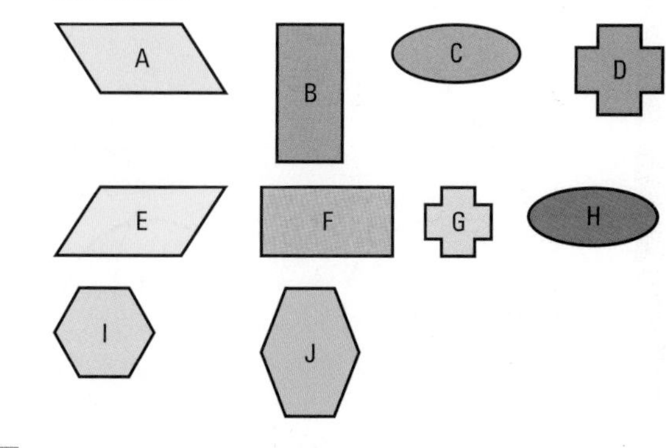

2 Which pairs of shapes are similar?

a

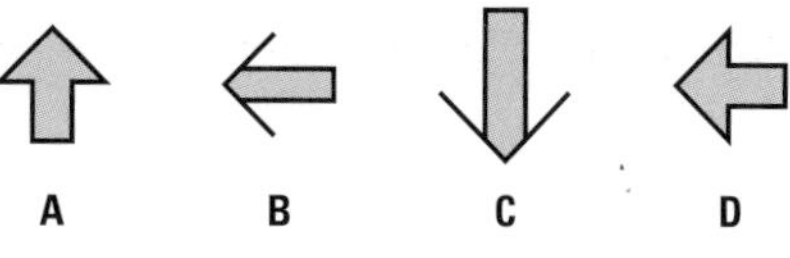

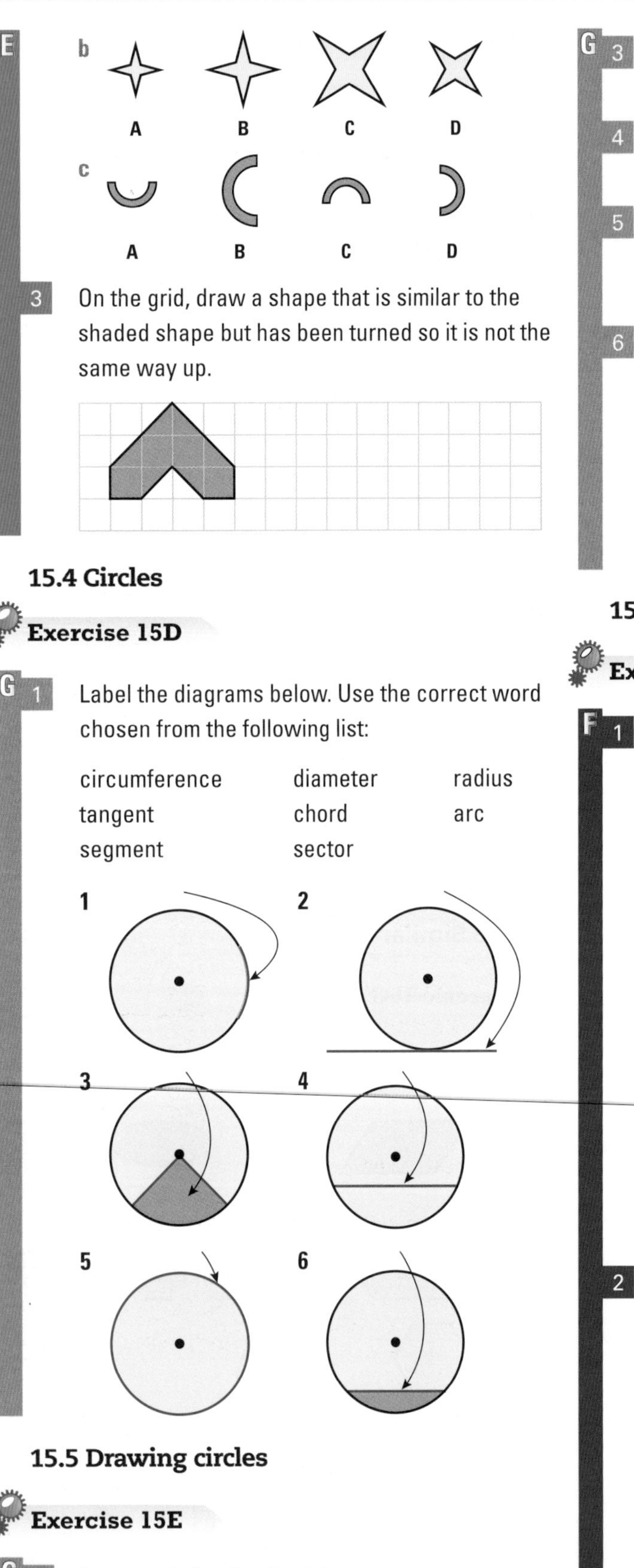

E

b A B C D

c A B C D

3 On the grid, draw a shape that is similar to the shaded shape but has been turned so it is not the same way up.

15.4 Circles

Exercise 15D

G

1 Label the diagrams below. Use the correct word chosen from the following list:

circumference	diameter	radius
tangent	chord	arc
segment	sector	

1 2 3 4 5 6

15.5 Drawing circles

Exercise 15E

G

1 Draw a circle of radius 4.5 cm.

2 Draw a circle with a diameter of 10 cm.

3 a Draw a circle of radius 5.6 cm.
b Shade a segment of your circle.

4 a Draw a circle of diameter 17.6 cm.
b Shade a sector of your circle.

5 a Draw a circle of diameter 8.8 cm.
b On your circle, draw and label
i a radius ii a chord iii a tangent.

6 Draw arcs with the following measurements.
a radius 5 cm, angle 30°
b radius 6 cm, angle 80°
c radius 4 cm, angle 60°
d radius 3 cm, angle 75°
e radius 6 cm, angle 40°
f radius 5.5 cm, angle 65°

15.6 Line symmetry

Exercise 15F

F

1 Copy the following shapes and draw all the lines of symmetry on each one.

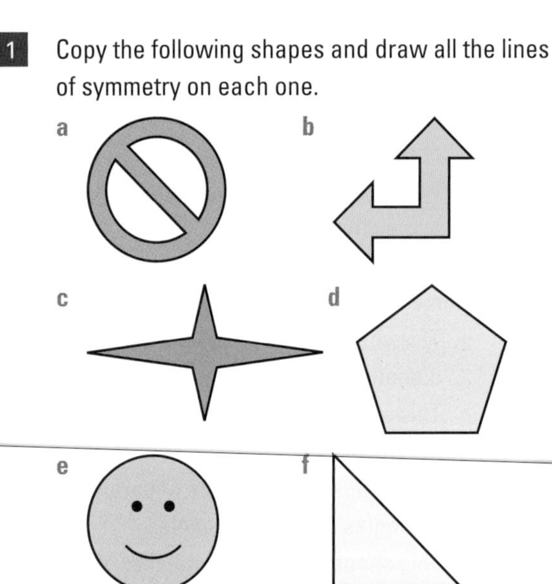

a b c d e f

2 For each shape, state whether or not it has line symmetry.
If it does, write down how many lines of symmetry it has.

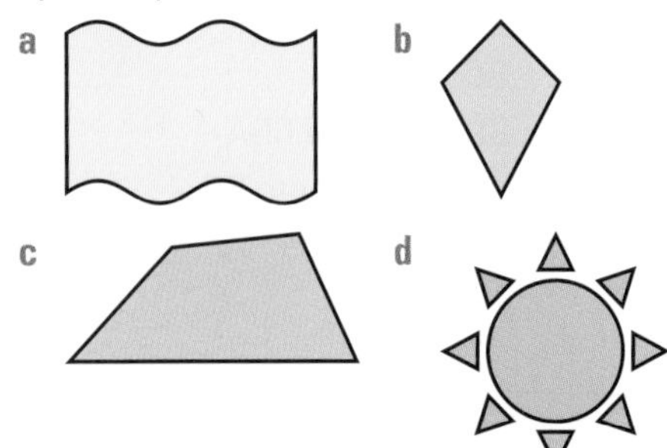

a b c d

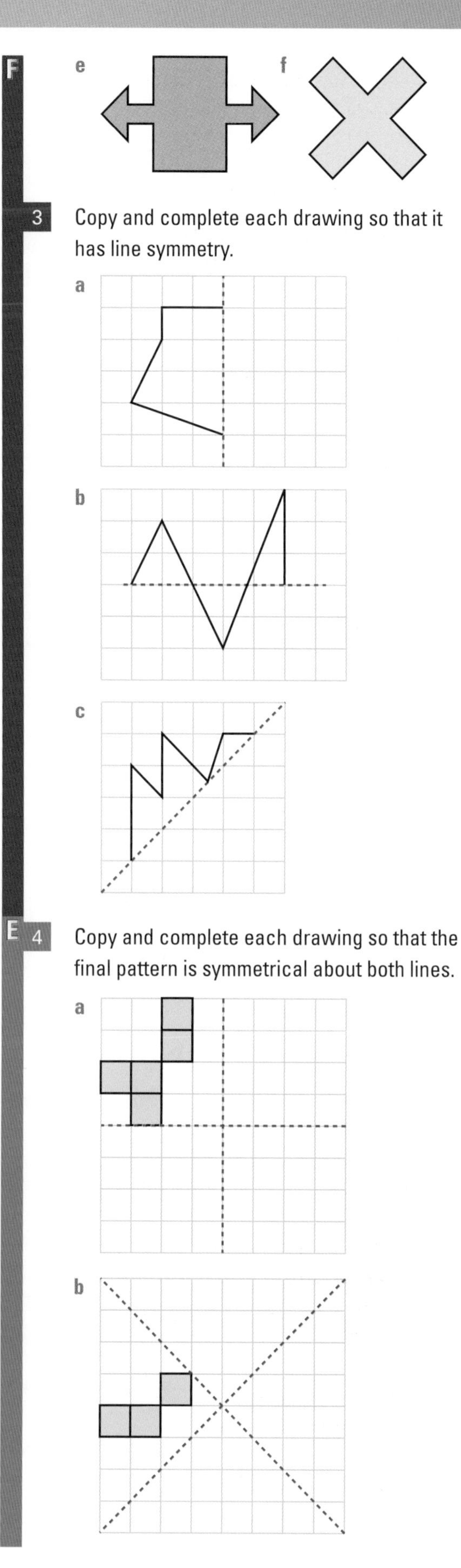

e f

3 Copy and complete each drawing so that it has line symmetry.

a

b

c

4 Copy and complete each drawing so that the final pattern is symmetrical about both lines.

a

b

15.7 Rotational symmetry

Exercise 15G

1 For each letter, write down if it has rotational symmetry or not. If it does, write down the order of rotational symmetry.

a S b M c A d N

2 Write down the order of rotational symmetry for each of the following shapes.

a b

c d

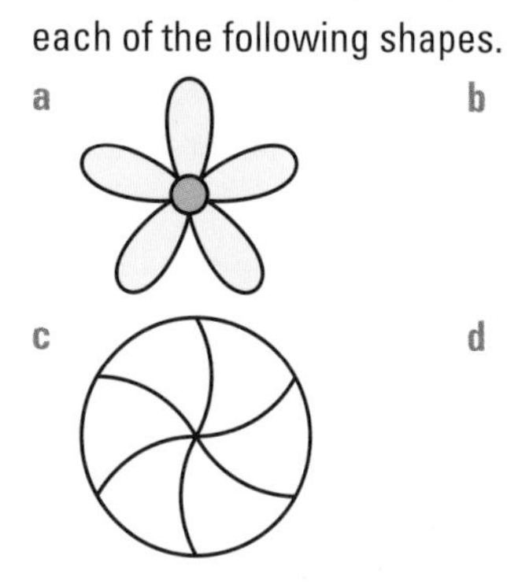

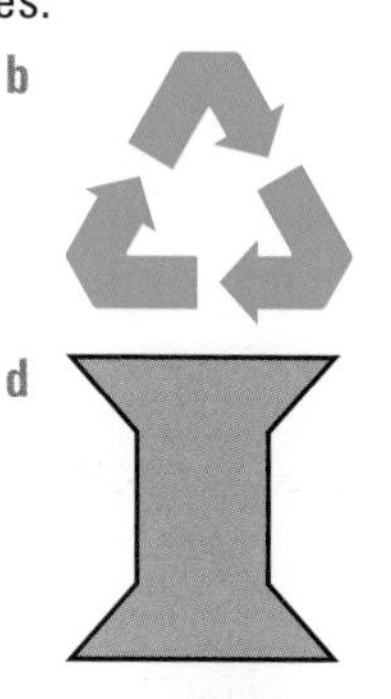

3 On a copy of this grid, add three squares to the shape so that it has rotational symmetry of order 2.

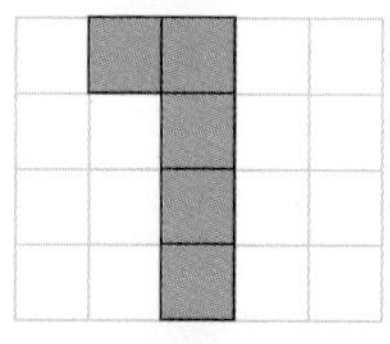

4 On a copy of this grid, add three squares to the shape so that it has rotational symmetry of order 4.

Exercise 15H

1 Copy and complete the following table.

Shape	Name of shape	Number of lines of symmetry	Order of rotational symmetry
(square)		4	
(triangle)	right-angled triangle		
(ellipse)			2
(pentagon)	pentagon		
(trapezium)			

16 Angles 2

Key Points

- **angles of a quadrilateral:**
  - **the sum of the interior angles of a quadrilateral** = 360°
  - **the sum of the exterior angles of a quadrilateral** = 360°
- **perpendicular lines:** lines that meet at an angle of 90°.
- **parallel lines:** lines that stay the same distance apart along their length and never meet.
- **angles on parallel lines:**
  - **corresponding angles** are equal

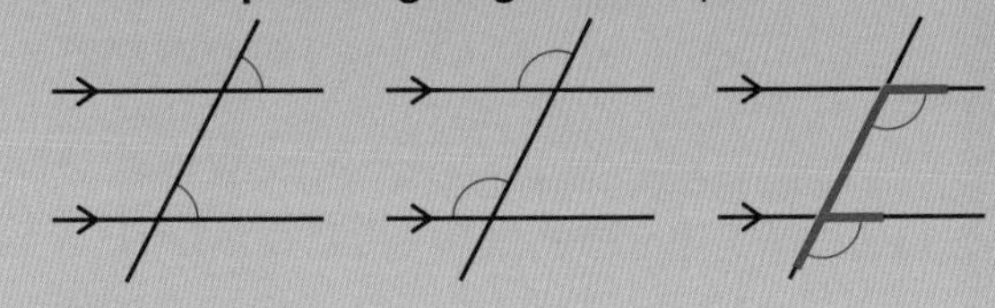

  - **alternate angles** are equal

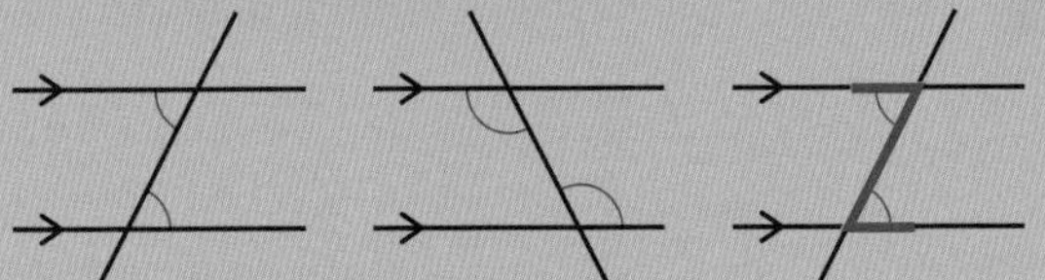

- **proving angle facts:** use known angle facts to prove that;
  - the exterior angle of a triangle = the sum of the 2 opposite interior angles
  - the angle sum of a triangle = 180°
  - opposite angles of a parallelogram are equal

16.1 Angles in quadrilaterals

Exercise 16A

Questions in this chapter are targeted at the grades indicated.

Find the missing angles in the following quadrilaterals.

E

1

2

3

4

5

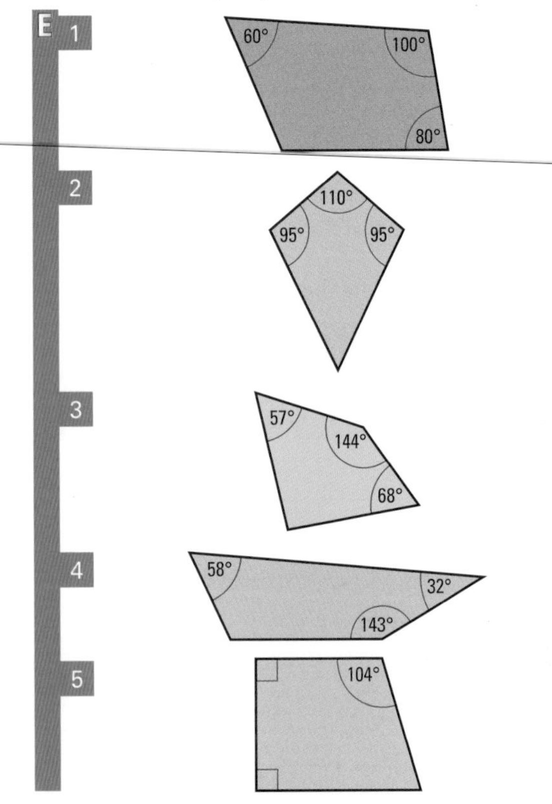

16.2 Perpendicular and parallel lines

Exercise 16B

G 1

A B C D E F G H I J K L

a Find and name as many pairs of parallel lines as you can in the diagram.

b Now find and name as many pairs of perpendicular lines as you can in the diagram.

16.3 Corresponding and alternate angles

Exercise 16C

D 1 a In the diagram, which pairs of angles are corresponding angles?

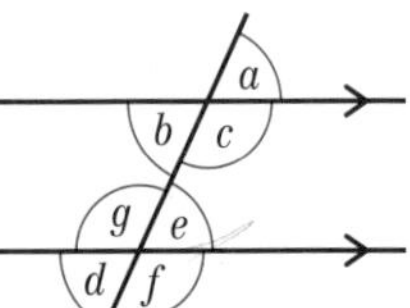

b Which pairs of angles in the diagram are alternate angles?

D 2 In the diagrams below, find the size of each angle marked with a letter.
Give reasons for your answers.

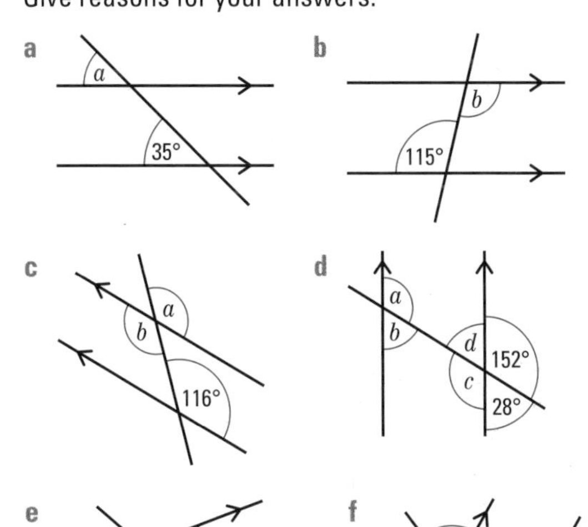

D 3 A02 Find the size of the angles marked with a letter in the diagrams below.

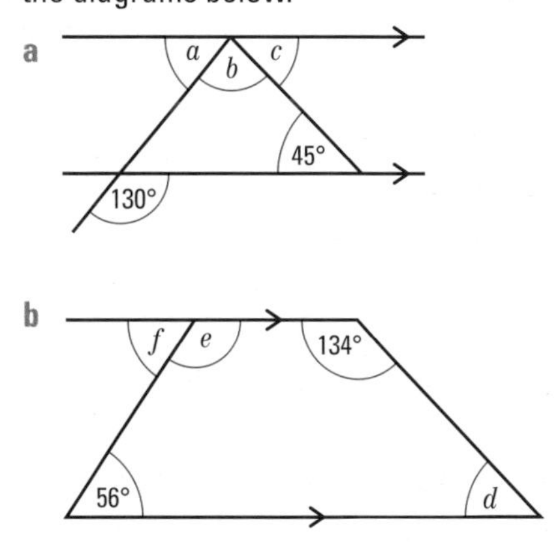

17 Measure

Key Points

- **12-hour clock:** uses am for times between midnight and midday and pm for times between midday and midnight.
- **24-hour clock:** goes from 0 to 24, where 0 and 24 are midnight.
- **speed, time and distance formula:**
  - speed $= \frac{\text{distance}}{\text{time}}$, usually measured in miles per hour (mph), kilometres per hour (km/h), metres per second (m/s)
  - time $= \frac{\text{distance}}{\text{speed}}$
  - distance = speed × time
  - average speed $= \frac{\text{total distance travelled}}{\text{total time taken}}$
- **converting time:**

60 seconds = 1 minute	366 days = 1 leap year
60 minutes = 1 hour	3 months = 1 quarter
24 hours = 1 day	12 months = 1 year
7 days = 1 week	52 weeks = 1 year
365 days = 1 year	

- **converting metric units:**

Length	Weight
10 mm = 1 cm	1000 mg = 1 g
100 cm = 1 m	1000 g = 1 kg
1000 mm = 1 m	1000 kg = 1 tonne
1000 m = 1 km	

Capacity
100 c*l* = 1 litre
1000 m*l* = 1 litre
1000 *l* = 1 cubic metre
1000 cm³ = 1 litre

- **converting imperial units:**
  12 inches = 1 foot
  3 feet = 1 yard
  16 ounces = 1 pound
  14 pounds = 1 stone
  8 pints = 1 gallon
- **converting metric units to imperial units:**

Metric		Imperial	Metric		Imperial
8 km	→	5 miles	1 kg	→	2.2 pounds
1 m	→	39 inches	25 g	→	1 ounce
30 cm	→	1 foot	4.5 litres	→	1 gallon
2.5 cm	→	1 inch	1 litre	→	1.75 pints

- **knowing the inaccuracy of measurements:** measurements given to the nearest unit may be inaccurate by up to one-half of a unit below and one-half of a unit above.

17.1 Reading scales

Exercise 17A

Questions in this chapter are targeted at the grades indicated.

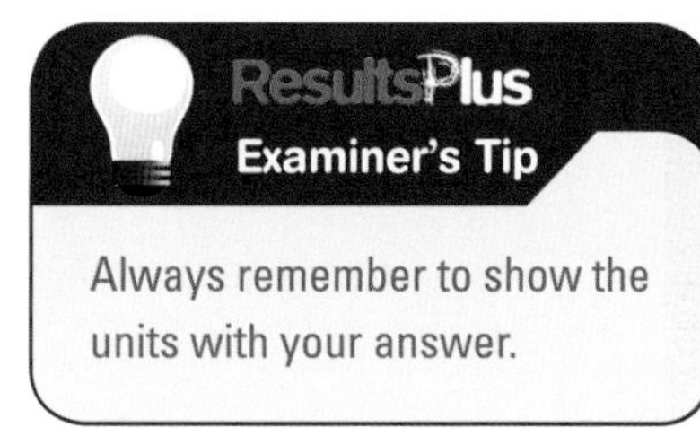

G 1 Measure and write down the lengths of these lines in cm and mm.

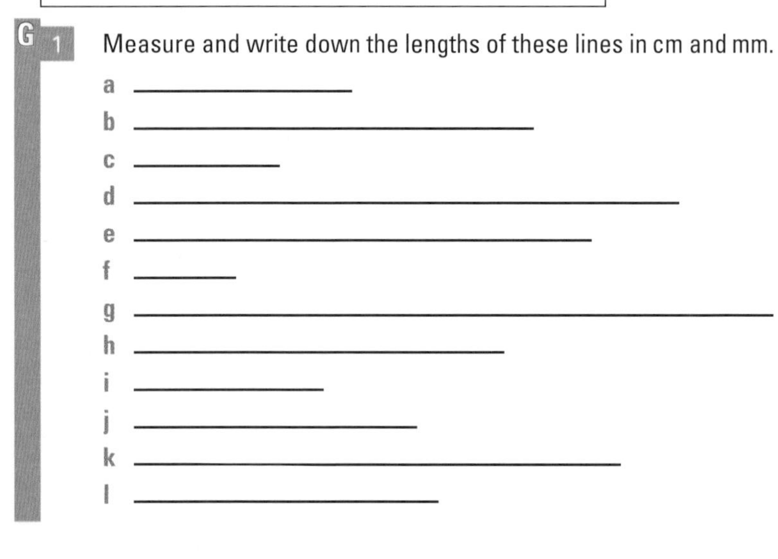

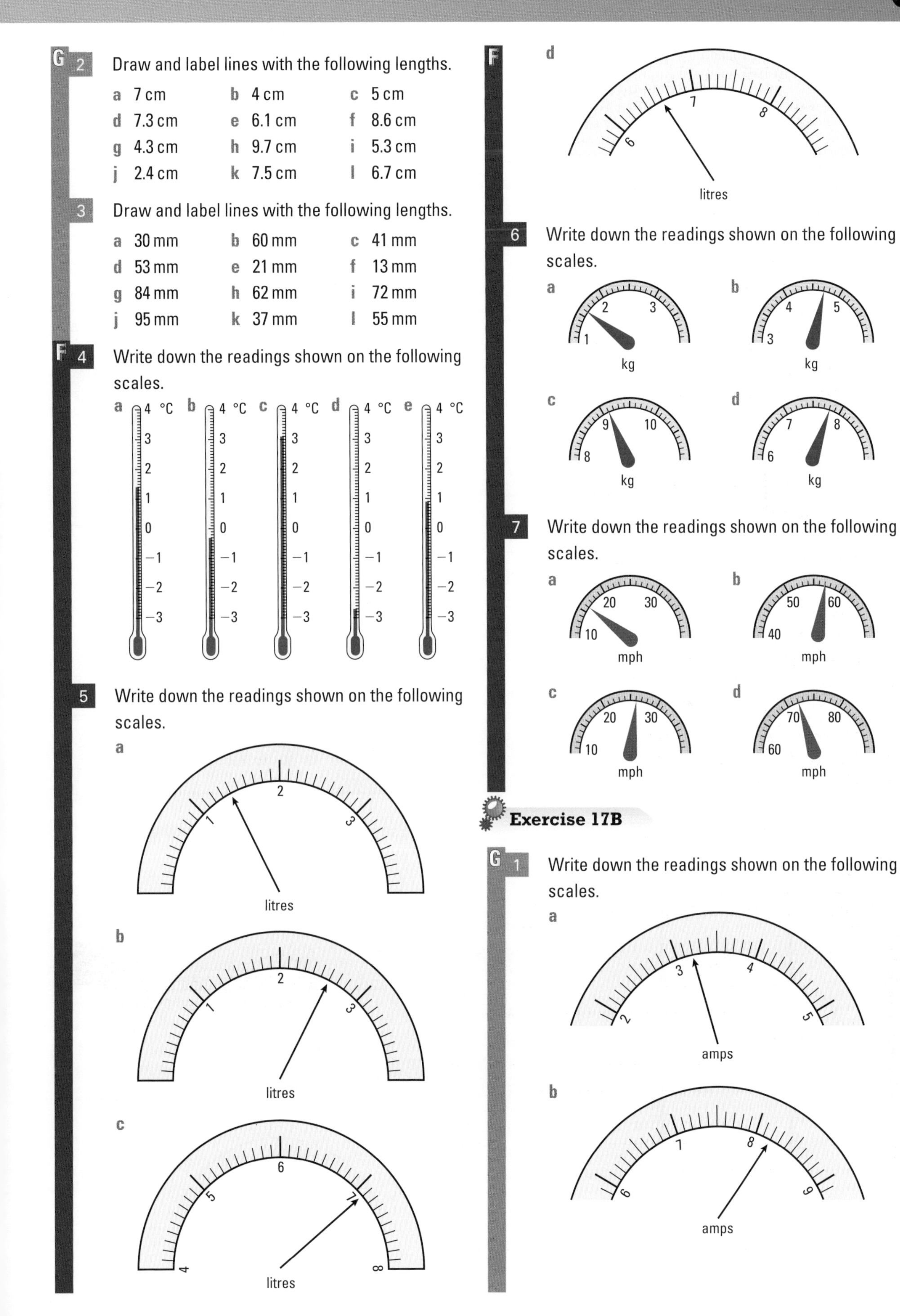

2 Draw and label lines with the following lengths.

a 7 cm	b 4 cm	c 5 cm
d 7.3 cm	e 6.1 cm	f 8.6 cm
g 4.3 cm	h 9.7 cm	i 5.3 cm
j 2.4 cm	k 7.5 cm	l 6.7 cm

3 Draw and label lines with the following lengths.

a 30 mm	b 60 mm	c 41 mm
d 53 mm	e 21 mm	f 13 mm
g 84 mm	h 62 mm	i 72 mm
j 95 mm	k 37 mm	l 55 mm

4 Write down the readings shown on the following scales.

a b c d e

5 Write down the readings shown on the following scales.

a

b

c

d

6 Write down the readings shown on the following scales.

a b c d

7 Write down the readings shown on the following scales.

a b c d

Exercise 17B

1 Write down the readings shown on the following scales.

a

b

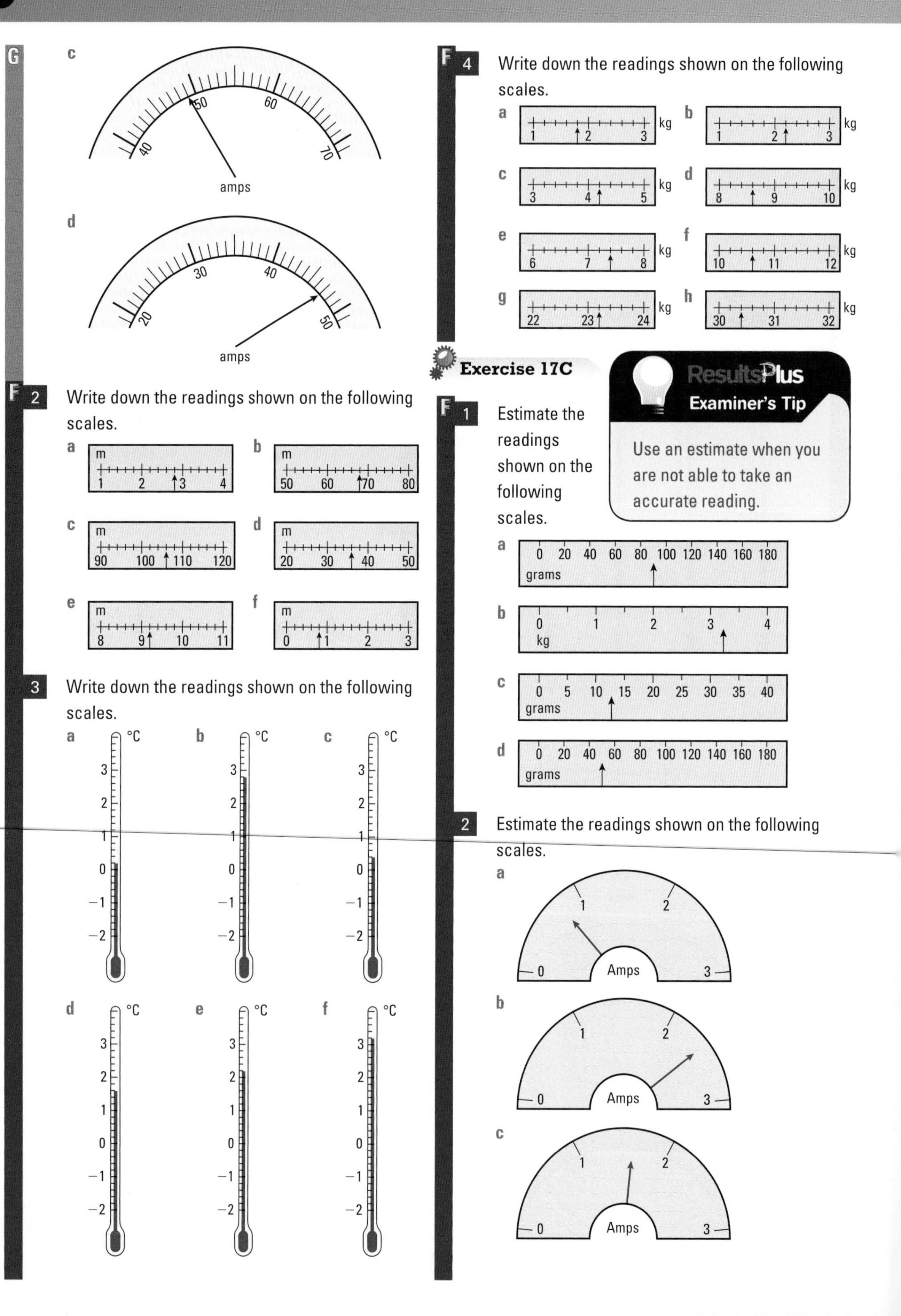

G

c

d

F **2** Write down the readings shown on the following scales.

a

b

c

d

e

f

3 Write down the readings shown on the following scales.

a b c d e f

F **4** Write down the readings shown on the following scales.

a

b

c

d

e

f

g

h

Exercise 17C

ResultsPlus

Examiner's Tip

Use an estimate when you are not able to take an accurate reading.

F **1** Estimate the readings shown on the following scales.

a

b

c

d

2 Estimate the readings shown on the following scales.

a

b

c

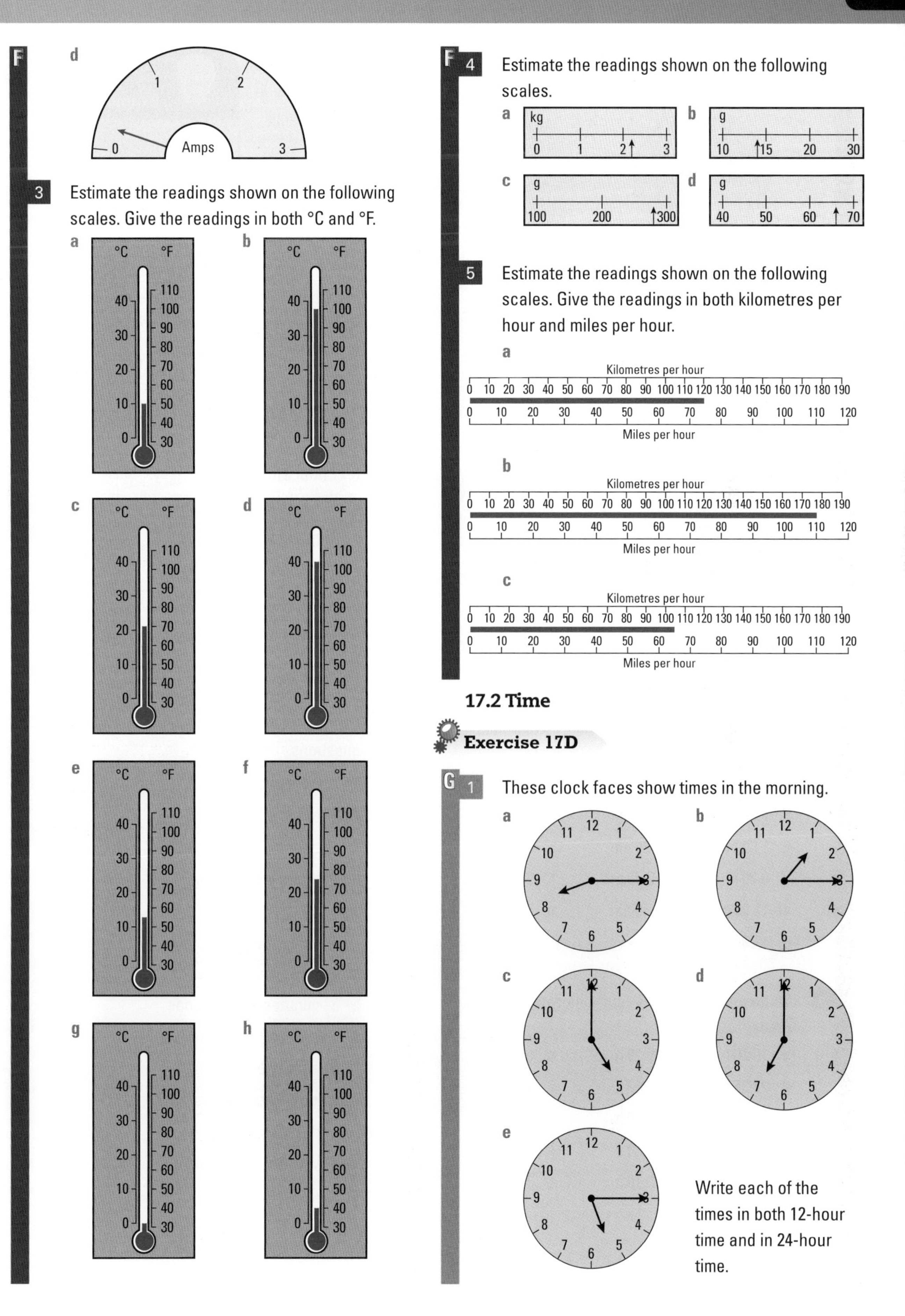

d

3 Estimate the readings shown on the following scales. Give the readings in both °C and °F.

a b c d e f g h

4 Estimate the readings shown on the following scales.

a b c d

5 Estimate the readings shown on the following scales. Give the readings in both kilometres per hour and miles per hour.

a

b

c

17.2 Time

Exercise 17D

1 These clock faces show times in the morning.

a b c d e

Write each of the times in both 12-hour time and in 24-hour time.

G 2 These clock faces show times in the afternoon and evening.

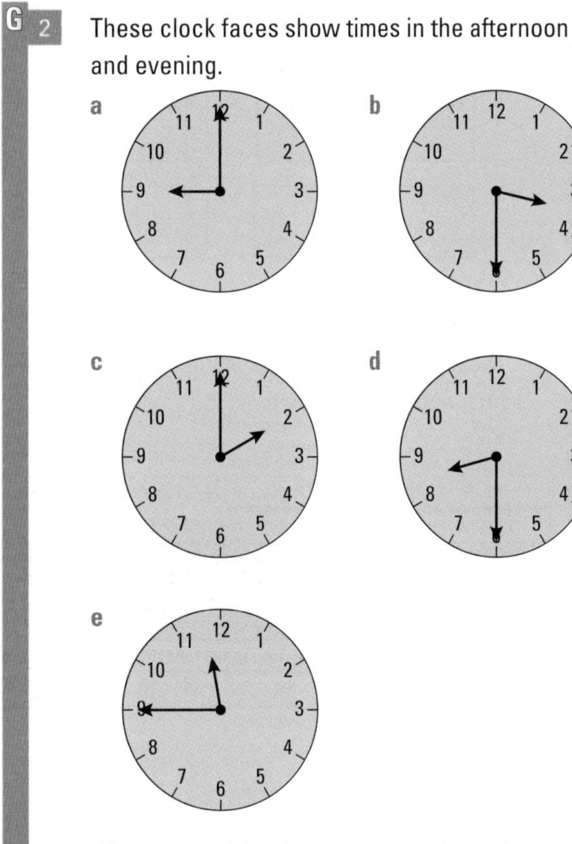

Write each of the times in both 12-hour time and 24-hour time.

3 Write the following 12-hour times as 24-hour times.

a 9 am
b 10.15 am
c 4.40 pm
d 7.20 am
e 9.55 pm
f 4.25 pm
g 1.30 am
h 6.25 pm
i 11.15 pm
j 6.20 am
k 10.45 am
l 2.15 pm
m 12.25 pm
n 3.50 am
o 2.50 pm
p 1.20 pm

4 Write the following 24-hour times as 12-hour times.

a 10:10 h
b 09:20 h
c 08:40 h
d 22:35 h
e 15:17 h
f 10:35 h
g 17:16 h
h 18:25 h
i 16:20 h
j 16:10 h
k 07:30 h
l 12:35 h
m 04:42 h
n 23:16 h
o 11:17 h
p 17:37 h

G 5 Write down the correct time in the following questions.

ResultsPlus
Examiner's Tip

It may be useful to use a clock face to help you answer this type of question. There will usually be a clock in the exam room that you can use.

a 5 hours before 3.15 pm
b $5\frac{1}{2}$ hours before 16:40 h
c $4\frac{1}{4}$ hours after 11.15 am
d $3\frac{1}{4}$ hours before 1 pm
e $2\frac{1}{2}$ hours before 13:00 h
f 1 hour after 2.40 am
g $4\frac{1}{2}$ hours after 11.20 am
h $3\frac{1}{4}$ hours before 14:45 h
i $6\frac{1}{2}$ hours after 10:55 h
j $5\frac{1}{2}$ hours before 15:10 h
k $4\frac{1}{4}$ hours before 3 am
l $3\frac{3}{4}$ hours after 22:00 h

6 Change the units of time in the following questions.

a 2 years into weeks
b $2\frac{1}{2}$ hours into minutes
c 15 minutes into seconds
d 15 years into months
e 5 days into hours
f 480 minutes into hours
g 416 weeks into years
h $3\frac{1}{2}$ minutes into seconds
i 150 seconds into minutes
j 5 years into weeks
k 6 hours into minutes
l $4\frac{1}{2}$ days into hours
m $2\frac{1}{2}$ years into weeks
n 3 years into days

F 7 A02

a A woman buys 2 magazines a month.
How many does she buy in a year?

b Fiona takes 2 mins to plant a bulb in the garden.
How many can she plant in an hour?

c Michael pays his paper bill once a month.
How many times does he pay in a year?

d The figure on the Town Hall clock hits the bell with his hammer every 15 mins from 9 am to 6 pm.
How many times does he hit the bell in a day?

Exercise 17E

F 1 Work out the time difference between each of the following times:

a 9.50 am to 11.15 am b 07:30 h to 09:20 h

c 12.15 pm to 3.25 pm d 12:45 h to 00:15 h

e 10 pm Monday to 9 am Tuesday

f 09:17 h to 12:27 h

g 10.37 am to 1.15 pm h 03:42 h to 21:14 h

i 8.30 am to 10.15 pm j 8.15 am to 1.05 pm

k 10:15 Tue to 08:05 Wed l 17:35 h to 21:50 h

F 2 Ruth leaves her house at 7.15 am. She travels by train to her parents' home, arriving at 10.35 pm. How long does her journey take?

3 A man arrives at work at 08:25 h and leaves at 17:15 h. How long is he at work?

4 A ferry sets sail from Portsmouth at 08:50 h and arrives in France at 14:40 h.
How long does the crossing take?

5 A02 Jessica has to catch a flight from Gatwick at 13:15 h. She wants to travel by train from London Victoria to Gatwick. She has to allow a minimum of 2 hours for a security check and checking her boarding card. Which train should she get?

London Victoria	09:47	10:47	11:47	12:47
Clapham Junction	09:53	10:53	11:53	12:53
East Croydon	10:07	11:07	12:07	13:07
Gatwick Airport	10:29	11:29	12:29	13:29

Exercise 17F

F 1 Use the bus timetable to answer the questions below.

Bus timetable: Ordsall to Bury

Ordsall, Salford Quays			07:30		08:30		18:30	19:00	20:00	21:00	22:00
Trafford Rd			07:35		08:35		18:35	19:05	20:05	21:05	22:05
Pendleton Precinct arr.			07:41		08:41		18:41				
Pendleton Precinct dep.	06:43	07:13	07:43	08:13	08:43	and	18:43	19:10	20:10	21:10	22:10
Lower Kersal	06:54	07:24	07:54	08:24	08:54	every	18:54	19:19	20:19	21:19	22:19
Agecroft	06:57	07:27	07:57	08:27	08:57	30	18:57	19:22	20:22	21:22	22:22
Butterstile Lane	07:04	07:34	08:04	08:34	09:04	mins	19:04				
Prestwich	07:10	07:40	08:10	08:40	09:10	until	19:10				
Besses o'th' Barn	07:14	07:44	08:14	08:44	09:14		19:14				
Unsworth	07:24	07:54	08:24	08:54	09:24		19:24				
Bury	07:40	08:10	08:40	09:10	09:40		19:40				

a How long does it take the 08:13 Pendleton bus to get to Unsworth?

b How frequently do buses leave Pendleton between 06:43 and 18:43?

c At what time does the last bus to Bury leave Ordsall?

d At what time does the first bus to call at Trafford Road reach Agecroft?

e How long does it take to travel from Lower Kersal to Prestwich?

f How long does it take to travel from Ordsall to Unsworth?

g How many buses call at Trafford Road before 11:00?

h How many buses call at Besses o'th' Barn during the day?

i Stella arrives at her bus stop at Agecroft at 7.45 am.
How long will she have to wait for a bus to Unsworth?

j Jack arrives at his bus stop in Trafford Road at 7.30 am.
How long will he have to wait for a bus to Agecroft?

A02 A03 k Kieron lives 5 minutes away from his bus stop in Butterstile Lane.
What is the latest time he can leave his house to get to Bury by 9 am?

l Pravin wants to catch a bus from Pendleton to Prestwich, to arrive in Prestwich no later than 12:30.
What is the departure time of the latest bus he can catch from Pendleton?

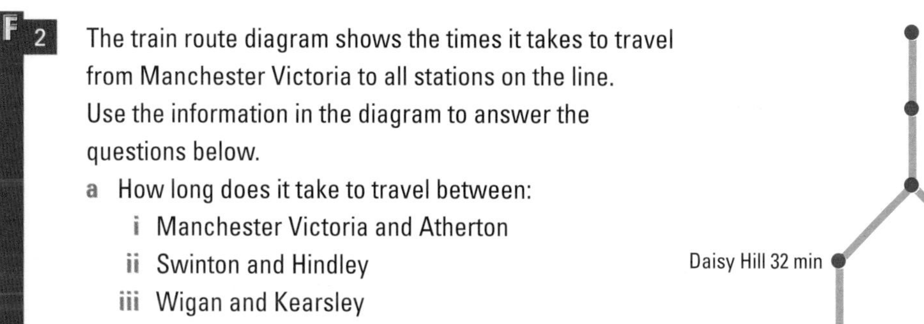

F **2** The train route diagram shows the times it takes to travel from Manchester Victoria to all stations on the line. Use the information in the diagram to answer the questions below.

a How long does it take to travel between:

i Manchester Victoria and Atherton
ii Swinton and Hindley
iii Wigan and Kearsley
iv Bolton and Ince?

AO2 AO3 **b** Faz is planning a trip from Swinton to Farnworth. She will have to wait 12 minutes at Salford Crescent to change trains. What will be her total journey time from Swinton to Farnworth?

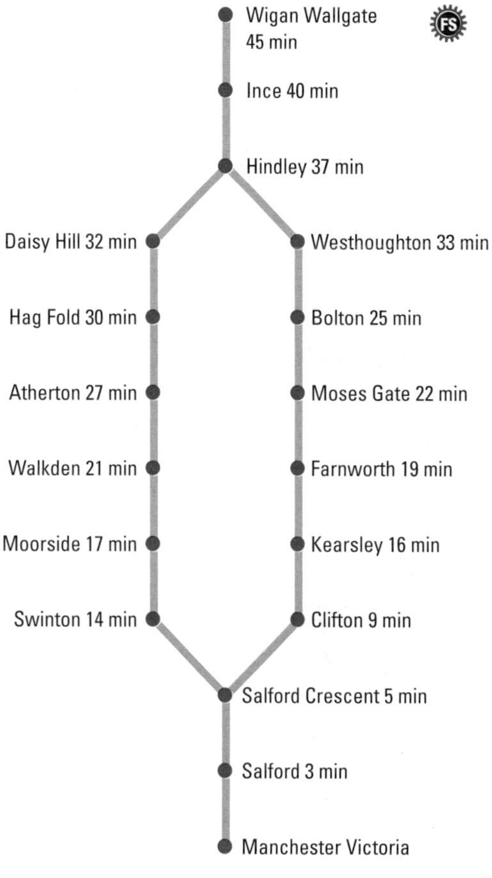

c Copy and complete the following timetables.

Manchester Victoria	09:35	11:05	Wigan Wallgate	10:05	11:30
Salford			Ince		
Salford Crescent			Hindley		
Swinton			Westhoughton		
Moorside			Bolton		
Walkden			Moses Gate		
Atherton			Farnworth		
Hag Fold			Kearsley		
Daisy Hill			Clifton		
Hindley			Salford Crescent		
Ince			Salford		
Wigan Wallgate			Manchester Victoria		

17.3 Metric units

Exercise 17G

Write down the metric units you would use to take the measurements listed below:

G
1. capacity of a swimming pool
2. dimensions of a room
3. the weight of a new-born kitten
4. the weight of a suitcase
5. the length of a pen
6. the weight of a packet of sugar
7. the length of a toe
8. the capacity of a carton of juice
9. the weight of a pill
10. the capacity of a car's fuel tank
11. the capacity of a bottle of lemonade
12. the height of a lorry
13. the distance from Calais to Paris
14. the weight of a lorry
15. the thickness of a book.

Exercise 17H

G

1. Convert the following lengths to centimetres.
   a 3 m b 150 mm c 9 m
   d 12 m e 300 mm f 45 m
   g 64 mm h 211 mm
2. Convert the following lengths to millimetres.
   a 4 cm b 8 cm c 24 cm
   d 30 cm e 300 cm f 6.4 cm
   g 17.3 cm h 6.15 cm
3. Convert the following lengths to metres.
   a 5 km b 400 cm c 30 km
   d 2000 cm e 55 km f 0.6 km
   g 1.6 km h 2.54 km
4. Convert the following weights to grams.
   a 3 kg b 20 kg c 300 kg
   d 150 kg e 3000 kg f 44 kg
   g 0.21 kg h 2.4 kg
5. Convert the following capacities to litres.
   a 3000 m*l* b 5000 m*l* c 40 000 m*l*
   d 45 000 m*l* e 3500 m*l* f 2700 m*l*
   g 5620 m*l* h 1330 m*l*
6. Convert the following capacities to millilitres.
   a 2 *l* b 30 *l* c 300 *l*
   d 150 *l* e 45 *l* f 5.7 *l*
   g 0.6 *l* h 1.34 *l*
7. Convert the following lengths to kilometres.
   a 4000 m b 7000 m c 20 000 m
   d 86 000 m e 3200 m f 6500 m
   g 4510 m h 4210 m
8. Convert the following weights to tonnes.
   a 3000 kg b 5000 kg c 20 000 kg
   d 75 000 kg e 2600 kg f 5400 kg
   g 6730 kg h 5420 kg
9. Convert the following weights to kilograms.
   a 2000 g b 2 tonnes c 30 000 g
   d 25 tonnes e 300 000 g f 3.5 tonnes
   g 5400 g h 3230 g

F

10. A02 A carton of sugar cubes weighs 0.5 kg. A sugar cube weighs 4 g.
    How many cubes are there in a box?
11. A02 Judy is making marmalade. Her preserving pan can fill 40 jars, each with a capacity of 200 m*l*.
    What is the capacity of the pan?

Exercise 17I

F

1. Put these lengths in order, with the smallest first.
   5 m 7 mm 2 cm 5 km 20 cm
   40 mm
2. Put these capacities in order, with the smallest first.
   800 m*l* 7 *l* 500 m*l* 2000 m*l* 3*l*
3. Put these weights in order, with the smallest first.
   400 g 540 g 0.6 kg 0.26 kg
4. Put these lengths in order, with the smallest first.
   30 cm 0.8 cm 730 mm 4.1 m
   200 mm 66 cm
5. Put these lengths in order, with the smallest first.
   5 cm 66 mm 42 cm 0.3 cm
   33 mm 0.5 cm 8 mm

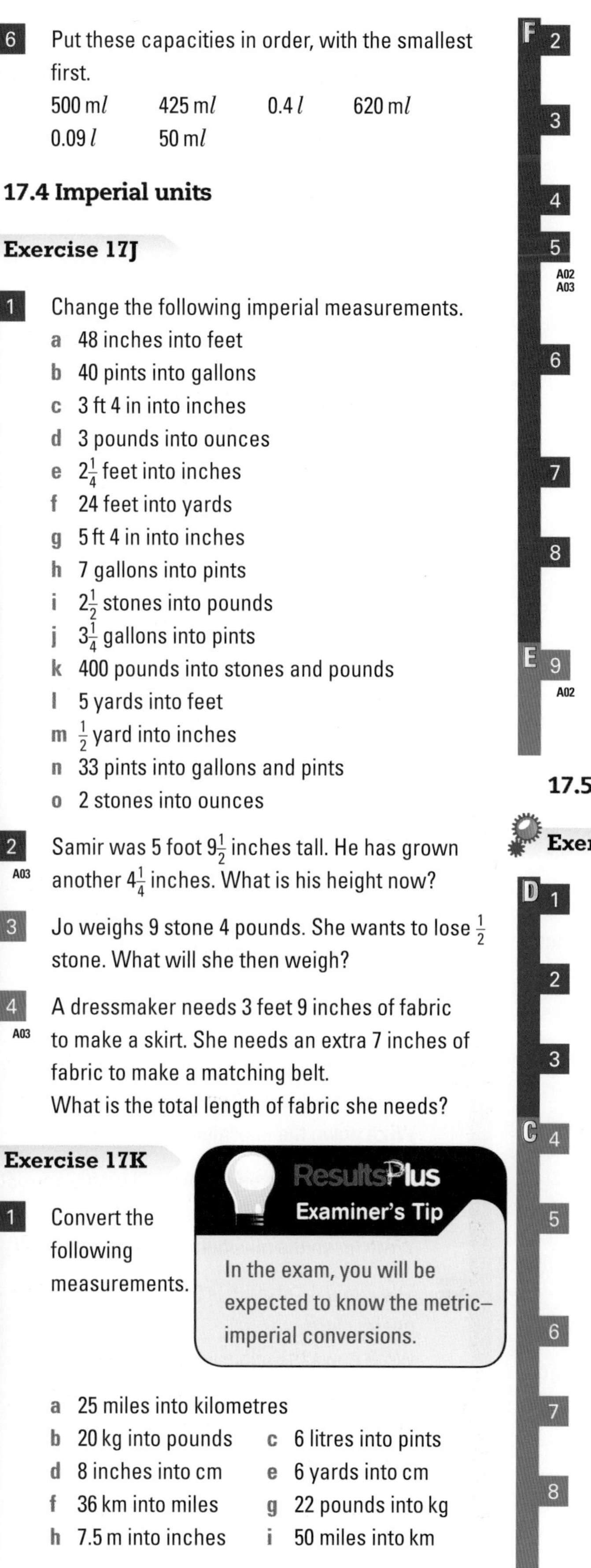

6 Put these capacities in order, with the smallest first.

500 ml 425 ml 0.4 l 620 ml
0.09 l 50 ml

17.4 Imperial units

Exercise 17J

1 Change the following imperial measurements.

a 48 inches into feet
b 40 pints into gallons
c 3 ft 4 in into inches
d 3 pounds into ounces
e $2\frac{1}{4}$ feet into inches
f 24 feet into yards
g 5 ft 4 in into inches
h 7 gallons into pints
i $2\frac{1}{2}$ stones into pounds
j $3\frac{1}{4}$ gallons into pints
k 400 pounds into stones and pounds
l 5 yards into feet
m $\frac{1}{2}$ yard into inches
n 33 pints into gallons and pints
o 2 stones into ounces

2 (A03) Samir was 5 foot $9\frac{1}{2}$ inches tall. He has grown another $4\frac{1}{4}$ inches. What is his height now?

3 Jo weighs 9 stone 4 pounds. She wants to lose $\frac{1}{2}$ stone. What will she then weigh?

4 (A03) A dressmaker needs 3 feet 9 inches of fabric to make a skirt. She needs an extra 7 inches of fabric to make a matching belt.
What is the total length of fabric she needs?

Exercise 17K

1 Convert the following measurements.

> **ResultsPlus Examiner's Tip**
> In the exam, you will be expected to know the metric–imperial conversions.

a 25 miles into kilometres
b 20 kg into pounds
c 6 litres into pints
d 8 inches into cm
e 6 yards into cm
f 36 km into miles
g 22 pounds into kg
h 7.5 m into inches
i 50 miles into km

2 A house is estimated to be 27 feet high.
What is this in metres?

3 A shirt collar measures 16 inches.
What is this in centimetres?

4 A tank holds 27 litres. How many gallons is this?

5 (A02 A03) There are 2 pints of milk in a container.
A glass will hold 250 ml.
How many children can have a full glass of milk?

6 Rasheen travels from London to Manchester.
The distance is 250 miles.
What is the distance in kilometres?

7 A container has a capacity of 4 litres.
What is this in pints?

8 A family on holiday in Majorca travel 175 km while touring the island.
How many miles do they travel?

9 (A02) A supermarket is selling 500 g of apples for 65p. A local shop is selling a 3-pound bag of apples for £2.10. Which is the better buy?

17.5 Speed

Exercise 17L

1 What is the average speed of a car that takes 3 hours to travel 120 miles?

2 Mike ran for 2 hours and covered 14 miles.
At what average speed was he running?

3 Dinesh was walking. He went 24 miles in 8 hours.
At what average speed was he walking?

4 Danielle swam 3 miles. It took her $1\frac{1}{2}$ hours.
What was her average speed?

5 Ying travelled 240 miles on a business trip. He left home at 9 am and arrived at his destination at 1 pm. What was his average speed?

6 What is the average speed of a train that takes $2\frac{1}{2}$ hours to travel 112 km?

7 After a $3\frac{1}{2}$ hour journey a car had travelled 203 miles. What was its average speed?

8 Jeff drives 150 miles in $2\frac{1}{2}$ hours.
Work out his average speed.

C

9 Ryan travels 40 km in 20 minutes.
Calculate his average speed in km/h.

10 An aeroplane flies 1225 km in $3\frac{1}{2}$ hours.
What is its average speed?

Exercise 17M

E

1 A car travels for 2 hours at 50 mph.
How far will the car have gone?

2 Find the time taken to travel 9 km at 3 km/h.

3 A delivery van takes 2 hours to complete a journey at a speed of 35 mph.
What distance will it have covered?

D

4 How long does it take to travel 45 miles at an average speed of 30 mph?

5 How long does it take to travel 75 km at 30 km/h?

6 Find the time taken to walk 5 miles at an average walking speed of $2\frac{1}{2}$ mph.

7 A man walks at an average speed of 5 km/h.
What distance will he cover in 1 hour 30 minutes?

8 Find the time taken to travel 21 miles at a speed of 6 mph.

9 An aeroplane flew at 550 mph. How far did it travel in $4\frac{1}{2}$ hours?

10 Rajesh travels for $3\frac{1}{2}$ hours at 42 mph. Calculate the distance he travelled.

Exercise 17N

C

1 How far will you have gone if you travel for 1 hour 45 minutes at 60 mph?

2 Fidel runs at an average speed of 5 mph. He goes running for 1 hour and 20 minutes.
What distance does he run in that time?

3 A train travelled a distance of 288 km in 3 hours 36 minutes.
What was the average speed of the train?

4 A cyclist rode at a speed of 10 mph. She covered a distance of 28 miles.
For how long was she cycling?
Give your answer in hours and minutes.

5 A speedboat has an average speed of 20 km/h. It travels around the coast for 3 hours 6 minutes.
What distance does it cover in that time?

6 A02 An aeroplane flew from London to Edinburgh.
It left London at 12:30 hours and arrived in Edinburgh at 13:45 hours. It travelled 344 miles.
What was the average speed of the aeroplane?

7 A lorry made a journey of 290 km at an average speed of 50 km/h. How long did it take?
Give your answer in hours and minutes.

8 A train was travelling on a track at a speed of 90 mph. It was travelling at this speed for 3 hours 36 minutes. How far did it go in this time?

9 A horse takes 8 minutes to gallop 7 km.
What is the average speed of the horse?

10 **a** A racing car travels at 75 m/s. Work out the distance the car travels in 0.4 seconds.
b Change a speed of 75 m/s into km/h.

17.6 Accuracy of measurements

Exercise 17O

C

1 The length of a purse is 18 cm correct to the nearest centimetre.
Write down the maximum length it could be.

2 The weight of an envelope is 35 grams correct to the nearest gram.
Write down the minimum weight it could be.

3 The capacity of a jug is 3 litres correct to the nearest litre.
Write down the minimum capacity of the jug.

4 The radius of a plate is 10.3 cm correct to the nearest millimetre.
Write down
a the least possible length it could be
b the greatest possible length it could be.

C

5 Fiona's height is 1.49 m correct to the nearest centimetre. Write down in metres

a the minimum possible height she could be

b the maximum possible height she could be.

6 The length of a pencil is 12 cm correct to the nearest cm.
The length of a pencil case is 122 mm correct to the nearest mm.
Explain why the pen might **not** fit in the case.

7 The width of a cupboard is measured to be 84 cm correct to the nearest centimetre.
There is a gap of 837 mm correct to the nearest mm in the wall.
Explain how the cupboard might fit in the wall.

18 Perimeter and area of 2D shapes

Key Points

- **perimeter:** the total distance around the edge of a 2D shape.
  - **perimeter of a rectangle** $= 2 \times \text{length} + 2 \times \text{width} = 2l + 2w$
- **area:** the amount of space inside a 2D shape.
  - **area of a square** = length × length $= l^2$
  - **area of a rectangle** = length × width $= l \times w$
  - **area of a triangle** $= \frac{1}{2} \times \text{base} \times \text{vertical height} = \frac{1}{2} \times b \times h$
  - **area of a parallelogram** = base × vertical height $= b \times h$
  - **area of a trapezium** $= \frac{1}{2} \times$ sum of parallel sides × distance between them $= \frac{1}{2}(a + b)h$
  - measured in square units, such as square millimetres (mm^2), square centimetres (cm^2), square metres (m^2), square kilometres (km^2)
- **finding the area of complicated shapes:** split the shape into a number of simpler shapes. The total area = the sum of the areas of each part.

18.1 Perimeter

Exercise 18A

Questions in this chapter are targeted at the grades indicated.

ResultsPlus Examiner's Tip

Always remember to include the units in your answer.

G 1 Here are three shapes drawn on a centimetre grid. Work out the perimeter of each shape.

F 2 Work out the perimeters of the following shapes.

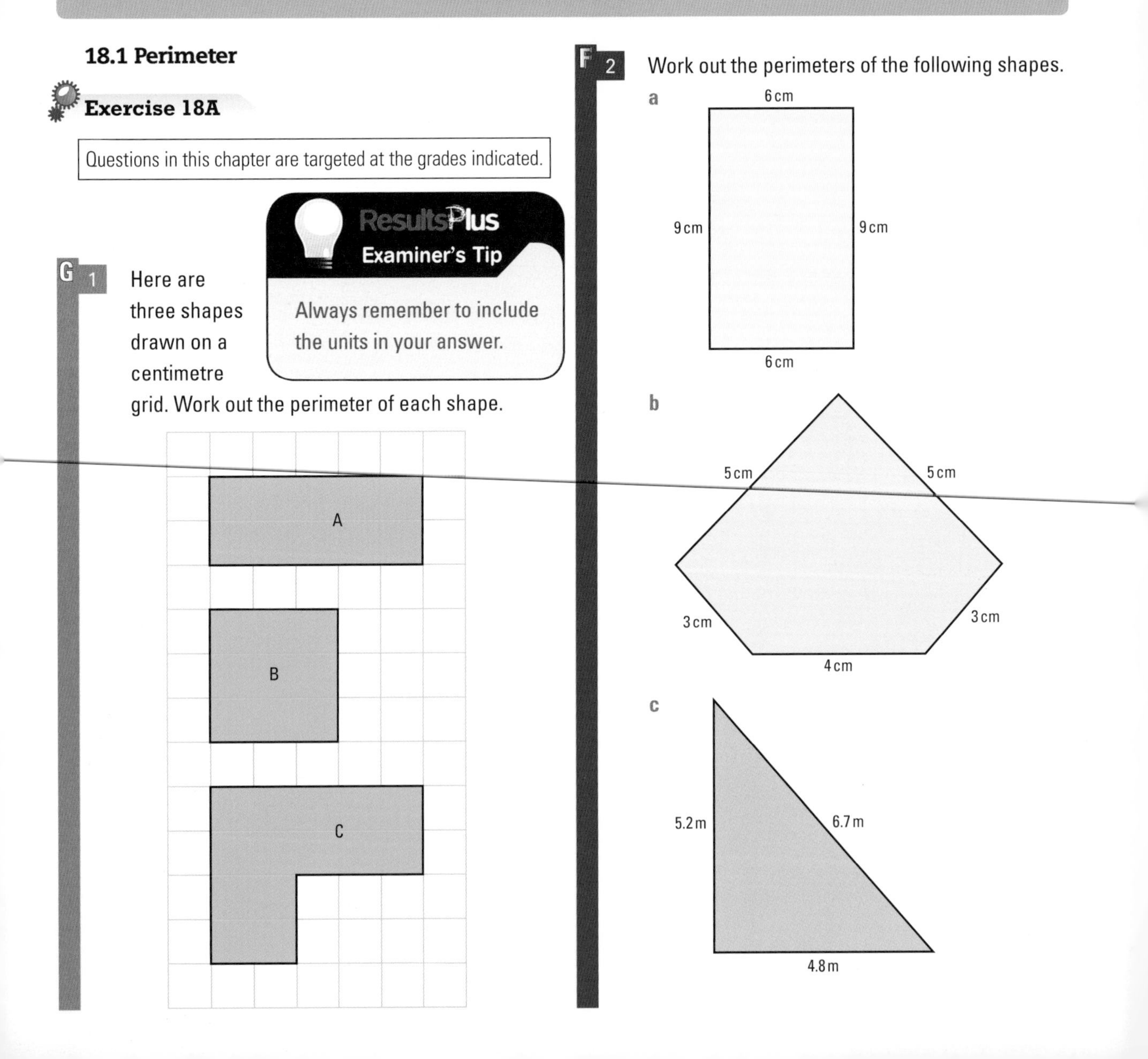

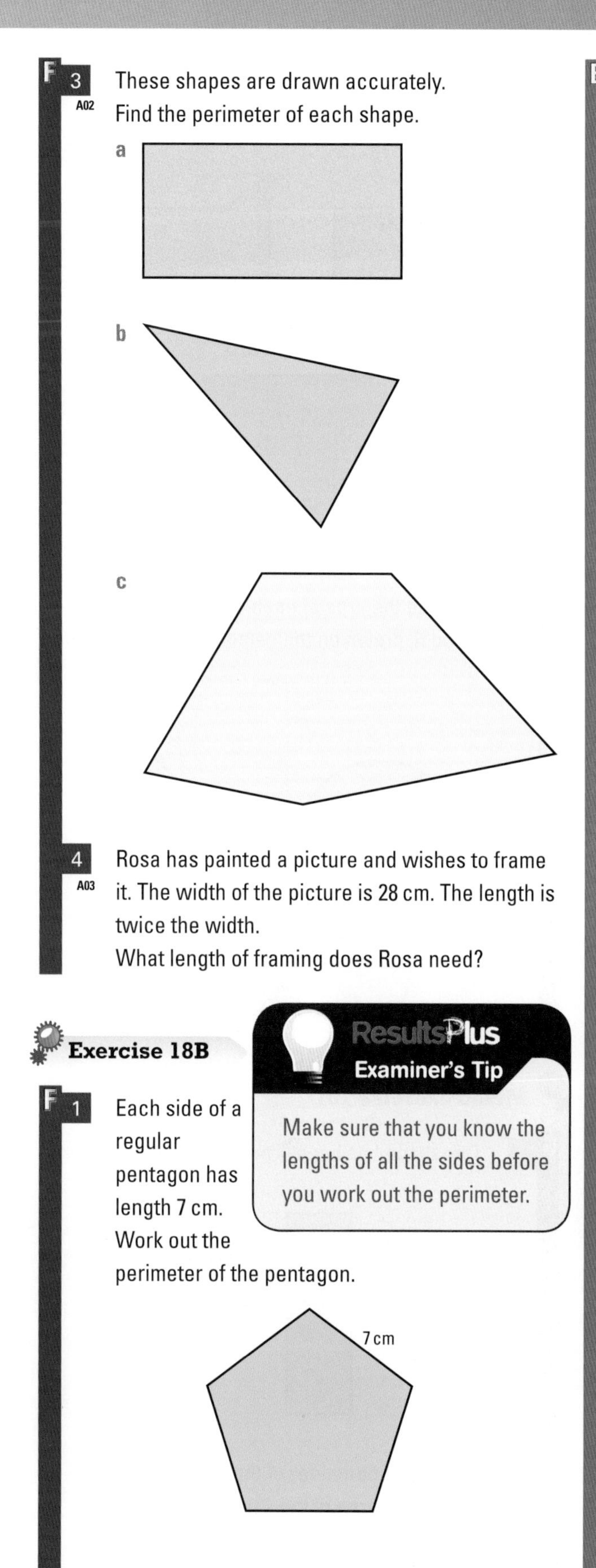

F 3 A02 These shapes are drawn accurately.
Find the perimeter of each shape.

4 A03 Rosa has painted a picture and wishes to frame it. The width of the picture is 28 cm. The length is twice the width.
What length of framing does Rosa need?

Exercise 18B

ResultsPlus Examiner's Tip

Make sure that you know the lengths of all the sides before you work out the perimeter.

F 1 Each side of a regular pentagon has length 7 cm.
Work out the perimeter of the pentagon.

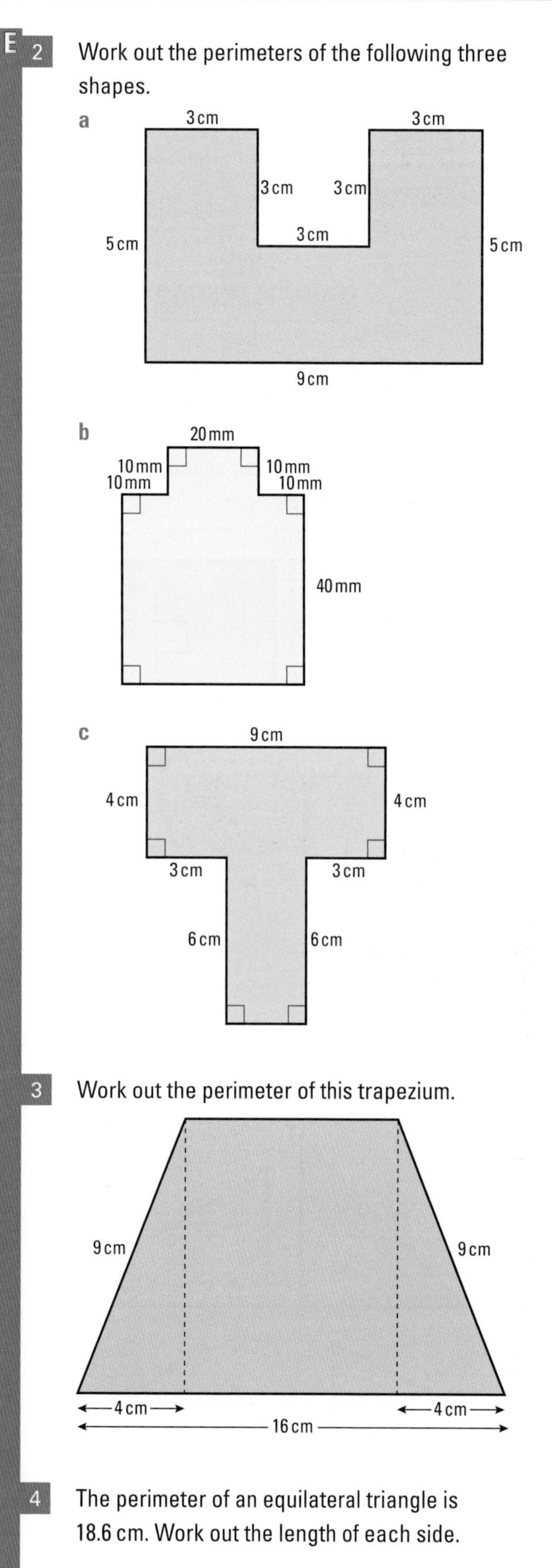

E 2 Work out the perimeters of the following three shapes.

3 Work out the perimeter of this trapezium.

4 The perimeter of an equilateral triangle is 18.6 cm. Work out the length of each side.

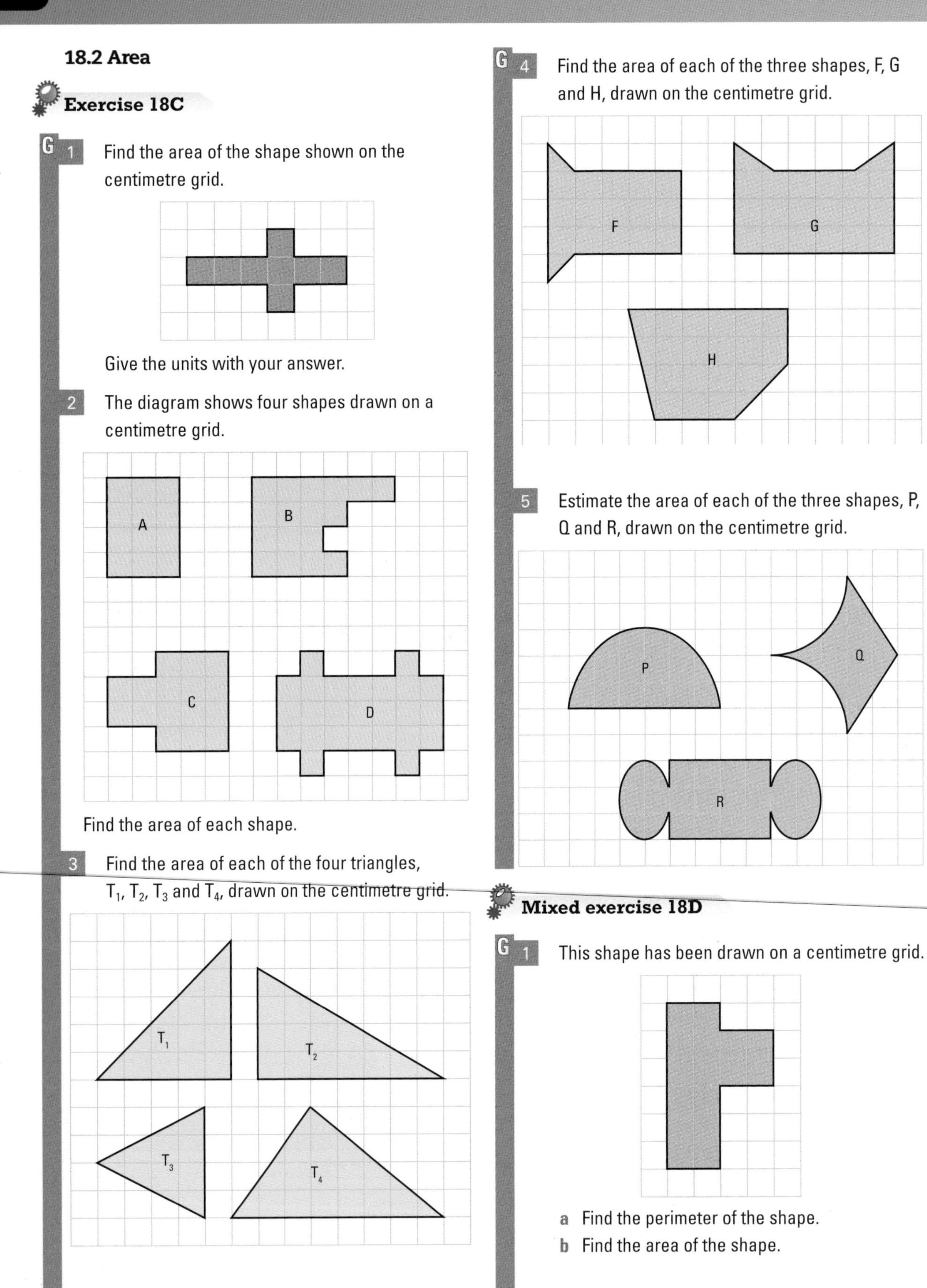

18.2 Area

Exercise 18C

G **1** Find the area of the shape shown on the centimetre grid.

Give the units with your answer.

2 The diagram shows four shapes drawn on a centimetre grid.

Find the area of each shape.

3 Find the area of each of the four triangles, T_1, T_2, T_3 and T_4, drawn on the centimetre grid.

G **4** Find the area of each of the three shapes, F, G and H, drawn on the centimetre grid.

5 Estimate the area of each of the three shapes, P, Q and R, drawn on the centimetre grid.

Mixed exercise 18D

G **1** This shape has been drawn on a centimetre grid.

a Find the perimeter of the shape.

b Find the area of the shape.

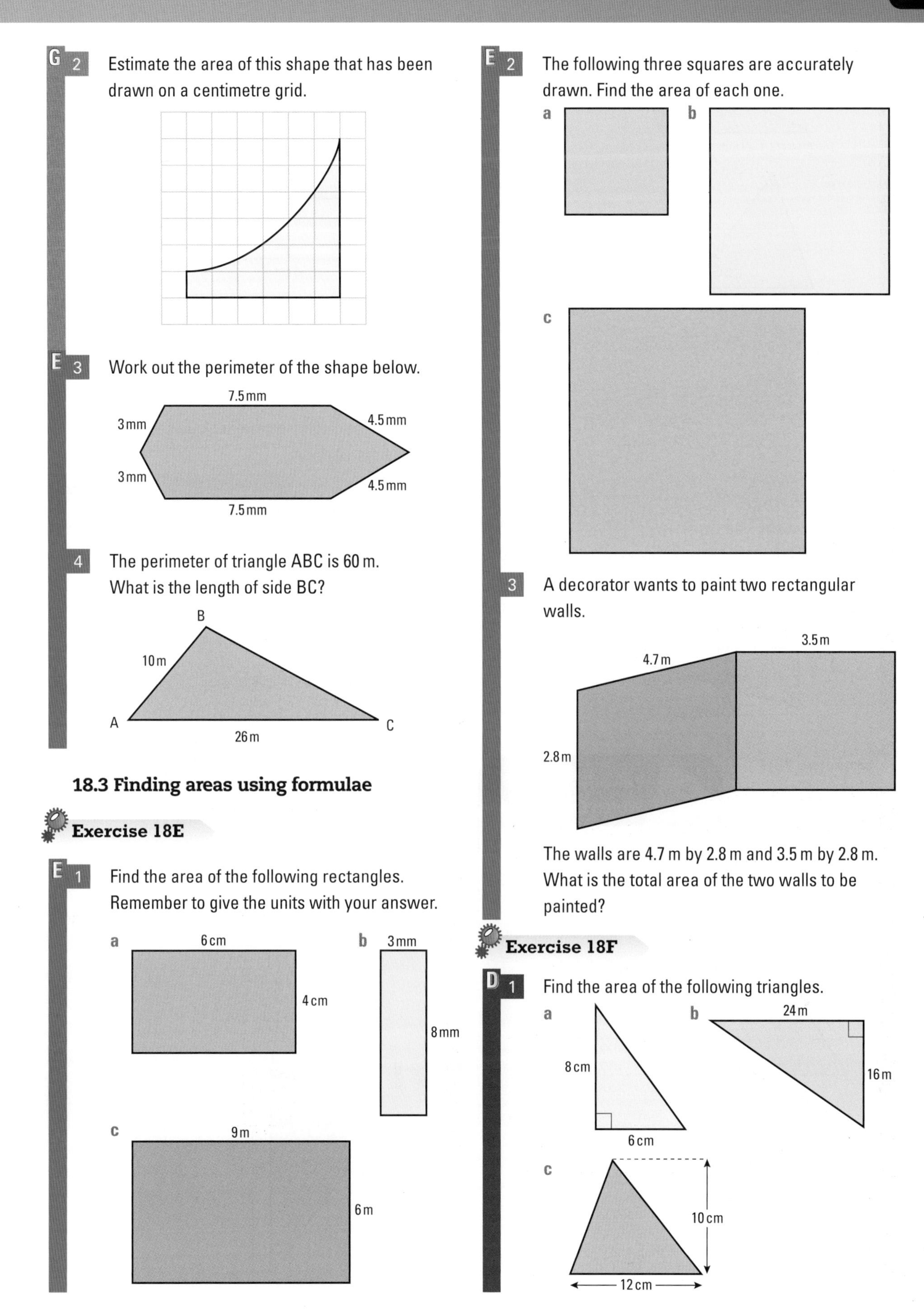

G **2** Estimate the area of this shape that has been drawn on a centimetre grid.

E **3** Work out the perimeter of the shape below.

7.5 mm
3 mm
4.5 mm
3 mm
4.5 mm
7.5 mm

4 The perimeter of triangle ABC is 60 m.
What is the length of side BC?

B
10 m
A
C
26 m

E **2** The following three squares are accurately drawn. Find the area of each one.

a **b** **c**

3 A decorator wants to paint two rectangular walls.

3.5 m
4.7 m
2.8 m

The walls are 4.7 m by 2.8 m and 3.5 m by 2.8 m.
What is the total area of the two walls to be painted?

18.3 Finding areas using formulae

Exercise 18E

E **1** Find the area of the following rectangles.
Remember to give the units with your answer.

a 6 cm, 4 cm

b 3 mm, 8 mm

c 9 m, 6 m

Exercise 18F

D **1** Find the area of the following triangles.

a 8 cm, 6 cm

b 24 m, 16 m

c 10 cm, 12 cm

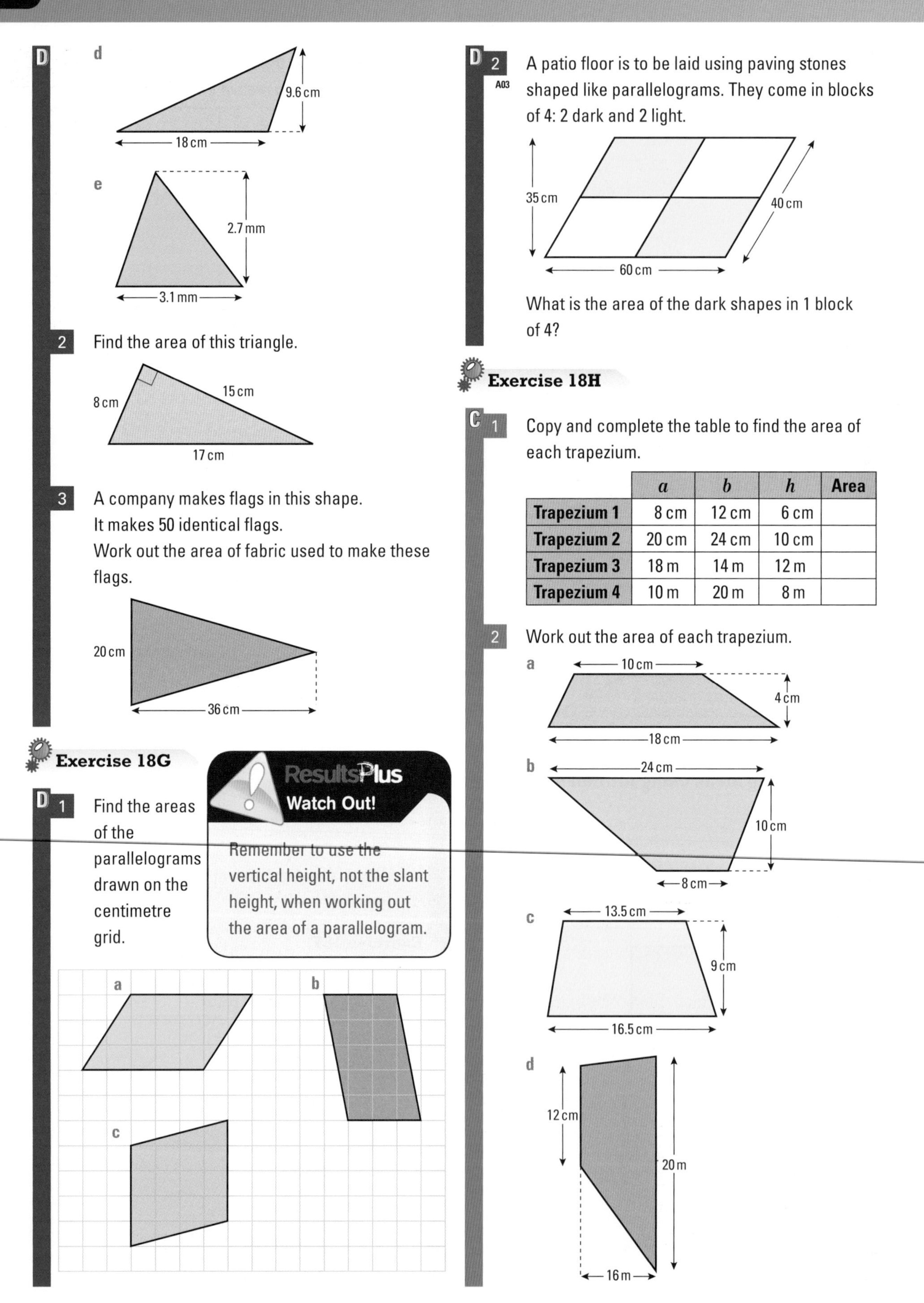

d

e

2 Find the area of this triangle.

3 A company makes flags in this shape.
It makes 50 identical flags.
Work out the area of fabric used to make these flags.

Exercise 18G

1 Find the areas of the parallelograms drawn on the centimetre grid.

ResultsPlus Watch Out!

Remember to use the vertical height, not the slant height, when working out the area of a parallelogram.

2 (A03) A patio floor is to be laid using paving stones shaped like parallelograms. They come in blocks of 4: 2 dark and 2 light.

What is the area of the dark shapes in 1 block of 4?

Exercise 18H

1 Copy and complete the table to find the area of each trapezium.

	a	b	h	Area
Trapezium 1	8 cm	12 cm	6 cm	
Trapezium 2	20 cm	24 cm	10 cm	
Trapezium 3	18 m	14 m	12 m	
Trapezium 4	10 m	20 m	8 m	

2 Work out the area of each trapezium.

a

b

c

d

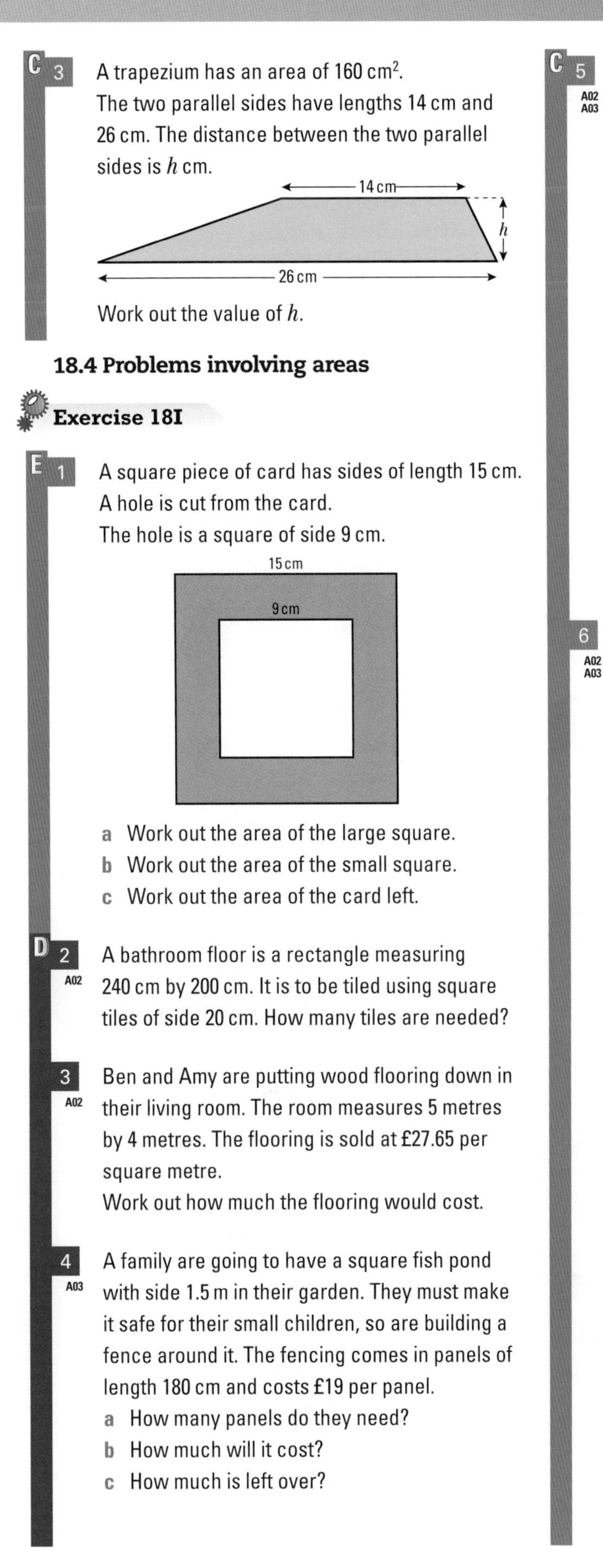

C 3 A trapezium has an area of 160 cm^2.
The two parallel sides have lengths 14 cm and 26 cm. The distance between the two parallel sides is h cm.

Work out the value of h.

18.4 Problems involving areas

Exercise 18I

E 1 A square piece of card has sides of length 15 cm.
A hole is cut from the card.
The hole is a square of side 9 cm.

a Work out the area of the large square.
b Work out the area of the small square.
c Work out the area of the card left.

D 2 A02 A bathroom floor is a rectangle measuring 240 cm by 200 cm. It is to be tiled using square tiles of side 20 cm. How many tiles are needed?

3 A02 Ben and Amy are putting wood flooring down in their living room. The room measures 5 metres by 4 metres. The flooring is sold at £27.65 per square metre.
Work out how much the flooring would cost.

4 A03 A family are going to have a square fish pond with side 1.5 m in their garden. They must make it safe for their small children, so are building a fence around it. The fencing comes in panels of length 180 cm and costs £19 per panel.
a How many panels do they need?
b How much will it cost?
c How much is left over?

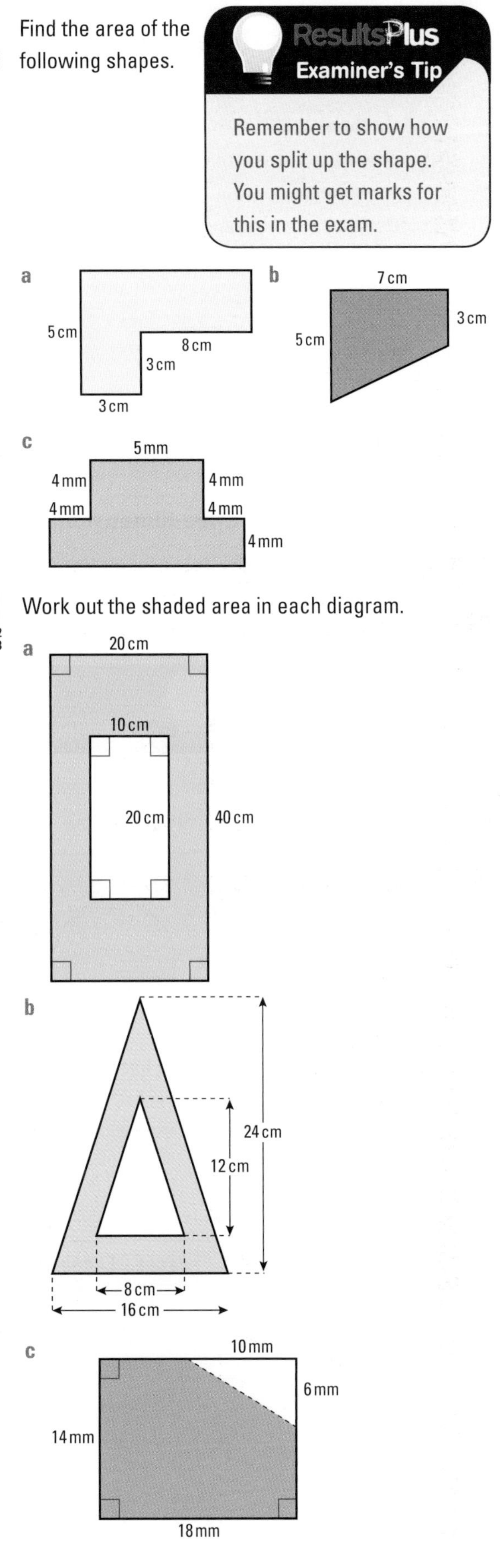

C 5 A02 A03 Find the area of the following shapes.

ResultsPlus
Examiner's Tip
Remember to show how you split up the shape. You might get marks for this in the exam.

6 A02 A03 Work out the shaded area in each diagram.

19 Three-dimensional shapes

Key Points

- **pyramid:** a 3D shape with a base of any 2D shape and sloping triangular sides.
- **prism:** a 3D shape with two parallel faces and rectangular sides joining them.
- **faces:** the flat surfaces of a 3D shape.
- **edges:** the lines where two faces meet.
- **vertex:** the point, or corner, at which edges meet.
- **vertices:** the plural of vertex.
- **net:** a pattern of 2D shapes that can be folded to make a hollow solid shape.
- **volume:** the amount of space a 3D shape takes up.
  - **volume of a prism** = **the area of the cross section × length**
- **surface area:** the area of the net that can be used to build the shape. Measured in square units.
- **drawing 3D objects:** use isometric paper with the vertical lines going down the page and no horizontal lines.

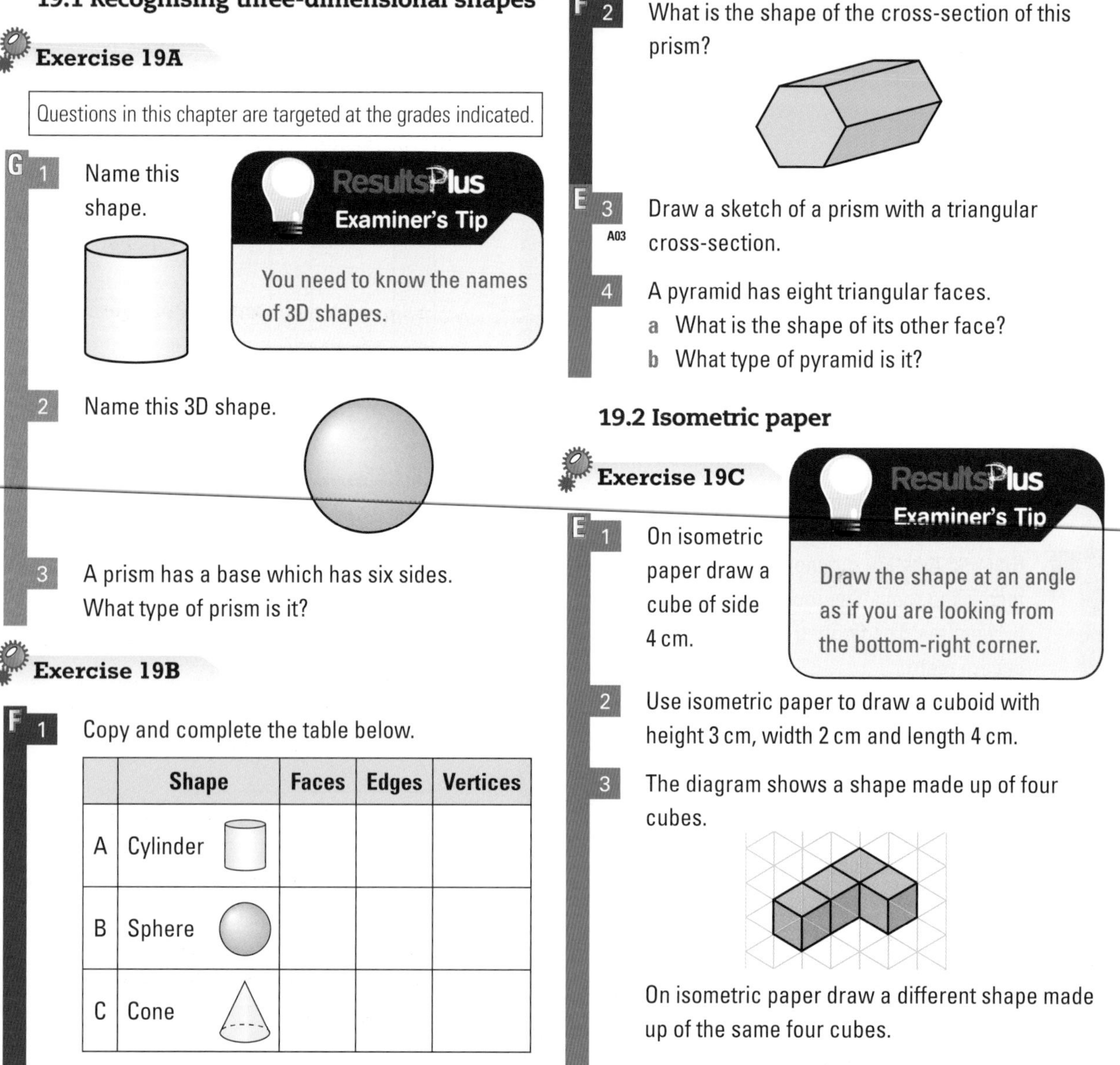

19.1 Recognising three-dimensional shapes

Exercise 19A

Questions in this chapter are targeted at the grades indicated.

G

1. Name this shape.

ResultsPlus Examiner's Tip

You need to know the names of 3D shapes.

2. Name this 3D shape.

3. A prism has a base which has six sides. What type of prism is it?

Exercise 19B

F

1. Copy and complete the table below.

	Shape	Faces	Edges	Vertices
A	Cylinder			
B	Sphere			
C	Cone			

2. What is the shape of the cross-section of this prism?

E

3. Draw a sketch of a prism with a triangular cross-section. A03

4. A pyramid has eight triangular faces.
   a What is the shape of its other face?
   b What type of pyramid is it?

19.2 Isometric paper

Exercise 19C

E

1. On isometric paper draw a cube of side 4 cm.

ResultsPlus Examiner's Tip

Draw the shape at an angle as if you are looking from the bottom-right corner.

2. Use isometric paper to draw a cuboid with height 3 cm, width 2 cm and length 4 cm.

3. The diagram shows a shape made up of four cubes.

On isometric paper draw a different shape made up of the same four cubes.

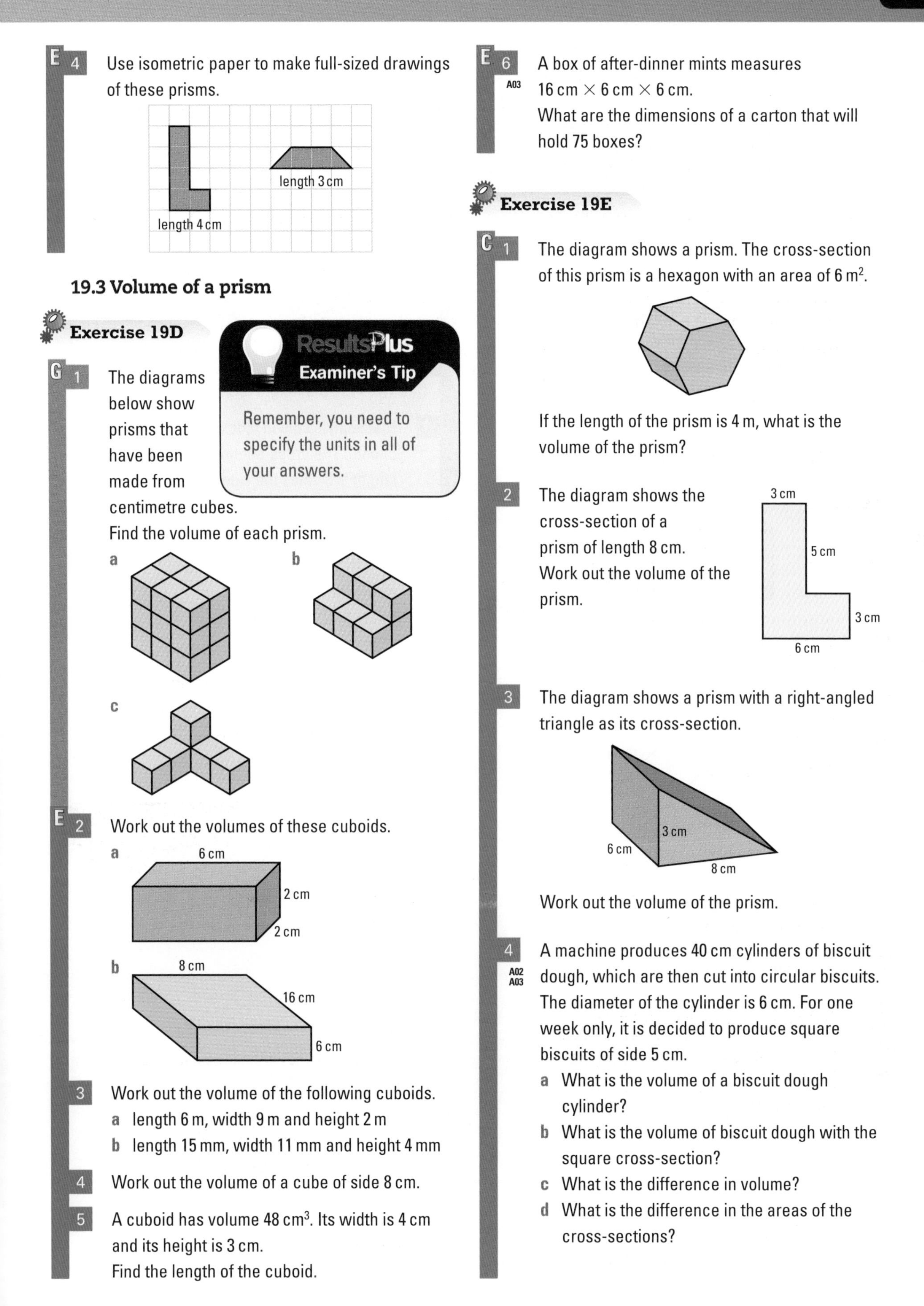

E 4 Use isometric paper to make full-sized drawings of these prisms.

19.3 Volume of a prism

Exercise 19D

ResultsPlus Examiner's Tip

Remember, you need to specify the units in all of your answers.

G 1 The diagrams below show prisms that have been made from centimetre cubes.
Find the volume of each prism.

a

b

c

E 2 Work out the volumes of these cuboids.

a

b

3 Work out the volume of the following cuboids.
a length 6 m, width 9 m and height 2 m
b length 15 mm, width 11 mm and height 4 mm

4 Work out the volume of a cube of side 8 cm.

5 A cuboid has volume 48 cm^3. Its width is 4 cm and its height is 3 cm.
Find the length of the cuboid.

E 6 A03 A box of after-dinner mints measures 16 cm × 6 cm × 6 cm.
What are the dimensions of a carton that will hold 75 boxes?

Exercise 19E

C 1 The diagram shows a prism. The cross-section of this prism is a hexagon with an area of 6 m^2.

If the length of the prism is 4 m, what is the volume of the prism?

2 The diagram shows the cross-section of a prism of length 8 cm.
Work out the volume of the prism.

3 The diagram shows a prism with a right-angled triangle as its cross-section.

Work out the volume of the prism.

4 A02 A03 A machine produces 40 cm cylinders of biscuit dough, which are then cut into circular biscuits. The diameter of the cylinder is 6 cm. For one week only, it is decided to produce square biscuits of side 5 cm.
a What is the volume of a biscuit dough cylinder?
b What is the volume of biscuit dough with the square cross-section?
c What is the difference in volume?
d What is the difference in the areas of the cross-sections?

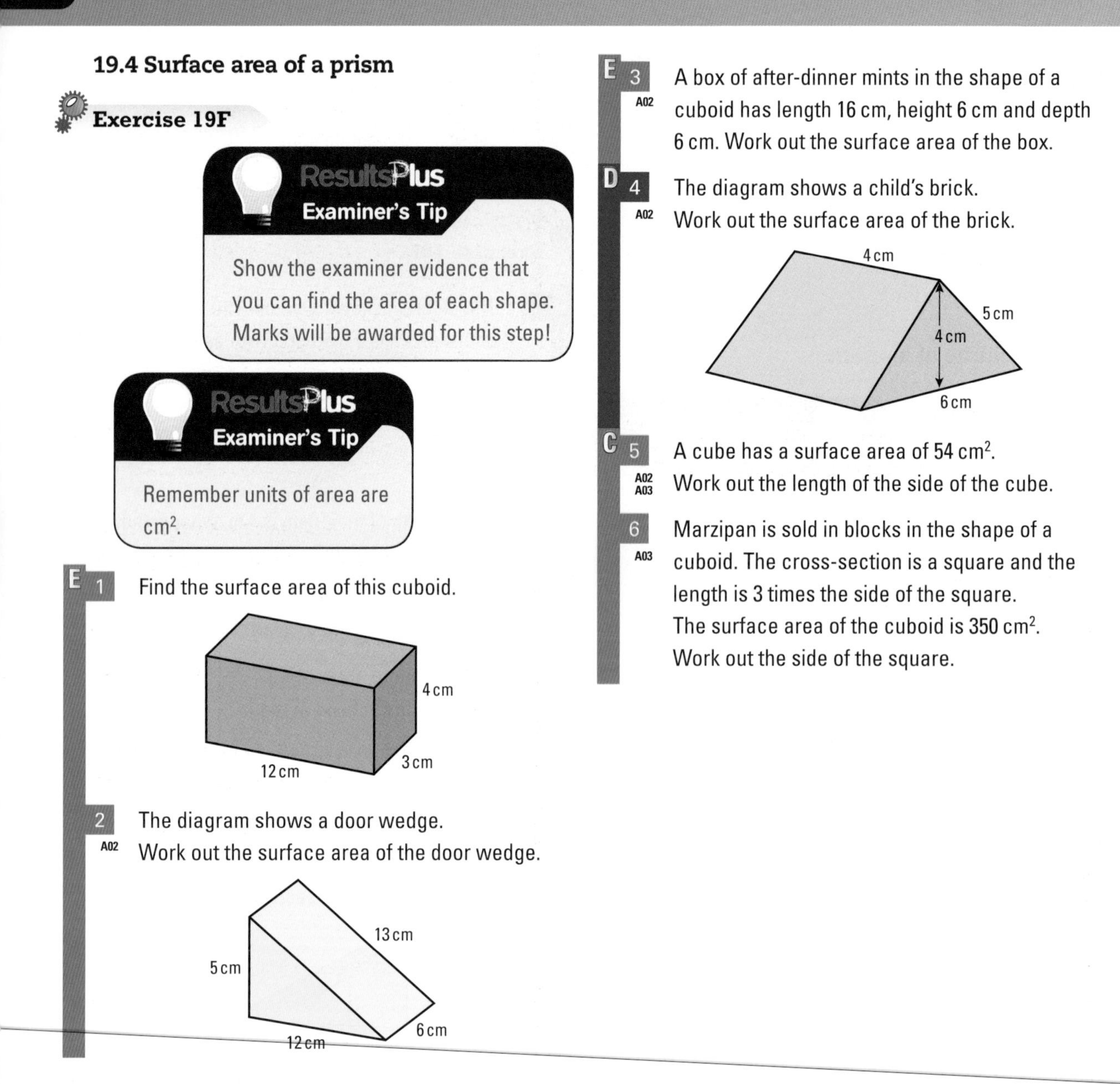

19.4 Surface area of a prism

Exercise 19F

ResultsPlus Examiner's Tip

Show the examiner evidence that you can find the area of each shape. Marks will be awarded for this step!

ResultsPlus Examiner's Tip

Remember units of area are cm^2.

E 1 Find the surface area of this cuboid.

2 (A02) The diagram shows a door wedge.
Work out the surface area of the door wedge.

E 3 (A02) A box of after-dinner mints in the shape of a cuboid has length 16 cm, height 6 cm and depth 6 cm. Work out the surface area of the box.

D 4 (A02) The diagram shows a child's brick.
Work out the surface area of the brick.

C 5 (A02, A03) A cube has a surface area of 54 cm^2.
Work out the length of the side of the cube.

6 (A03) Marzipan is sold in blocks in the shape of a cuboid. The cross-section is a square and the length is 3 times the side of the square.
The surface area of the cuboid is 350 cm^2.
Work out the side of the square.

1 Using a calculator

Key Points

- **reciprocal of a number:** 1 divided by the number.
  - any number multiplied by its reciprocal = 1
  - zero has no reciprocal
- **reciprocal of a fraction:** the fraction turned upside down.
- **using a calculator:**
  - $1/x$ or x^{-1} works out a reciprocal
  - x^2 works out squares
  - x^y or $x^{\blacksquare}$ works out powers
  - $\sqrt{}$ works out square roots
  - $\sqrt[3]{}$ works out cube roots
  - remember BIDMAS

1.1 Finding reciprocals

Exercise 1A

D

1 Find the reciprocals of these numbers.

a 8 b 5 c 20
d 2 e 4

2 Find the reciprocals of these fractions.

a $\frac{1}{5}$ b $\frac{1}{2}$ c $\frac{3}{4}$
d $\frac{4}{6}$ e $\frac{7}{10}$

3 Use your calculator to find the reciprocals of these numbers.

a 3.5 b 40 c 18
d 60 e 0.4 f 0.6
g 0.06 h 0.250 i 0.05
j 0.02

4 The reciprocal of 100 is 0.01.
What is the reciprocal of 0.01?

5 a Find the reciprocal of 60.
b Multiply 60 by its reciprocal.

6 a Find the reciprocal of 1000.
b Multiply 1000 by its reciprocal.

7 Find the reciprocal of 50.

8 Find the reciprocal of 0.4.

1.2 Interpreting a calculator display

Exercise 1B

E

1 The total cost of five adult theme park tickets is £53.40. Work out the cost of one adult theme park ticket.

2 Alfie buys five bars of chocolate costing £1.25 each and one packet of sweets costing £1.35. Work out the total cost.

3 Evie buys two bottles of shampoo costing £3.58 each. Work out how much change she should get from £10.

4 A garden centre sells plants for £1.90 each. Hakim buys 24 plants. Work out the total cost.

5 Dave's company pays him 35p for each mile that he drives his car.
Work out how much money Dave's company pays him when he drives 226 miles.

6 Petrol costs 115.9 pence per litre. Tom buys 42 litres of petrol. How much should Tom pay?

7 Shelley needs 170 tiles for a room. FS
Tiles are sold in boxes. There are 12 tiles in each box. Work out the least number of boxes of tiles that Shelley needs.

8 345 students and teachers are going on a coach trip. Each coach holds 52 passengers.
Work out the smallest number of coaches needed.

9 A cup holds 225 m*l* of apple juice.
How many of these cups can be completely filled using 2000 m*l* of apple juice?

10 It takes 35 seconds to fill a bucket. How many buckets can be completely filled in 30 minutes?

11 The battery life of a torch is 380 hours.
Work out the battery life in days and hours.

E 12 A pen costs 47p. Sam has £5. He buys as many pens as he can.
Work out how much change Sam should get from £5.

D 13* A02 A03 Gary uses his mobile phone for a monthly call average of 140 minutes and 400 texts.
He compares two different tariffs.
a Contract: 18 months giving 500 minutes and 500 texts for £13.50 per month.
b Pay-as-you-go: £15 gives 600 texts and paying 20p per minute for each call.
Which should he use?

1.3 Working out powers and roots

Exercise 1C

E 1 Work out
a 3.5^2 b 2.2^2 c 37^2
d 2.8^2 e 27.9^2

2 Work out
a 5^3 b 21^3 c 3.1^3
d 4.4^3 e 2.5^3

3 Work out
a 3^5 b 12^4 c 8^5
d 11^4 e 6^6

4 Work out
a $17^2 + 20$ b $2.4^2 + 25$
c $29^2 - 322$ d $3.7^2 + 5.42$

5 Work out
a $3.1^3 + 1.96$ b $2.9^3 + 5.29$
c $3.5^3 - 7.29$ d $2.4^3 - 1.544$

Exercise 1D

E 1 Work out
a $\sqrt{625}$
b $\sqrt{324}$
c $\sqrt{1296}$
d $\sqrt{529}$

2 Work out
a $\sqrt{2.89}$
b $\sqrt{20.25}$
c $\sqrt{15.21}$
d $\sqrt{84.64}$

ResultsPlus Examiner's Tip

Write down all the figures on your calculator display before you round your answer to one decimal place. (See Section 5.7 on Rounding.)

E 3 Work these out, giving your answers correct to one decimal place.
a $\sqrt{140}$ b $\sqrt{85}$ c $\sqrt{134}$ d $\sqrt{220}$

4 Work out
a $\sqrt[3]{343}$ b $\sqrt[3]{512}$ c $\sqrt[3]{1331}$ d $\sqrt[3]{2197}$

5 Work these out, giving your answers correct to one decimal place.
a $\sqrt[3]{50}$ b $\sqrt[3]{300}$ c $\sqrt[3]{220}$ d $\sqrt[3]{85}$

6 Work out
a $\sqrt{51.84} + 8.4$ b $\sqrt{841} - 23.1$
c $\sqrt[3]{9.261} - 1.3$ d $\sqrt[3]{1.728} + 8.1$

1.4 Using a calculator to work out complex calculations

Exercise 1E

E 1 Work out
a $(5.2 + 2.7)^3$ b $(12.4 - 9.71)^3$
c $(2.43 + 1.87)^2$ d $(5.1 - 3.7)^2$

2 Work out
a $12^3 + 13^3$ b $34^3 + 6^2$
c $12^2 - 23^2$ d $38^3 - 18^3$

3 Work out
a $\sqrt{18.7 + 15.63}$ b $\sqrt{514 - 195}$
c $\sqrt[3]{127 - 63}$ d $\sqrt[3]{2.85 + 2.525}$

D 4 Work out the value of each of these.
Write down all the figures on your calculator display.
a $\frac{4.78 - 1.42}{0.72}$
b $\frac{84.38}{3.62 + 5.78}$
c $\frac{12.24 \times 2.5}{8.6}$ d $\frac{36.35}{12.6 - 5.8}$

ResultsPlus Examiner's Tip

If you work out the numerator and the denominator separately, make sure you write down the value of each.

5 Work out the value of each of these.
Write down all the figures on your calculator display.
a $\frac{23.2 - 16.84}{2.8 + 3.41}$ b $\frac{5.6 \times 8.1}{22.5 - 13.9}$
c $\frac{6.37 \times 4.52}{2.8 + 7.19}$ d $\frac{27.6 + 19.82}{23.6 - 5.94}$

2 Percentages

Key Points

- **per cent:** out of 100.
- **percentage (%):** a quantity out of 100. Can also be written as a decimal or a fraction.
- **knowing basic percentage equivalents:** know these percentages.

Percentage	1%	10%	25%	50%	75%
Decimal	0.01	0.1	0.25	0.5	0.75
Fraction	$\frac{1}{100}$	$\frac{1}{10}$	$\frac{1}{4}$	$\frac{1}{2}$	$\frac{3}{4}$

- **finding percentages of quantities:** write the percentage as a fraction, and then multiply the fraction by the quantity.
- **increasing a quantity by a percentage:** work out the increase and add it to the original quantity.
- **decreasing a quantity by a percentage:** work out the decrease and subtract it from the original quantity.
- **using the multiplier method:** work out the multiplier for an increase or decrease. Then multiply the original amount by the multiplier to find the new amount.
- **writing one quantity as a percentage of another quantity:** write the first quantity as a fraction of the second quantity, then convert the fraction to a percentage.

2.1 Finding percentages of quantities

Exercise 2A

D

1 Work out

a 14% of £40 b 68% of 45 kg
c 45% of £370 d 73% of 640 km
e 84% of 330 ml f 32% of \$90
g 6% of £170 h 29% of 1500 m

2 There are 240 employees in a company. 55% of the employees are women.
How many of the employees are women?

3 Stuart invested £1500 in a savings account. At the end of the year he received 4% interest.
Work out how much interest he received.

4 There are 225 students in Year 11.
20% of these students study geography.
How many of the students study geography?

5 Sarah's salary is £46 000. Her employer agrees to increase her salary in line with inflation.
The rate of inflation this year is 3%.
Work out her new salary.

6 In a restaurant a service charge of 10.5% is added to the cost of the meal.
Work out the service charge when the cost of the meal is £60.

7 VAT is charged at the rate of $17\frac{1}{2}$%.
Work out how much VAT will be charged on:
a a ladder costing £76
b a garage bill of £140.

8 The rate of simple interest is 3% per year. FS
Work out the simple interest paid on £400 in one year.

9* A02 A03 The cash price of a sofa is £540. FS
A Credit Plan requires a deposit of 5% of the cash price and 24 monthly payments of £24.
Which is the cheaper way to buy the sofa, paying cash or using the credit plan?
Explain your answer.

C

10 A02 A 100 g tub of margarine has the following nutrional content. FS

fat	35 g
sodium	1.4 g
carbohydrate	2.6 g
protein	0.2 g

a What percentage of the margarine is carbohydrate?
b How many grams of fat and sodium would there be in a 250 g tub?

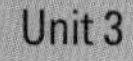

C 11 (A03) Ruby bought 50 melons at 90p each.
She sold all the melons.
On each of the first 36 melons she made a 35% profit.
On each of the remaining melons she made a 40% loss.
Work out the overall profit or loss that Ruby made.

2.2 Using percentages

Exercise 2B

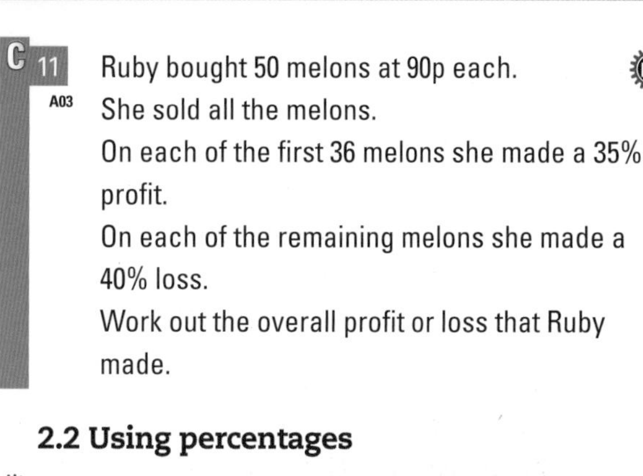

E 1 A packet contains 750 g of cereal plus 20% extra cereal.
Work out the weight of cereal in the packet.

2 Jennie's salary is £27 500. Her salary is increased by 3%. Work out her new salary.

3 The price of rail fares increased by 11%. Before the increase the price of a ticket was £77.
Work out the price of the ticket after the increase.

4 A travel company reduced the prices of its holidays by 12%. What is the new price of a holiday which was originally priced at £795?

5 A car battery costs £58 plus VAT at $17\frac{1}{2}$%.
Work out the total cost of the car battery.

6 VAT at $17\frac{1}{2}$% is added to a telephone bill of £66.
Work out the total bill.

7 Lisa invests £3500. The interest rate is 3.5% per year. How much will Lisa have in her account at the end of one year?

8 Seryn invests £2500 for 3 years at 4% simple interest. Work out the value of her investment after one year.

D 9 Asif bought a car for £9000. In one year the value of the car depreciated by 10%.
Work out the value of the car one year after he bought it.

10 The normal price of a pack of cakes is £1.85.
The normal price is reduced by 25%.
Work out the price after the reduction.

D 11 A store reduced all normal prices by 15% in a two-day sale. Work out the sale price of:

a a drill with a normal price of £60
b a lawnmower with a normal price of £190
c a tin of paint with a normal price of £12.

12 (A03) In a super-sale a shop reduces its sale prices by a further 15%.

In the super-sale, Sally buys a television with a normal price of £320.
How much does she pay?

13 (A03) Overdean Garage has a loyalty scheme for businesses who buy their vans from the garage.
The scheme gives customers a discount on the cost of labour and the cost of parts.
The percentage discount depends on the age of the van.

Age of vehicle (years)	Labour discount	Parts discount
4	10%	5%
5	12.5%	5%
6	15%	5%
7	17.5%	10%
8	20%	10%
9	25%	10%
10 or older	30%	10%

Alan bought a new van from Overdean Garage in September 2005.
Today, Overdean Garage carried out some repairs on the van.
Copy and complete the bill for the repairs.

Item	Cost before discount	% discount	Cost after discount
Labour	£240	...............%	£..............
Parts	£64	...............%	£..............
Overdean Garage		Total before VAT	£..............
		VAT at $17\frac{1}{2}$%	£..............
		Total with VAT	£..............

2.3 Writing one quantity as a percentage of another

Exercise 2C

G

1. a Write 9 out of 10 as a percentage.
   b Write £27 out of £50 as a percentage.
   c Write 17 kg out of 20 kg as a percentage.
   d Write 16p out of £1 as a percentage.
   e Write 140 m*l* out of 200 m*l* as a percentage.
2. There are 25 trees in a park.
   Of these trees, 16 are chestnut trees.
   What percentage of the trees are chestnut trees?
3. Adam planted 400 flower seeds.
   Of these seeds, 320 germinated.
   What percentage of the seeds germinated?
4. A glass contains 500 m*l* of drink.
   300 m*l* of the drink is water.
   What percentage of the drink is water?
5. Danni scored 36 out of 60 in a test.
   Write 36 out of 60 as a percentage.
6. A02 There were 60 members of a film society.
   They need to elect an events organiser.
   Three members decide to stand for the position.
   Jack gets 15 votes, Andy gets 38% of the vote and Faz gets the remaining votes.
   What percentage of the votes did Faz get?

Exercise 2D

F

1. a Write 360 g as a percentage of 750 g.
   b Write £9.60 as a percentage of £120.
   c Write 140 m*l* as a percentage of 350 m*l*.
   d Write 321 km as a percentage of 480 km.
   e Write 70p as a percentage of £2.50.
2. A rugby team played 55 matches. The team won 22 of these matches.
   What percentage of the matches did the team win?
3. There are 32 students in a class. On Friday, four of these students were absent.
   What percentage of the students were absent on Friday?
4. There are 1650 students in a school. 264 of the students are in Year 7.
   What percentage of the students are in Year 7?
5. 120 g of cheese contains 19.4 g of carbohydrates and 6.2 g of protein.
   What percentage of the cheese is:
   a carbohydrates b protein?
6. A03 The audience at a comedy club consisted of 212 men and 190 women.
   What percentage of the audience was male?
7. A02 FS A mixture of gravel and sand is being mixed by Gary ready for icy conditions.
   He knows the best mixture is with 50% gravel.
   He has made 12 bucketfuls, but it is only 25% gravel. How many more bucketfuls of gravel must be added for the best mixture?

3 Equations

Key Points

- **an equation:** has an equals sign and a symbol or letter that represents an unknown number.
- **coefficient:** a number in front of an unknown.
- **trial and improvement:** a method used to find an approximate solution to an equation, if all other methods cannot be used.
- **solving an equation:** rearrange the equation so the unknown or unknowns appear on one side of the equation only. Solutions can be whole numbers, fractions, decimals, or negative numbers.
- **rearranging an equation:** use the balance method.
  - add the same number to both sides
  - subtract the same number from both sides
  - multiply both sides by the same number
  - divide both sides by the same number

3.1 Using simple equations

Exercise 3A

Questions in this chapter are targeted at the grades indicated.

In questions 1–10, write each equation using a letter.

G

1 $\square - 5 = 9$

2 $\square + 4 = 11$

3 $7 \times \square = 32$

4 $11 + \square = 20$

5 $4 \times \square - 3 = 32$

6 $\square \times 5 = 21$

7 $5 \times (\square + 1) = 24$

8 $\square \times 4 + 6 = 13$

9 $(4 + \square) \times 2 = 16$

10 $8 + 5 \times \square = 27$

In questions 11–15

a express the problem as an equation

b find the number.

G

11 Jon thinks of a number and adds 7 to it. The answer is 12.

12 Kimberley thinks of a number and multiplies it by 5. The answer is 55.

13 Victoria thinks of a number. She adds 2 to it and multiplies the result by 4. The answer is 28.

14 Adem thinks of a number. He multiplies it by 2 and subtracts 5 from the result. The answer is 15.

15 Ryan thinks of a number. He multiplies it by 6 and adds 7 to the result. The answer is 35.

3.2 Solving equations with one operation

Exercise 3B

Solve these equations.

ResultsPlus Examiner's Tip

'Solve' means find the value of the letter.

F

1 $q - 3 = 8$

2 $y + 3 = 7$

3 $7 = a + 2$

4 $p - 5 = 6$

5 $k + 3 = 4$

6 $d - 6 = 3$

7 $x + 2 = 2$

8 $t - 3 = 0$

9 $r + 8 = 10$

10 $21 + x = 22$

11 $n + 1 = 3$

12 $x - 2 = 2$

13 $m + 8 = 12$

14 $y - 7 = 8$

15 $a + 6 = 9$

16 $q - 10 = 3$

17 $5 + p = 9$

18 $5 + t = 5$

19 $a + 17 = 31$

20 $h + 2 = 8$

21 $p - 15 = 26$

22 $w + 4 = 4$

23 $6 = b + 6$

24 $11 = v + 11$

Exercise 3C

Solve these equations.

F

1 $22 + q = 22$

2 $x + 8 = 15$

3 $5 = b + 4$

4 $y + 5 = 17$

5 $q - 3 = 7$

6 $a + 4 = 10$

F

7 $x - 8 = 15$ 8 $y - 5 = 17$
9 $s - 13 = 15$ 10 $a + 5 = 7$
11 $p - 5 = 7$ 12 $c + 19 = 21$
13 $4 + a = 6$ 14 $11 + p = 17$
15 $p - 4 = 9$ 16 $12 = p - 14$
17 $s + 12 = 16$ 18 $12 = c - 5$
19 $10 = p + 3$ 20 $12 = y - 10$
21 $17 = t + 10$ 22 $13 = p + 13$
23 $4 = a + 3$ 24 $p + 15 = 15$

Exercise 3D

Solve these equations.

F

1 $3a = 9$ 2 $4p = 16$
3 $5p = 30$ 4 $6s = 24$
5 $2k = 20$ 6 $7u = 21$
7 $2g = 18$ 8 $5l = 45$
9 $6j = 18$ 10 $8f = 24$
11 $3r = 33$ 12 $5v = 35$

Exercise 3E

Solve these equations.

E

1 $\frac{a}{2} = 4$ 2 $\frac{b}{5} = 3$
3 $\frac{s}{4} = 6$ 4 $\frac{c}{6} = 7$
5 $\frac{t}{4} = 2$ 6 $\frac{s}{8} = 7$
7 $\frac{h}{6} = 10$ 8 $\frac{f}{4} = 5$
9 $\frac{d}{3} = 13$ 10 $\frac{a}{3} = 12$
11 $\frac{b}{5} = 7$ 12 $\frac{r}{4} = 15$
13 $\frac{a}{12} = 6$ 14 $\frac{b}{2} = 12$
15 $\frac{k}{3} = 14$

Mixed exercise 3F

Solve these equations.

F

1 $d - 6 = 3$ 2 $4 + r = 9$
3 $q - 3 = 3$ 4 $s - 2 = 5$
5 $a + 4 = 7$ 6 $b + 3 = 8$
7 $8 + p = 8$ 8 $p - 3 = 5$
9 $4r = 16$ 10 $c + 6 = 9$
11 $2p = 4$ 12 $6 + e = 8$

E

13 $\frac{a}{2} = 3$ 14 $\frac{b}{3} = 12$
15 $\frac{s}{5} = 9$

3.3 Solving equations with two operations

Exercise 3G

Solve these equations.

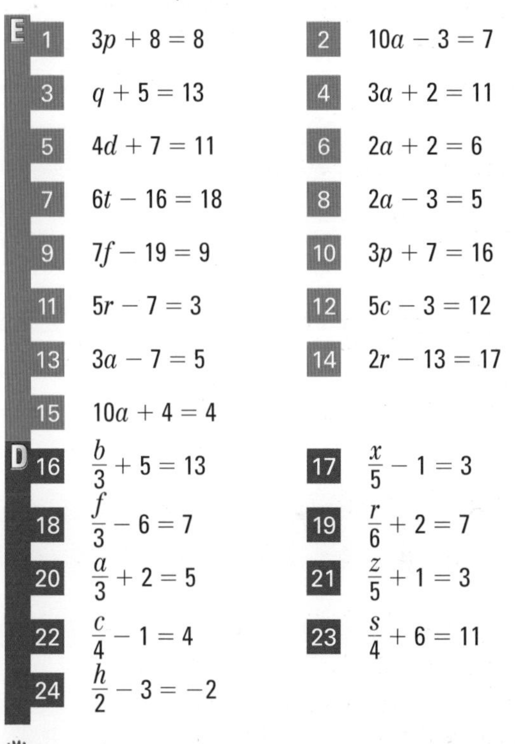

E

1 $3p + 8 = 8$ 2 $10a - 3 = 7$
3 $q + 5 = 13$ 4 $3a + 2 = 11$
5 $4d + 7 = 11$ 6 $2a + 2 = 6$
7 $6t - 16 = 18$ 8 $2a - 3 = 5$
9 $7f - 19 = 9$ 10 $3p + 7 = 16$
11 $5r - 7 = 3$ 12 $5c - 3 = 12$
13 $3a - 7 = 5$ 14 $2r - 13 = 17$
15 $10a + 4 = 4$

D

16 $\frac{b}{3} + 5 = 13$ 17 $\frac{x}{5} - 1 = 3$
18 $\frac{f}{3} - 6 = 7$ 19 $\frac{r}{6} + 2 = 7$
20 $\frac{a}{3} + 2 = 5$ 21 $\frac{z}{5} + 1 = 3$
22 $\frac{c}{4} - 1 = 4$ 23 $\frac{s}{4} + 6 = 11$
24 $\frac{h}{2} - 3 = -2$

Exercise 3H

Solve these equations.

E

1 $5p - 8 = 15$
2 $8k + 2 = 5$
3 $3a - 7 = 6$
4 $2a - 3 = 4$
5 $9u + 5 = 9$
6 $3a + 8 = 15$
7 $2a + 6 = 9$ 8 $5p + 8 = 15$
9 $5e + 2 = 2$ 10 $4t + 2 = 9$
11 $8j - 8 = 5$ 12 $7c - 7 = 4$

ResultsPlus
Examiner's Tip

Unless the question says give your answer in its simplest form, a mixed fraction such as $\frac{11}{5}$ is ok.

E

13 $7y + 9 = 15$

14 $3d - 8 = 3$

15 $4q - 5 = 4$

Exercise 3I

Solve these equations.

E

1 $5p - 3 = -13$

2 $2a + 7 = 3$

3 $2s + 6 = -4$

4 $5p + 13 = 3$

5 $2a + 6 = 2$

6 $3a + 9 = 6$

7 $2a + 7 = 1$

8 $3a + 10 = 1$

9 $4y + 8 = -8$

10 $2k - 5 = -9$

11 $3w + 7 = 1$

12 $8h + 9 = 1$

13 $13a + 8 = 8$

14 $9e + 47 = 20$

15 $6t - 13 = -13$

16 $4c + 15 = 11$

Mixed exercise 3J

Solve these equations.

E

1 $4f + 4 = 19$

2 $6k + 7 = 43$

3 $2s + 6 = 10$

4 $6y + 7 = 13$

5 $3s - 6 = 7$

6 $3d + 3 = 18$

7 $4h - 6 = 14$

8 $5p + 2 = 11$

9 $5m - 7 = 33$

10 $-7g - 3 = 12$

11 $4f - 7 = 12$

12 $5k - 12 = 7$

13 $-3s - 15 = 4$

14 $5g + 18 = 15$

15 $3e - 6 = -7$

16 $-2r + 12 = 3$

17 $5t + 15 = -4$

18 $7y - 14 = -20$

19 $9b + 6 = 1$

20 $-4f - 9 = -2$

21 $6j - 2 = 19$

22 $3h + 3 = 0$

23 $-3c - 7 = 0$

24 $8s + 7 = 4$

D

25 $\frac{z}{2} + 2 = 5$

26 $\frac{x}{5} - 3 = 4$

27 $\frac{p}{2} - 7 = -3$

28 $\frac{c}{3} + 5 = -2$

29 $\frac{a}{8} - 1 = 3$

30 $-\frac{e}{3} + 3 = 10$

3.4 Solving equations with brackets

Exercise 3K

Solve the equations.

D

1 $4g + 5 = 37$

2 $\frac{c}{6} = 5$

3 $\frac{f + 4}{5} = 3$

4 $5(b + 6) = 30$

5 $5(a - 5) = 60$

6 $5(e + 2) = 45$

7 $3(d - 5) = 12$

8 $4(m - 4) = 16$

9 $5(q + 6) = 30$

10 $\frac{h}{3} - 5 = 3$

11 $3c + 5 = 3$

12 $\frac{x}{3} + 7 = 6$

13 $6p - 1 = 2$

14 $3(2d - 7) = 27$

15 $2(b - 3) = 5$

16 $3(y - 1) = 1$

17 $5v + 3 = 9$

C

18 $\frac{n - 3}{6} = 3$

19 $\frac{t + 10}{4} = 1$

20 $\frac{3c + 4}{3} = 5$

3.5 Solving equations with letters on both sides

Exercise 3L

Solve the equations.

C

1 $7y = 2y + 20$

2 $x + 14 = 5x + 2$

3 $2a + 8 = a + 4$

4 $4d + 16 = 8d - 2$

5 $5p - 10 = 2p + 11$

6 $9q - 7 = 2q + 14$

7 $8b + 10 = 3b + 15$

8 $3c - 2 = c + 10$

9 $7r - 4 = 2r + 10$

10 $6v - 8 = 3v + 8$

11 $7m - 3 = 3m + 9$

12 $5b + 7 = 7b + 6$

13 $3n + 16 = 5n$

14 $4u + 4 = 2u + 9$

15 $5k + 2 = 2k + 2$

16 $3e = 7e - 17$

17 $9t + 4 = 4t + 8$

18 $9f = 3f + 5$

19 $3g + 4 = 9g - 3$

20 $2h + 7 = 8h - 2$

3.6 Solving equations with negative coefficients

Exercise 3M

Solve these equations.

C

1. $6 - 2x = x$
2. $4(x + 1) = 18 - 3x$
3. $3x + 4 = 12 - x$
4. $8 - x = 5$
5. $6 - 6x = 10 - 8x$
6. $9 - 4x = 1$
7. $40 - 13x = 1$
8. $2 - 6x = 9 - 7x$
9. $4 - x = x$
10. $12 - 6x = 5 - 3x$
11. $17 - 5x = 3x + 1$
12. $7 - 4x = 15$
13. $3(4 - x) = 5 + 4x$
14. $9 + 3x = 2 - 4x$
15. $4 - 4x = 9 - 9x$
16. $5 - 6x = 5$
17. $3 - 9x = 7 - 6x$
18. $9 - 2x = 4$
19. $5 + 2x = 7 - 3x$
20. $12 - 2x = 2 - 7x$

3.7 Using equations to solve problems

Exercise 3N

E

1. AO3 I think of a number. I multiply it by 5 and add 4. The result is 29. Find the number.

2. AO3 I think of a number. I multiply it by 7 and subtract the result from 53.
The answer is 11. Find the number.

D

3. AO3 The sizes of the angles of a quadrilateral are a, $a + 10°$, $a + 20°$ and $a + 30°$.
Find the size of the largest angle.

C

4. AO3 The diagram shows three angles at a point.

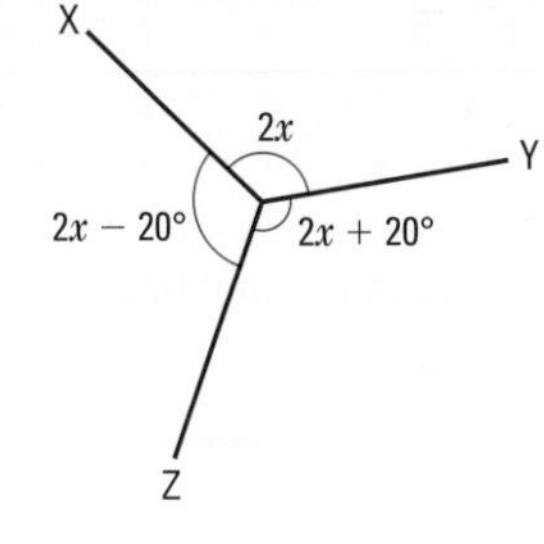

Find the size of each angle.

5. The lengths, in millimetres, of the sides of a triangle are $3x - 4$, $x + 5$ and $15 - 2x$.
The perimeter of the triangle is 28 mm.
Find the length of each side.

C

6. AO3 I think of a number. I multiply it by 6 and subtract 17 from the result.
The answer is the same as when I multiply the number by 4 and add 11 to the result.
Find the number.

7. AO3 The length of each side of a square is $3s - 2$ centimetres. The perimeter of the square is 28 cm.
Find the value of s.

8. AO3 Gill is 3 times as old as her daughter. She is also 28 years older than her. Find Gill's age.

9. AO3 The width of a rectangle is 5 cm less than its length. The perimeter of the rectangle is 66 cm.
Find its length.

10. AO3 The diagram shows a rectangle.

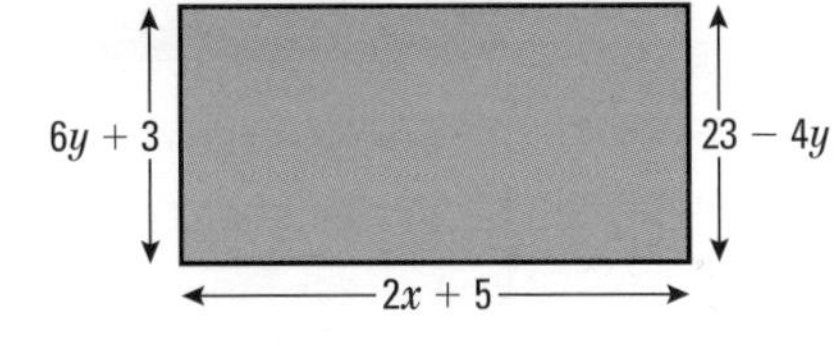

Find the values of x and y.

3.8 Solving equations by trial and improvement

Exercise 3O

C

1. Use a trial and improvement method to solve $x^3 + x = 6$, giving your answer correct to 2 d.p.

2. Use trial and improvement to solve $\dfrac{x^2 + x}{5} = 16$, giving your answer correct to 2 d.p.

3. Use a trial and improvement method to solve $y^3 + y = 50$, giving your answer correct to 1 d.p.

4. Use trial and improvement to solve $\dfrac{x^3}{2 + x} = 40$, giving your answer correct to 1 d.p.

5. Use trial and improvement to solve $x^2 + 2x = 40$, giving your answer correct to 2 d.p.

6. Use trial and improvement to solve $2x^3 + 2x = 48$, giving your answer correct to 2 d.p.

7. The equation $x^3 - 4x = 28$ has a solution between 3 and 4.
Use a trial and improvement method to find this solution. Give your answer correct to 1 d.p.
You must show all your working.

4 Inequalities

Key Points

- **inequality signs:**
  - $>$ means **greater than**
  - $\geqslant$ means **greater than or equal to**
  - $<$ means **less than**
  - $\leqslant$ means **less than or equal to**
- **drawing inequalities on a number line:** use an empty circle if the value is not included, and use a filled circle for a value that is included.

0 1 2 3 4 5 6 7 8 9 10

- **solving inequalities:** use the same method as linear equations, but remember not to multiply or divide both sides by a negative quantity.

4.1 Introducing inequalities

Exercise 4A

ResultsPlus
Examiner's Tip

Sometimes in the examination the question will ask for integers. These are the same as positive and negative whole numbers.

D

1 Put the correct sign ($<$ or $>$) between each pair of numbers to make a true statement.

a 3, 7 b 9, 4 c 7, 11
d 2, 3 e 6, 17 f 12, 11
g 21, 19 h 0.1, 0.2 i 3, 0.3
j 1.2, 2.2 k 0.1, 1 l 3.2, 3.29

2 Write down whether each statement is true or false. If it is false, write down the pair of numbers with the correct sign.

a $7 < 9$ b $6 > 4$ c $5 < 2$
d $7 < 7$ e $2 < 1$ f $21 = 12$
g $8 > 8.99$ h $3 < 2.99$ i $0 > 1$
j $2 > 12$ k $3 = 2$ l $2 > 0.22$

C

3 Write down the values of x that are whole numbers and satisfy these inequalities.

a $4 < x < 7$ b $3 < x < 6$
c $0 \leqslant x < 5$ d $3 < x < 7$
e $1 < x \leqslant 5$ f $2 < x < 5$
g $4 \leqslant x < 6$ h $-2 \leqslant x < 3$
i $-1 < x < 4$ j $-2 < x \leqslant 5$
k $-3 \leqslant x < 2$ l $-3 \leqslant x \leqslant 2$
m $0 < x < 3$ n $-1 < x \leqslant 3$
o $-4 \leqslant x < 0$ p $-2 \leqslant x \leqslant 2$

4.2 Representing inequalities on a number line

Exercise 4B

C

1 Draw six number lines from 0 to 10. Show these inequalities.

a $x > 5$ b $x > 4$ c $x < 3$
d $x > 2$ e $x < 5$ f $x > 6$

2 Draw ten number lines from 0 to 10. Show these inequalities.

a $2 < x < 7$ b $4 < x < 8$
c $4 \leqslant x < 8$ d $6 < x \leqslant 9$
e $3 \leqslant x \leqslant 6$ f $1 < x \leqslant 8$
g $2 \leqslant x < 5$ h $3 < x < 7$
i $4 \leqslant x < 6$ j $1 < x \leqslant 5$

3 Draw ten number lines from -5 to 5. Show these inequalities.

a $-3 \leqslant x < 5$ b $-2 < x < 4$
c $-1 < x \leqslant 4$ d $-5 \leqslant x \leqslant 0$
e $0 < x < 5$ f $-3 < x \leqslant 1$
g $-4 \leqslant x < 2$ h $0 \leqslant x \leqslant 4$
i $-5 \leqslant x < 1$ j $-2 \leqslant x < 2$

4 Write down the inequalities represented on these number lines.

a 0 1 2 3 4 5 6 7 8 9 10

b 0 1 2 3 4 5 6 7 8 9 10

c 0 1 2 3 4 5 6 7 8 9 10

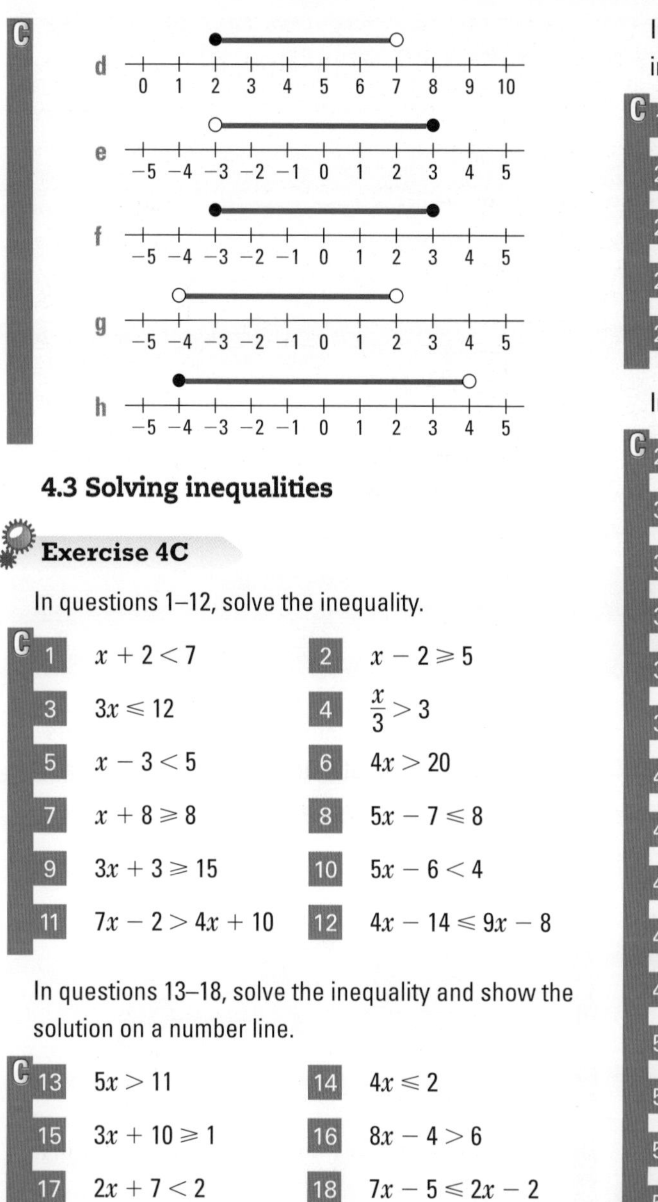

4.3 Solving inequalities

Exercise 4C

In questions 1–12, solve the inequality.

1. $x + 2 < 7$
2. $x - 2 \geqslant 5$
3. $3x \leqslant 12$
4. $\dfrac{x}{3} > 3$
5. $x - 3 < 5$
6. $4x > 20$
7. $x + 8 \geqslant 8$
8. $5x - 7 \leqslant 8$
9. $3x + 3 \geqslant 15$
10. $5x - 6 < 4$
11. $7x - 2 > 4x + 10$
12. $4x - 14 \leqslant 9x - 8$

In questions 13–18, solve the inequality and show the solution on a number line.

13. $5x > 11$
14. $4x \leqslant 2$
15. $3x + 10 \geqslant 1$
16. $8x - 4 > 6$
17. $2x + 7 < 2$
18. $7x - 5 \leqslant 2x - 2$

In questions 19–27, find all the integers that satisfy the inequality.

19. $4 \leqslant 2x \leqslant 6$
20. $-6 \leqslant 3x < 12$
21. $-10 < 5x \leqslant 5$
22. $0 \leqslant 6x < 18$
23. $-12 < 4x \leqslant 0$
24. $4 \leqslant 3x < 5$
25. $-7 < 5x \leqslant 20$
26. $-7 < 2x < 7$
27. $-5 < 3x < 0$

In questions 28–54, solve the inequality.

28. $8x < 22$
29. $5x \geqslant 3$
30. $5x > -20$
31. $3x \geqslant -7$
32. $\dfrac{x}{4} > -3$
33. $27 < 6x$
34. $4x - 7 \geqslant 2$
35. $6x + 9 \leqslant 3$
36. $8x - 3 > 6$
37. $10 < 7x + 3$
38. $5x + 4 \geqslant 2x + 10$
39. $7x + 3 \leqslant 3x - 3$
40. $8x - 2 > 5x - 7$
41. $9x - 8 < 5x + 4$
42. $2x + 10 \geqslant 7x - 5$
43. $4(x - 3) \geqslant 8$
44. $5(x + 2) > 15$
45. $3(x + 1) < x + 8$
46. $7 - x \leqslant 2$
47. $8 - 3x > 3$
48. $2 - 5x < 5$
49. $8 - 2x \geqslant 3x + 3$
50. $2(x - 3) \leqslant 3 - x$
51. $10 - 3x > 2x - 2$
52. $7 - 5x \leqslant 3 - 3x$
53. $4 - 5x \geqslant 5 - 7x$
54. $12 - 2x < 3 - 5x$
55. Solve the inequality $6x + 5 > 3x - 9$.
    Write down the smallest integer that satisfies it.
56. Solve the inequality $3x + 5 \leqslant 2 - 2x$.
    Write down the largest integer that satisfies it.

5 Quadratic graphs

Key Points

- **filling rate graph:** a graph showing the height of the water as a container is filled at a constant rate.
  - containers with straight vertical sides fill at a constant rate
  - containers with sides that bulge out fill quickly at first, slower at the bulge and then speed up again
  - containers with sides that curve in fill slowly at first, faster in the middle and then slow down again
  - thin containers fill faster than fat containers
- **drawing a quadratic graph:** make a table of values for x. Calculate the value of y for each value of x and plot the points on a grid.
- **understanding quadratic graphs such as $y = ax^2 + b$:**
  - b moves the graph up or down
  - a brings the graph closer to the y-axis
  - if there is a minus sign in front of the x^2 then the graph turns upside down

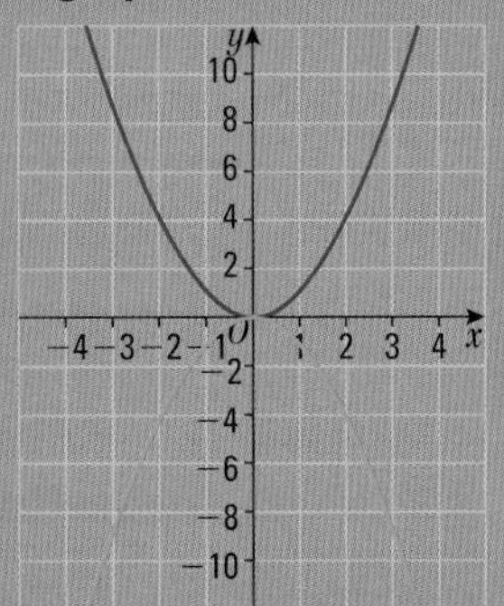

- **solving quadratic equations:** draw the graph of the equation. The solutions are where it crosses the x-axis.

5.1 Interpreting real-life graphs

Exercise 5A

C **1** Liquid is poured at a constant rate into the containers. The height of the liquid in the container h in cm is plotted against the passage of time t in seconds. Match these containers with their graphs.

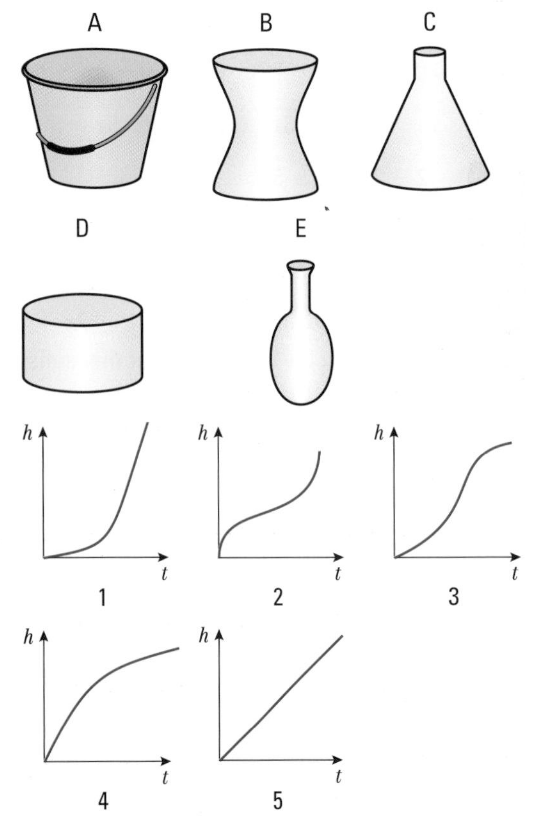

C **2** Liquid is poured into each of these containers at a constant rate. Draw, on the same graph, the height of the liquid h against the time t in seconds.

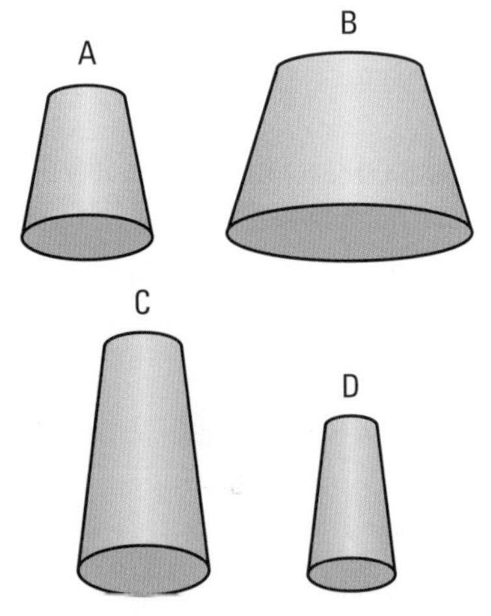

3* Here is a graph that shows the height of water in a washing-up bowl. For each part of the graph, describe what may have happened.

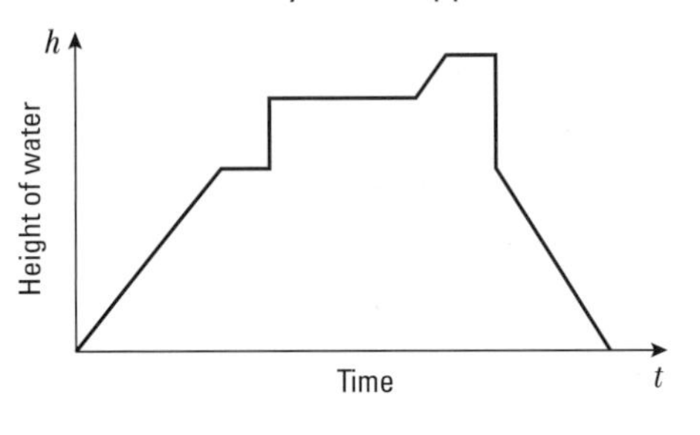

C 4* Here is a graph that shows the height, in cm, of the water in a paddling pool.
Explain, giving the heights and the times, what happened at each stage of the process.

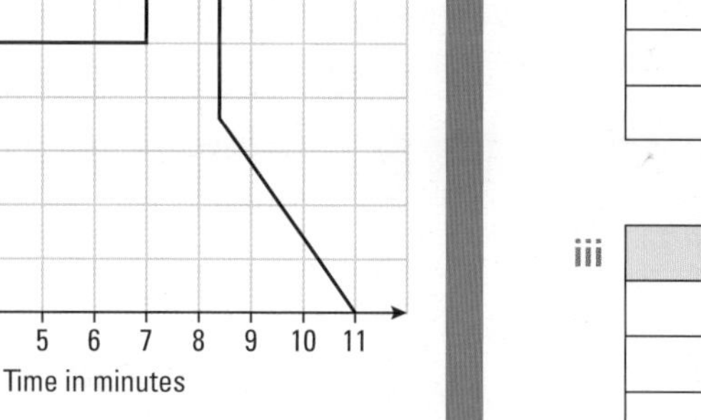

5* Here is a graph that shows the height, in metres, of a glider during a flight.

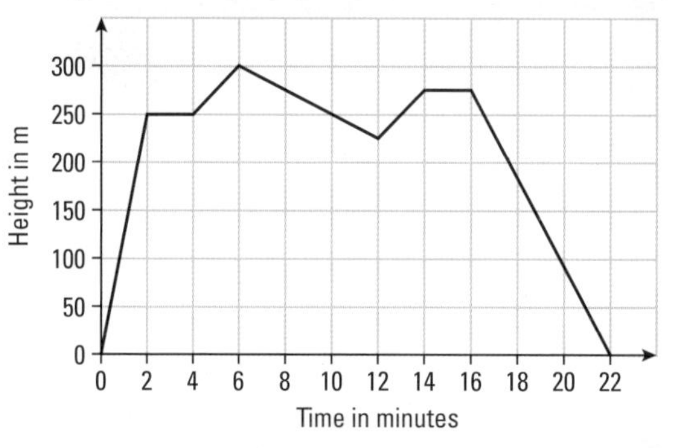

Describe the flight of the glider.

5.2 Drawing quadratic graphs

Exercise 5B

C 1 **a** Copy and complete the tables of values for these quadratic graphs.

b Draw the graphs on a coordinate grid with x-axis drawn from -3 to $+3$ and y-axis drawn from -12 to $+12$.

i

x	$y = x^2 + 1$
-3	
-2	5
-1	
0	1
1	
2	5
3	

ii

x	$y = -x^2 - 1$
-3	
-2	-5
-1	
0	-1
1	
2	-5
3	

iii

x	$y = -x^2 + 2$
-3	-7
-2	
-1	
0	2
1	
2	
3	-7

iv

x	$y = -x^2 + 3$
-3	
-2	-1
-1	
0	3
1	
2	
3	-6

v

x	$y = x^2 + 4$
-3	
-2	8
-1	
0	
1	
2	8
3	

2 **a** Copy and complete the tables of values for these quadratic graphs.

b Draw the graphs on a coordinate grid with x-axis drawn from -3 to $+3$ and y-axis drawn from -20 to $+20$.

C

i

x	$y = 2x^2 + 2$
−3	
−2	10
−1	
0	2
1	
2	10
3	

ii

x	$y = -2x^2 + 2$
−3	
−2	−6
−1	
0	2
1	
2	−6
3	

iii

x	$y = -3x^2 + 2$
−2	−10
−1	
0	2
1	
2	

iv

x	$y = 2x^2 - 2$
−3	
−2	6
−1	
0	−2
1	
2	
3	

v

x	$y = -2x^2 - 2$
−3	
−2	−10
−1	
0	
1	−4
2	
3	

C 3 Draw these quadratic graphs on a coordinate grid with x-axis drawn from −3 to +3 and y-axis drawn from −15 to +15.

a $y = 2x^2$
b $y = 2x^2 + 3$
c $y = 2x^2 - 3$
d $y = -2x^2 + 3$
e $y = -2x^2 - 3$

4 Draw these quadratic graphs on a coordinate grid with x-axis drawn from −3 to +3 and y-axis drawn from −20 to +20.

a $y = 3x^2$
b $y = -3x^2$
c $y = 2x^2$
d $y = -2x^2$
e $y = -(x + 2)^2$

5 Draw these quadratic graphs on a coordinate grid with x-axis drawn from −3 to +3 and y-axis drawn from −30 to +30.

a $y = 2x^2$
b $y = 2x^2 - 2$
c $y = -2x^2 - 2$
d $y = -2x^2$
e $y = 2x^2 + 3$

Exercise 5C

C 1 a Copy and complete the tables of values for these quadratic graphs.

b Draw the graphs on a coordinate grid with x-axis drawn from −3 to +3 and y-axis drawn from −20 to +20.

i

x	x^2	$2x$	+2	$y = x^2 + 2x + 2$
−3	+9	−6	+2	5
−2				
−1				
0	0	0	+2	2
1				
2				
3				

ii

x	x^2	$3x$	+3	$y = x^2 + 3x + 3$
−3				
−2	+4	−6	+3	1
−1				
0	0	0	+3	3
1				
2				
3				

C

iii

x	x^2	$2x$	-3	$y = x^2 + 2x - 3$
-3	$+9$	-6	-3	0
-2				
-1				
0	0	0	-3	-3
1				
2				
3				

iv

x	x^2	$-2x$	$+2$	$y = x^2 - 2x + 2$
-3	$+9$	$+6$	$+2$	17
-2				
-1				
0				
1	$+1$	-2	$+2$	1
2				
3				

v

x	x^2	$-2x$	-2	$y = x^2 - 2x - 2$
-3	$+9$	$+6$	-2	13
-2				
-1				
0	0	0	-2	-2
1				
2				
3				

2 Draw these quadratic graphs on a coordinate grid with x-axis drawn from -3 to $+3$ and y-axis drawn from -15 to $+15$.

a $y = x^2$ b $y = x^2 + 2x$
c $y = x^2 - 2x$ d $y = -x^2 + 2x$
e $y = -x^2 - 2x$

3 Draw these quadratic graphs on a coordinate grid with x-axis drawn from -3 to $+3$ and y-axis drawn from -20 to $+20$.

a $y = x^2 + 2x + 1$ b $y = -x^2 + 2x + 2$
c $y = 2x^2 + x - 2$ d $y = -2x^2 + x + 2$
e $y = (x - 2)^2$ f $y = (x + 2)(x - 2)$

C 4 Draw these quadratic graphs on a coordinate grid with x-axis drawn from -3 to $+3$ and y-axis drawn from -30 to $+30$.

a $y = x^2 - 2x + 3$ b $y = 4x^2 - 3x$
c $y = -4x^2 + 3x$ d $y = -x^2 + 3x$
e $y = 3x^2 - 2x$ f $(x + 1)(x - 2)$

5.3 Using graphs of quadratic functions to solve equations

Exercise 5D

C 1 a Draw the graph of $y = 2x^2 - 3x - 1$ for values of x from -2 to $+4$.
b Use your graph to solve the equations.
i $2x^2 - 3x - 1 = 0$ ii $2x^2 - 3x - 1 = 10$

2 a Draw the graph of $y = x^2 - 3x - 1$ for values of x from -2 to $+5$.
b Use your graph to solve the equations.
i $x^2 - 3x - 1 = 0$ ii $x^2 - 3x - 1 = 5$

3 a Draw the graph of $y = 2x^2 - 3x$ for values of x from -2 to $+3$.
b Use your graph to solve the equations.
i $2x^2 - 3x = 0$ ii $2x^2 - 3x = 2$

4 a Draw the graph of $y = x^2 - 2x$ for values of x from -2 to $+4$.
b Use your graph to solve the equations.
i $x^2 - 2x = 0$ ii $x^2 - 2x = 4$

5 a Draw the graph of $y = x^2 - 4x + 2$ for values of x from -1 to $+5$.
b Use your graph to solve the equations.
i $x^2 - 4x + 2 = 0$ ii $x^2 - 4x + 2 = 2$

6 Formulae

Key Points

- **subject of the formula:** the variable that appears on its own on one side of the = sign. E.g. A is the subject of the formula $A = lw$.
- **substituting numbers into expressions:** replace the variables in the expression with the numeric values given.
- **changing the subject of a formula:** carry out the same operations on both sides of the equal sign to isolate the subject.

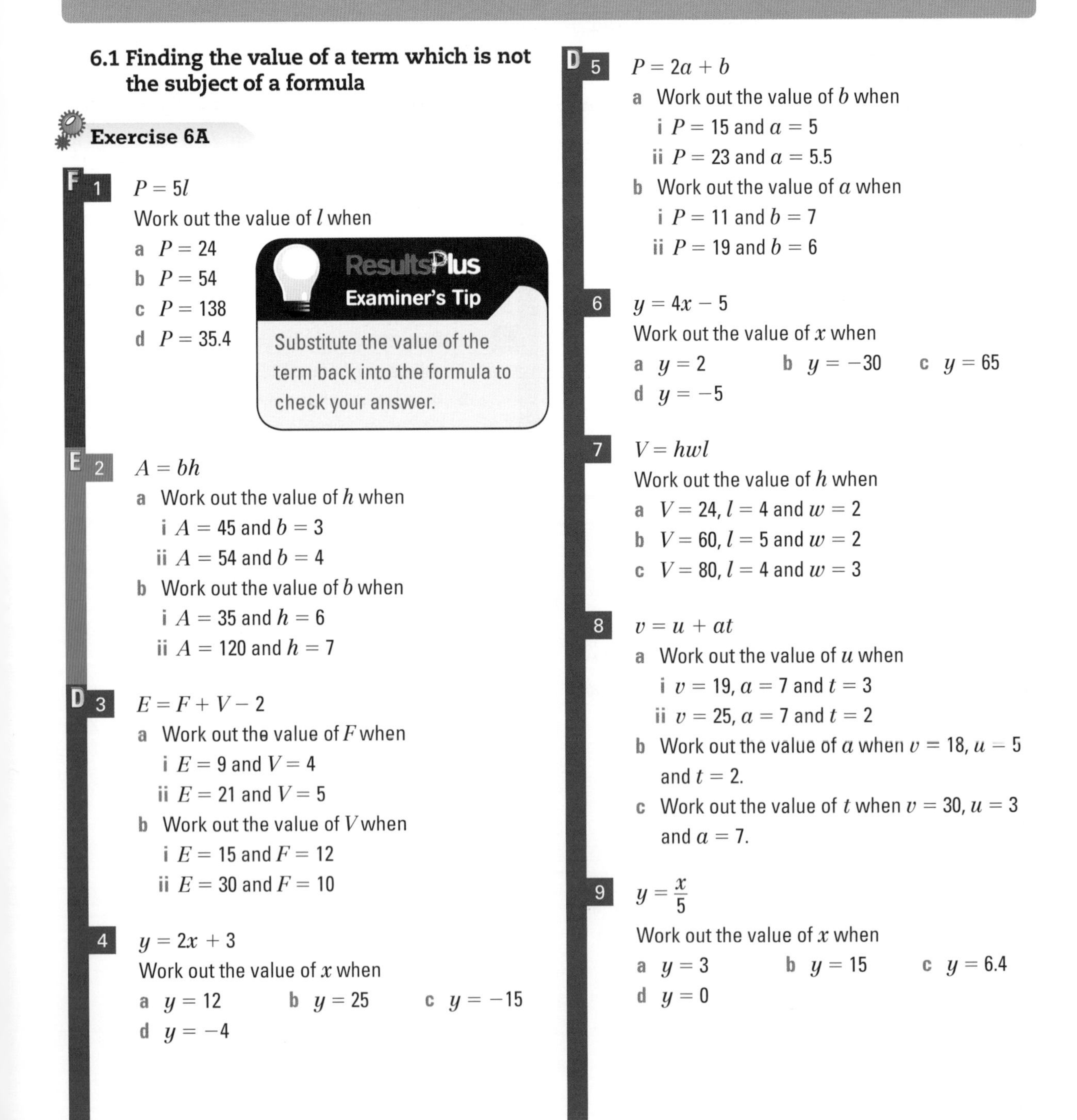

6.1 Finding the value of a term which is not the subject of a formula

Exercise 6A

F

1 $P = 5l$

Work out the value of l when

a $P = 24$

b $P = 54$

c $P = 138$

d $P = 35.4$

ResultsPlus
Examiner's Tip

Substitute the value of the term back into the formula to check your answer.

E

2 $A = bh$

a Work out the value of h when

i $A = 45$ and $b = 3$

ii $A = 54$ and $b = 4$

b Work out the value of b when

i $A = 35$ and $h = 6$

ii $A = 120$ and $h = 7$

D

3 $E = F + V - 2$

a Work out the value of F when

i $E = 9$ and $V = 4$

ii $E = 21$ and $V = 5$

b Work out the value of V when

i $E = 15$ and $F = 12$

ii $E = 30$ and $F = 10$

4 $y = 2x + 3$

Work out the value of x when

a $y = 12$ b $y = 25$ c $y = -15$

d $y = -4$

D

5 $P = 2a + b$

a Work out the value of b when

i $P = 15$ and $a = 5$

ii $P = 23$ and $a = 5.5$

b Work out the value of a when

i $P = 11$ and $b = 7$

ii $P = 19$ and $b = 6$

6 $y = 4x - 5$

Work out the value of x when

a $y = 2$ b $y = -30$ c $y = 65$

d $y = -5$

7 $V = hwl$

Work out the value of h when

a $V = 24$, $l = 4$ and $w = 2$

b $V = 60$, $l = 5$ and $w = 2$

c $V = 80$, $l = 4$ and $w = 3$

8 $v = u + at$

a Work out the value of u when

i $v = 19$, $a = 7$ and $t = 3$

ii $v = 25$, $a = 7$ and $t = 2$

b Work out the value of a when $v = 18$, $u = 5$ and $t = 2$.

c Work out the value of t when $v = 30$, $u = 3$ and $a = 7$.

9 $y = \frac{x}{5}$

Work out the value of x when

a $y = 3$ b $y = 15$ c $y = 6.4$

d $y = 0$

D 10 $t = \frac{d}{s}$

Work out the value of d when

a $t = 4$ and $s = 5$

b $t = 6$ and $s = -8$

c $t = 7.5$ and $s = 5$

d $t = 4.6$ and $s = -10.4$

6.2 Changing the subject of a formula

Exercise 6B

Rearrange each formula to make the letter in square brackets the subject.

D

1 $P = 5m$ [m]

2 $P = IV$ [V]

3 $A = lw$ [l]

4 $C = 2\pi r$ [r]

5 $V = lwh$ [l]

6 $A = \pi rl$ [l]

C

7 $y = 4x - 5$ [x]

8 $t = 3n + 4$ [n]

9 $P = 3x + y$ [y]

10 $y = mx + c$ [c]

11 $v = u - gt$ [g]

12 $v = u - gt$ [u]

13 $A = \frac{1}{2}bh$ [h]

14 $I = \frac{PRT}{100}$ [P]

15 $T = \frac{D}{V}$ [D]

16 $\frac{PV}{T} = k$ [T]

17 $\frac{PV}{T} = k$ [V]

18 $I = m(v - u)$ [m]

19 $A = \frac{1}{2}(a + b)h$ [a]

20 $y = \frac{1}{4}x - 2$ [x]

21 $y = 3(x - 1)$ [x]

22 $x = 4(y + 3)$ [y]

23 $H = 15 - \frac{A}{2}$ [A]

24 $3x - 2y = 4$ [x]

25 $5x - 2y = 6$ [y]

26 $P = 4(q - 7) - 6(q - 6)$ [q]

27 $6y^2 - 4x = 8(x - 8y)$ [x]

7 Angles and two-dimensional shapes

Key Points

- **polygon:** a 2D shape with straight lines.
- **regular polygon:** a polygon with all sides equal and all interior angles equal.
- **angles of a polygon:**
  - **interior angle of a polygon + exterior angle of a polygon** = 180°
  - **the sum of the exterior angles of a polygon** = 360°
  - **the sum of the interior angles of a polygon** = $(n - 2) \times 180°$ where n is the number of sides
- **congruent shapes:** shapes that have exactly the same size and shape.
- **tessellation:** when a shape can be repeatedly drawn to cover an area without any gaps.
- **bearing:** a way to describe direction as an angle measured clockwise from North, and written as a 3-figure number.
- **scale:** a ratio that shows the relationship between a drawn length and an actual length.
- **drawing a triangle given 3 sides:** use compasses only.

7.1 Polygons

Exercise 7A

F 1 Name the following polygons.

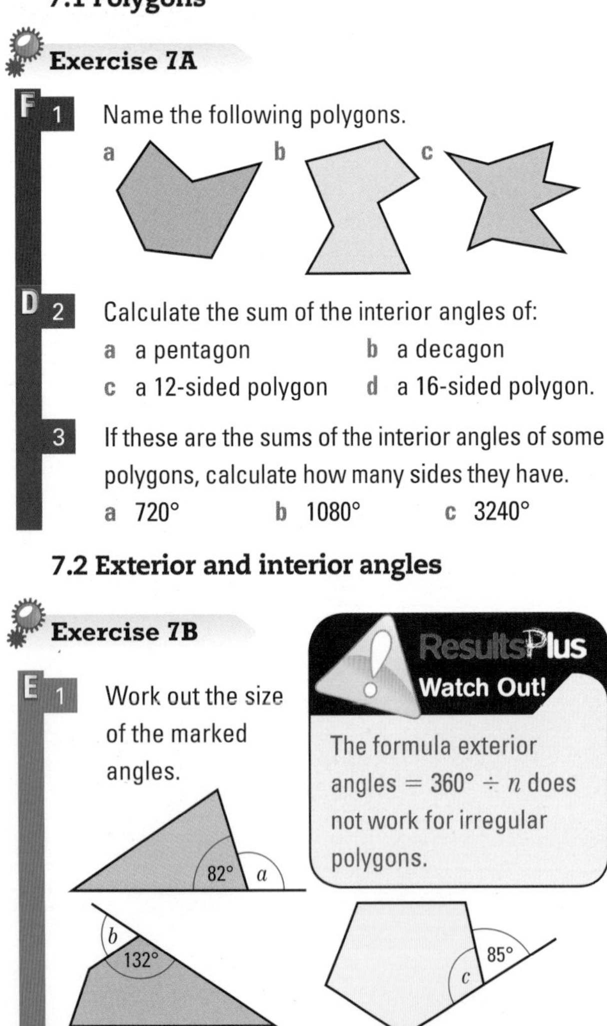

D 2 Calculate the sum of the interior angles of:
a a pentagon b a decagon
c a 12-sided polygon d a 16-sided polygon.

3 If these are the sums of the interior angles of some polygons, calculate how many sides they have.
a 720° b 1080° c 3240°

7.2 Exterior and interior angles

Exercise 7B

E 1 Work out the size of the marked angles.

82° a

b 132°

85° c

44° 56° d e

g f 83° 42° 100°

ResultsPlus Watch Out!

The formula exterior angles = 360° ÷ n does not work for irregular polygons.

D 2 Work out the exterior angle of a regular: (A03)
a hexagon b octagon c decagon.

3 Use your answers to question 2 to work out the interior angle of each polygon.

C 4 A regular polygon has an exterior angle of 24°. Work out how many sides it has. (A03)

5 A regular polygon has an interior angle of 160°. Work out: (A03)
a its exterior angle
b how many sides it has.

7.3 Congruent shapes

Exercise 7C

E 1 Write down the letters of three pairs of shapes that are congruent.

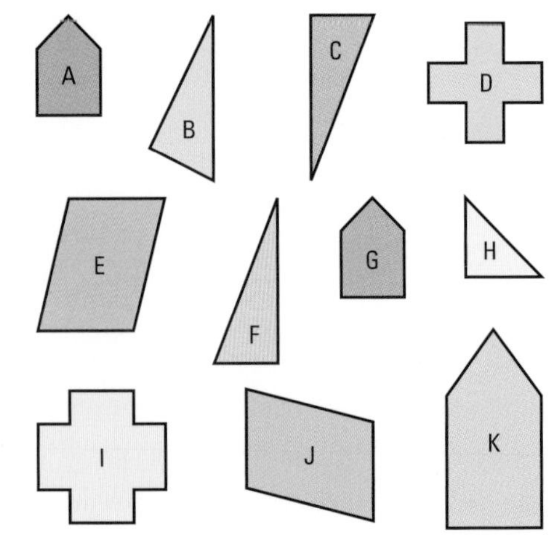

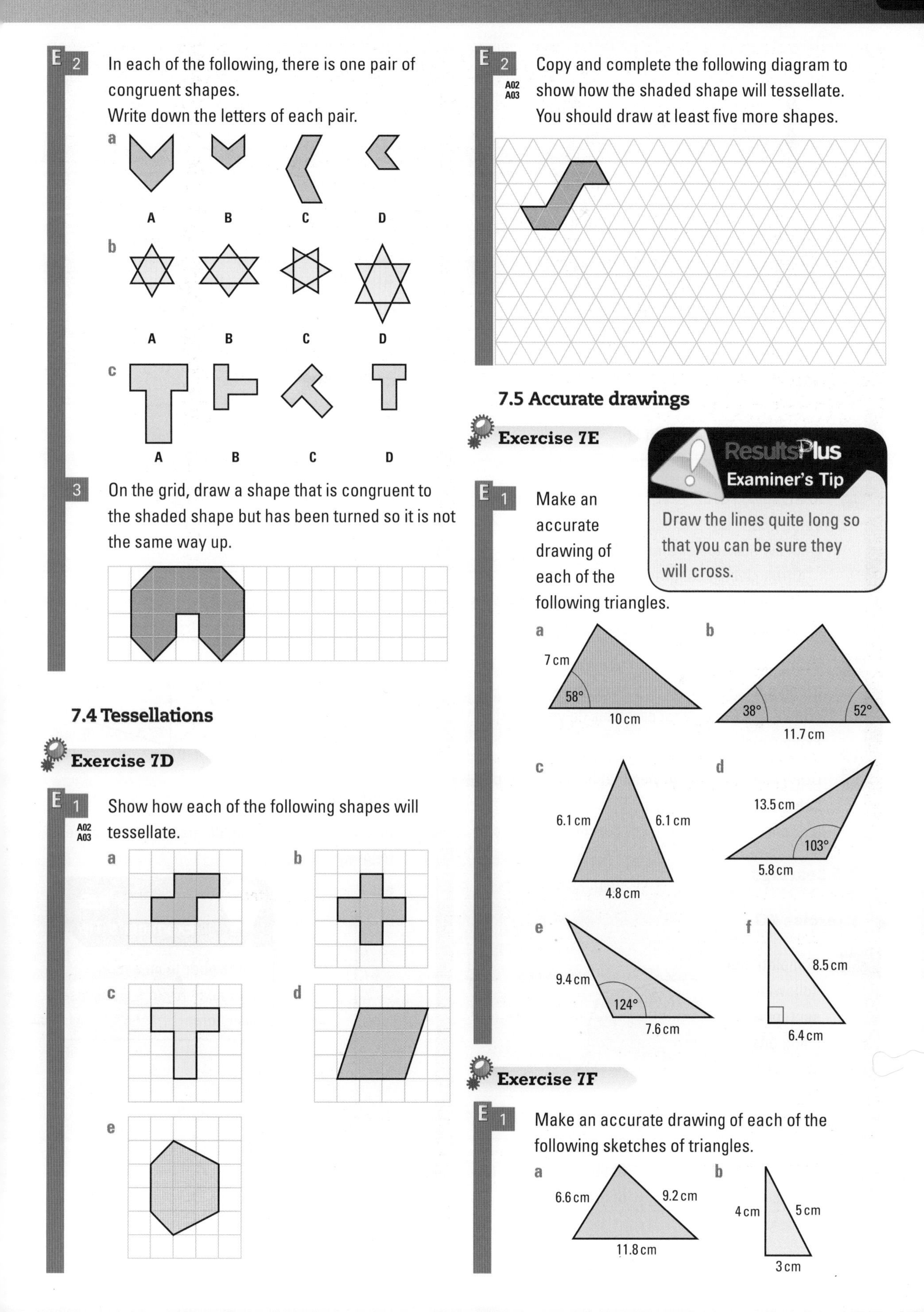

2 In each of the following, there is one pair of congruent shapes.
Write down the letters of each pair.

3 On the grid, draw a shape that is congruent to the shaded shape but has been turned so it is not the same way up.

7.4 Tessellations

Exercise 7D

1 Show how each of the following shapes will tessellate.

2 Copy and complete the following diagram to show how the shaded shape will tessellate.
You should draw at least five more shapes.

7.5 Accurate drawings

Exercise 7E

ResultsPlus Examiner's Tip
Draw the lines quite long so that you can be sure they will cross.

1 Make an accurate drawing of each of the following triangles.

Exercise 7F

1 Make an accurate drawing of each of the following sketches of triangles.

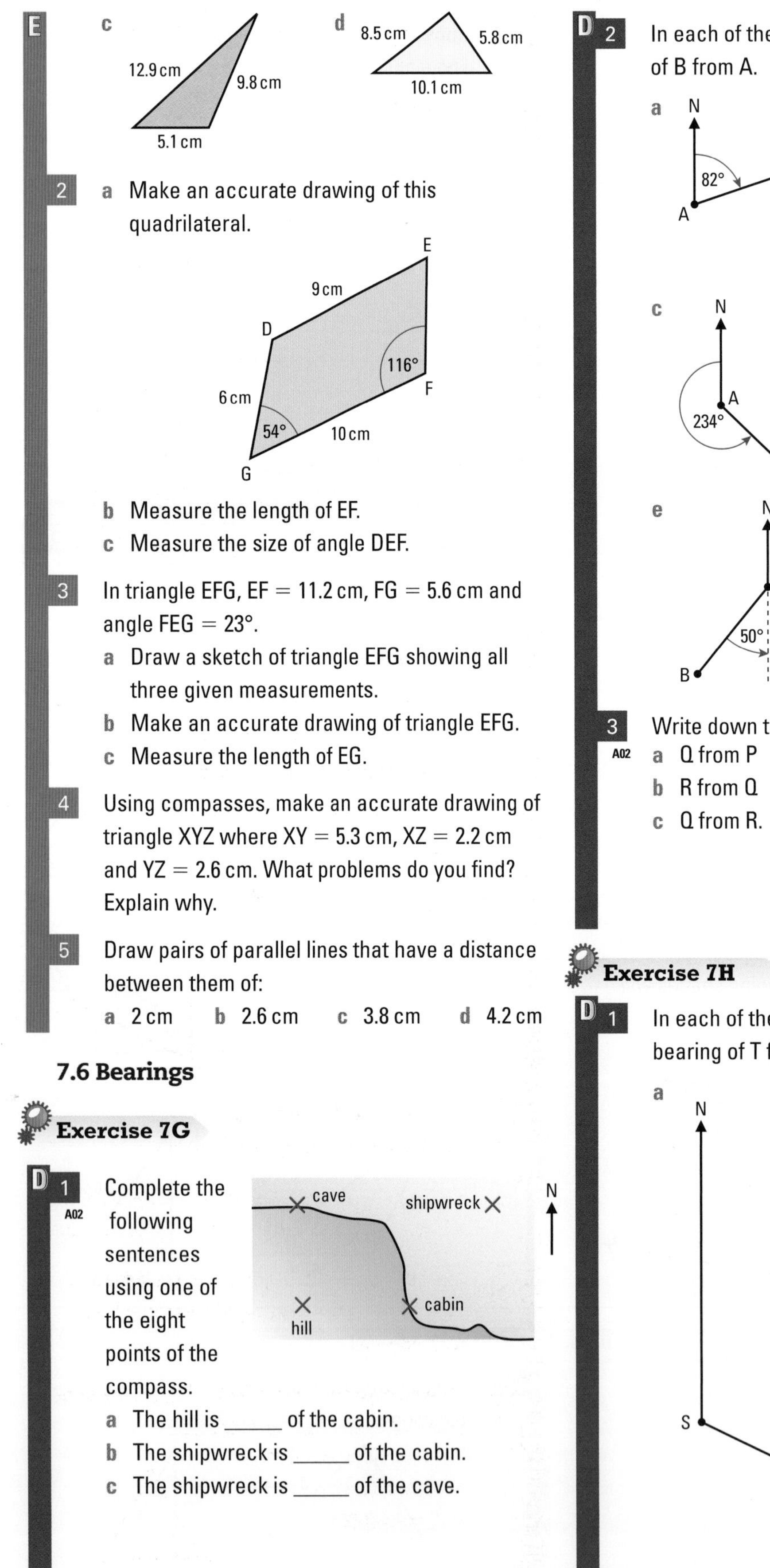

E

c

d

2 a Make an accurate drawing of this quadrilateral.

b Measure the length of EF.

c Measure the size of angle DEF.

3 In triangle EFG, EF = 11.2 cm, FG = 5.6 cm and angle FEG = 23°.

a Draw a sketch of triangle EFG showing all three given measurements.

b Make an accurate drawing of triangle EFG.

c Measure the length of EG.

4 Using compasses, make an accurate drawing of triangle XYZ where XY = 5.3 cm, XZ = 2.2 cm and YZ = 2.6 cm. What problems do you find? Explain why.

5 Draw pairs of parallel lines that have a distance between them of:

a 2 cm b 2.6 cm c 3.8 cm d 4.2 cm

7.6 Bearings

Exercise 7G

D 1 A02 Complete the following sentences using one of the eight points of the compass.

a The hill is _____ of the cabin.

b The shipwreck is _____ of the cabin.

c The shipwreck is _____ of the cave.

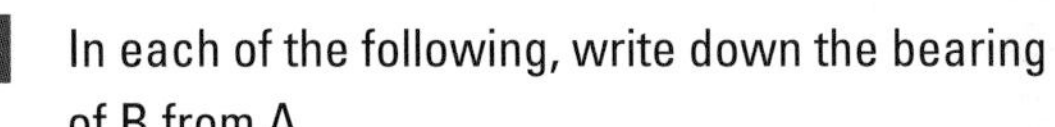

D 2 In each of the following, write down the bearing of B from A.

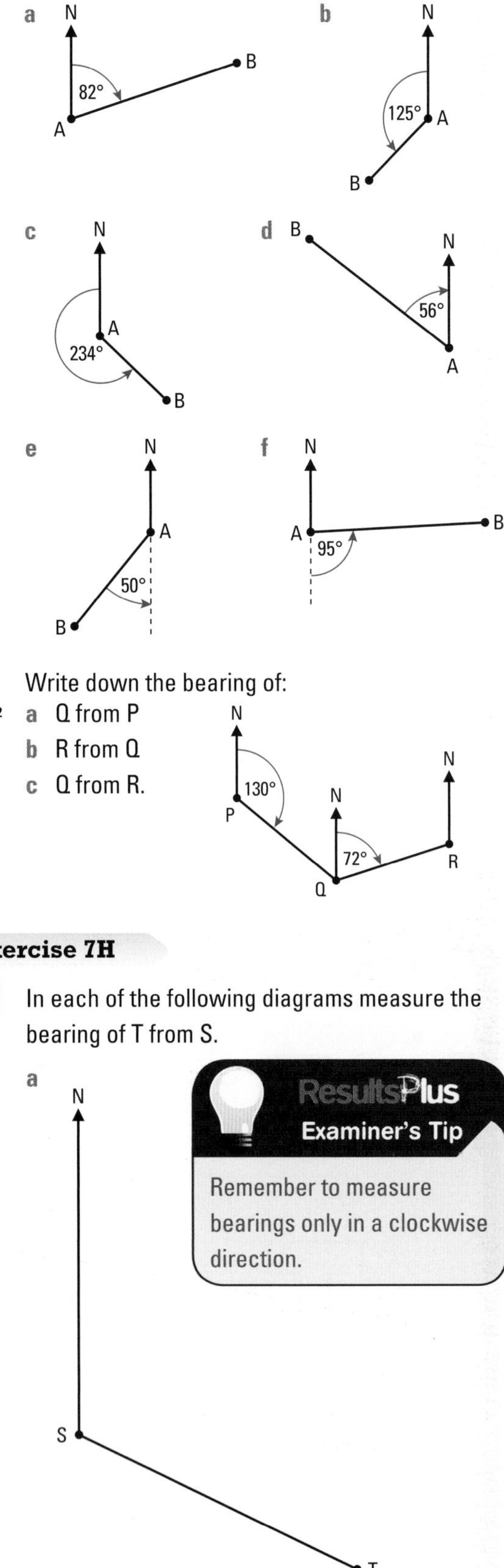

a b c d e f

3 A02 Write down the bearing of:

a Q from P

b R from Q

c Q from R.

Exercise 7H

D 1 In each of the following diagrams measure the bearing of T from S.

a

ResultsPlus Examiner's Tip

Remember to measure bearings only in a clockwise direction.

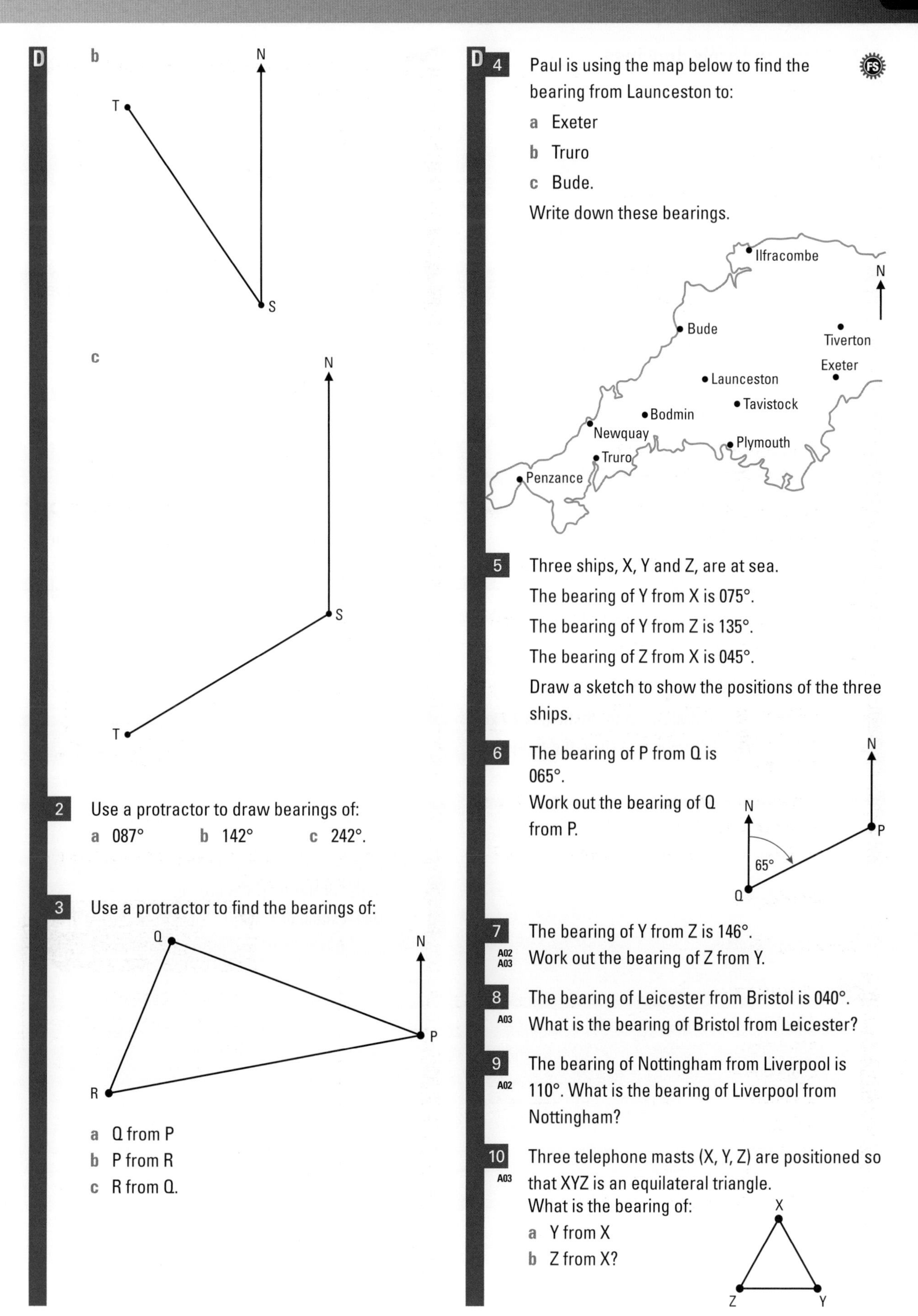

D

b

c

2 Use a protractor to draw bearings of:

a 087° **b** 142° **c** 242°.

3 Use a protractor to find the bearings of:

a Q from P

b P from R

c R from Q.

D **4** Paul is using the map below to find the bearing from Launceston to:

a Exeter

b Truro

c Bude.

Write down these bearings.

5 Three ships, X, Y and Z, are at sea.

The bearing of Y from X is 075°.

The bearing of Y from Z is 135°.

The bearing of Z from X is 045°.

Draw a sketch to show the positions of the three ships.

6 The bearing of P from Q is 065°.

Work out the bearing of Q from P.

7 (A02, A03) The bearing of Y from Z is 146°.

Work out the bearing of Z from Y.

8 (A03) The bearing of Leicester from Bristol is 040°.

What is the bearing of Bristol from Leicester?

9 (A02) The bearing of Nottingham from Liverpool is 110°. What is the bearing of Liverpool from Nottingham?

10 (A03) Three telephone masts (X, Y, Z) are positioned so that XYZ is an equilateral triangle.

What is the bearing of:

a Y from X

b Z from X?

7.7 Maps and scale drawings

Exercise 7I

D

1. The scale on a map is 1 cm to 2 km.
   The distance between Banton and Starwell is 6.5 cm on the map.
   How many kilometres apart are Banton and Starwell in real life?

2. A map is drawn on a scale of 3 cm to 1 km.
   a Work out the real length of a reservoir, which is 5.2 cm long on the map.
   b The distance between the Post Office in Angate and the memorial in Denway is 6.9 km. Work out the distance between them on the map.

3. Sabrina walks for 8 miles on a bearing of 070°. Use a scale of 1 cm to represent 1 mile to show the journey.

4. Pete runs for 5 km on a bearing of 130°. Use a scale of 1 cm to represent 2 km to show the journey.

5. Tayford is 2 km due south of Mickham. The bearing of Winfield from Mickham is 135° and the distance from Mickham to Winfield is 2.7 km.
   a Make a scale drawing to show the three villages. Use a scale of 1 : 25 000.
   b Use your drawing to find
      i the distance of Winfield from Tayford
      ii the bearing of Winfield from Tayford.

D

6. Ken sails his boat from the Isle of Wight for 24 km on a bearing of 140°. (FS)
   He then sails on a bearing of 260° for 12 km.
   How far is Ian from his starting point?
   What bearing does he need to sail on to get back to the start?
   Use a scale of 1 cm to represent 2 km.

7. Peg Leg the pirate buried his treasure 80 yards from the big tree on a bearing of 055°. One-Eyed Rick dug up the treasure and moved it 60 yards on a bearing of 320° from where it had been buried.
   How far is the treasure from the big tree now?
   What bearing is the new hiding place of the treasure?
   Use a scale of 1 cm to represent 10 yards.

8. Alex flew his plane on a bearing of 320° for 240 km. He then changed direction and flew on a bearing of 160° for 120 km. (FS)
   What bearing must Alex fly on to get back to the start?
   How far is he away from the start?

9. This is a sketch of Hardeep's bedroom. It is *not* drawn to scale.

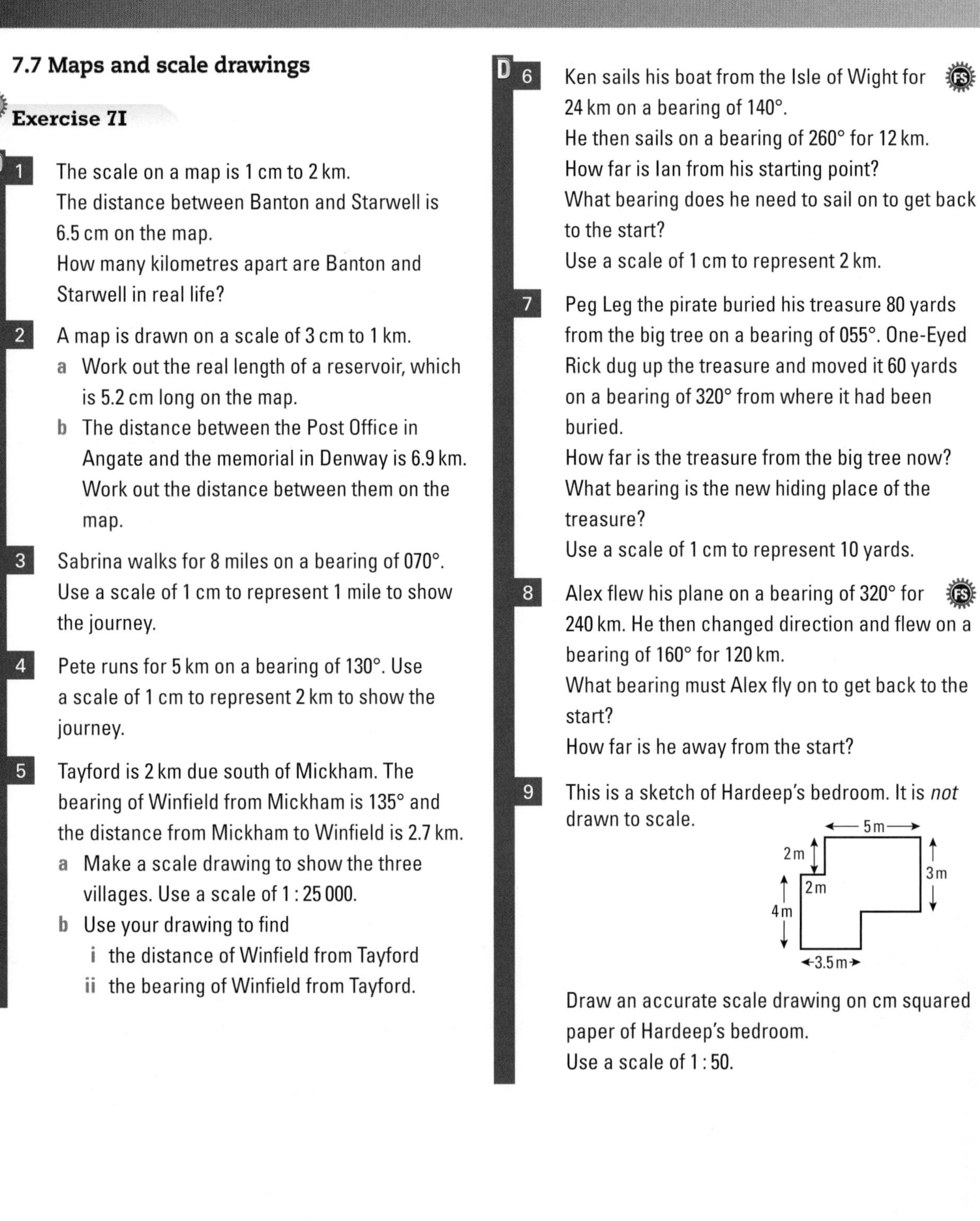

   Draw an accurate scale drawing on cm squared paper of Hardeep's bedroom.
   Use a scale of 1 : 50.

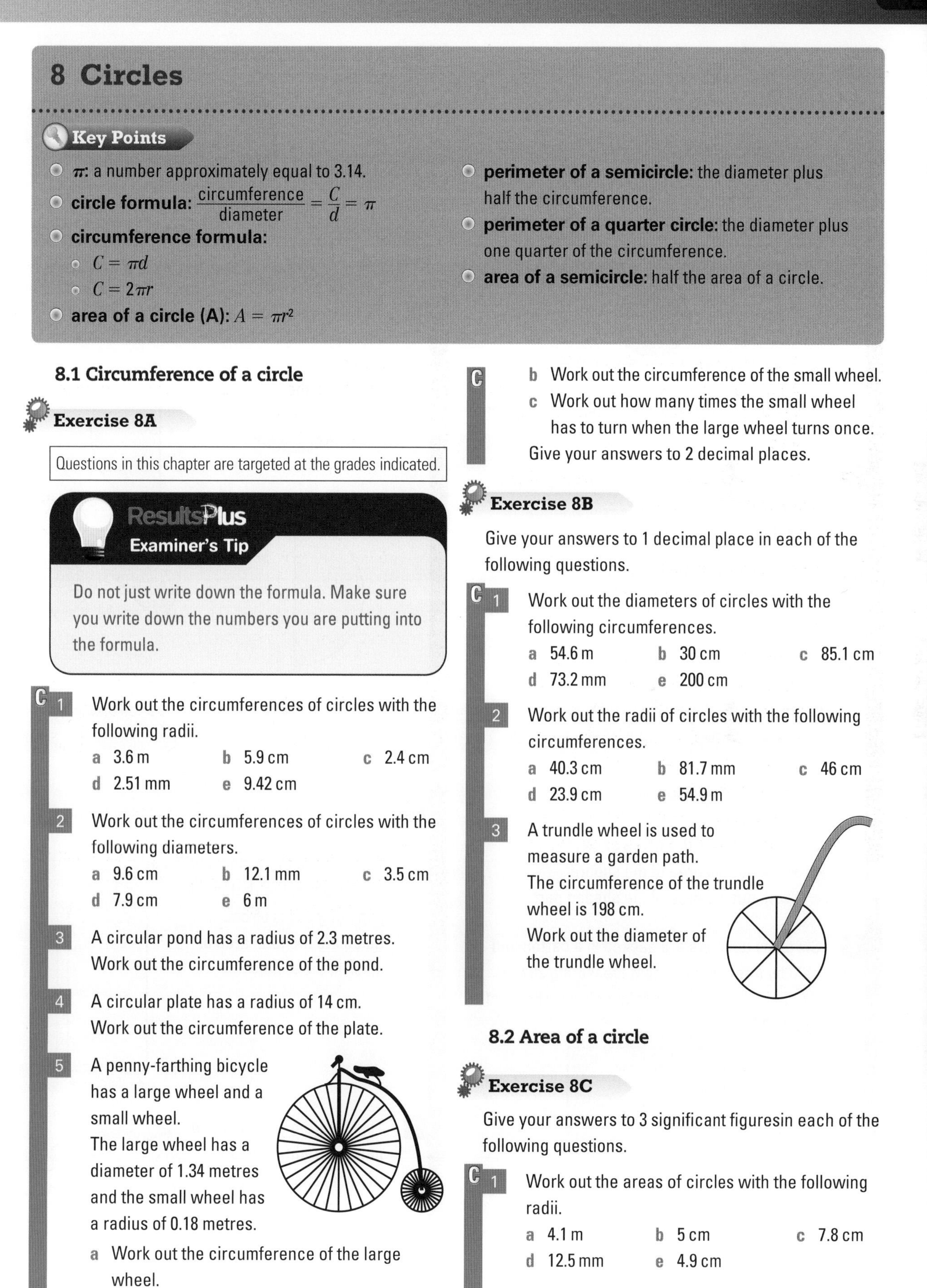

8 Circles

Key Points

- π: a number approximately equal to 3.14.
- **circle formula:** $\frac{\text{circumference}}{\text{diameter}} = \frac{C}{d} = \pi$
- **circumference formula:**
  - $C = \pi d$
  - $C = 2\pi r$
- **area of a circle (A):** $A = \pi r^2$
- **perimeter of a semicircle:** the diameter plus half the circumference.
- **perimeter of a quarter circle:** the diameter plus one quarter of the circumference.
- **area of a semicircle:** half the area of a circle.

8.1 Circumference of a circle

Exercise 8A

Questions in this chapter are targeted at the grades indicated.

ResultsPlus
Examiner's Tip

Do not just write down the formula. Make sure you write down the numbers you are putting into the formula.

C

1 Work out the circumferences of circles with the following radii.

a 3.6 m b 5.9 cm c 2.4 cm
d 2.51 mm e 9.42 cm

2 Work out the circumferences of circles with the following diameters.

a 9.6 cm b 12.1 mm c 3.5 cm
d 7.9 cm e 6 m

3 A circular pond has a radius of 2.3 metres.
Work out the circumference of the pond.

4 A circular plate has a radius of 14 cm.
Work out the circumference of the plate.

5 A penny-farthing bicycle has a large wheel and a small wheel.
The large wheel has a diameter of 1.34 metres and the small wheel has a radius of 0.18 metres.

a Work out the circumference of the large wheel.

b Work out the circumference of the small wheel.

c Work out how many times the small wheel has to turn when the large wheel turns once.

Give your answers to 2 decimal places.

Exercise 8B

Give your answers to 1 decimal place in each of the following questions.

C

1 Work out the diameters of circles with the following circumferences.

a 54.6 m b 30 cm c 85.1 cm
d 73.2 mm e 200 cm

2 Work out the radii of circles with the following circumferences.

a 40.3 cm b 81.7 mm c 46 cm
d 23.9 cm e 54.9 m

3 A trundle wheel is used to measure a garden path.
The circumference of the trundle wheel is 198 cm.
Work out the diameter of the trundle wheel.

8.2 Area of a circle

Exercise 8C

Give your answers to 3 significant figuresin each of the following questions.

C

1 Work out the areas of circles with the following radii.

a 4.1 m b 5 cm c 7.8 cm
d 12.5 mm e 4.9 cm

C

2 Work out the areas of circles with the following diameters.

a 20 cm b 40.4 mm c 4.7 cm
d 31.2 cm e 6 m

3 A goat is tied to a post in the middle of a field covered in grass. He is tied so that he can eat the grass within 7.6 m of the post.

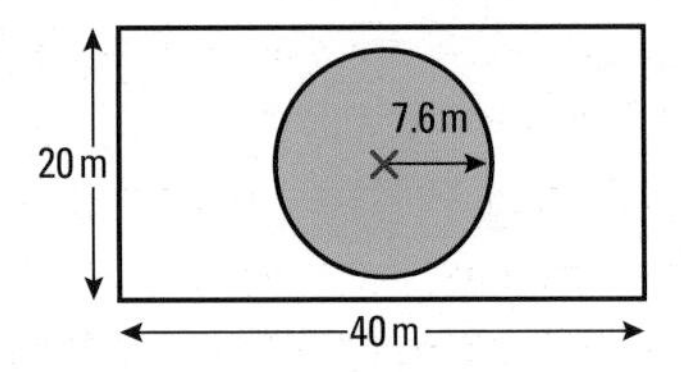

Work out the area of grass from which he cannot eat.

4 Mandy makes some jam. She covers the tops of the jars with circular pieces of material of diameter 5.6 cm.
Work out the area of material covering one jar.

5 The diagram shows a square of side 7 cm inside a circle of radius 9 cm.

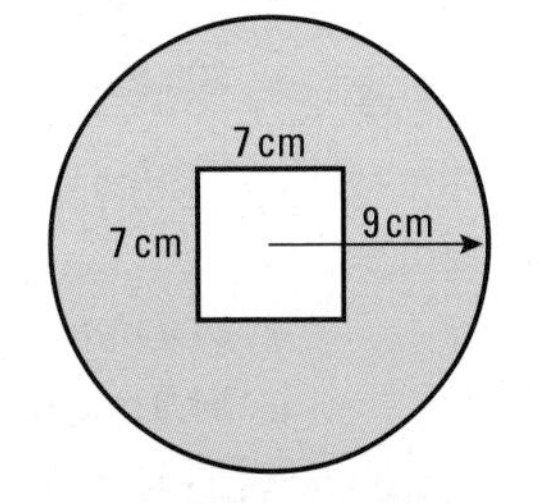

a Work out the area of the circle.
b Work out the area of the square.
c Work out the area of the shaded part.

6 Six glass tumblers are packed in a carton and held by a rectangular card measuring 29 cm by 21 cm. The tumblers fit into circular holes of radius 3.5 cm.

A03

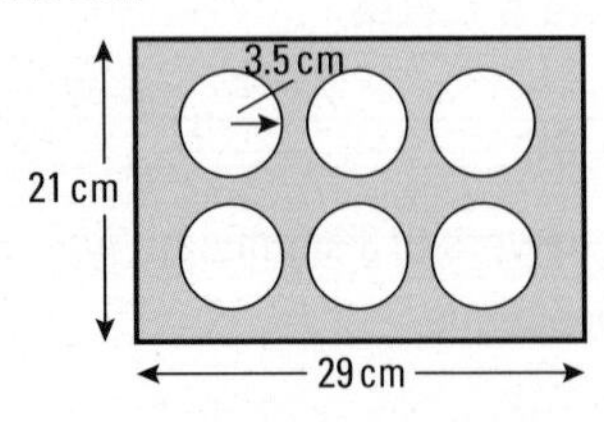

Work out the area of card that is left.

8.3 Area and perimeter of half and quarter circles

Exercise 8D

Give your answers to 3 significant figures in each of the following questions.

When finding the perimeter of a semicircle remember that it is all the way round the shape, not just the arc length.

C

1 Calculate the perimeter and the area of each sector.

A02

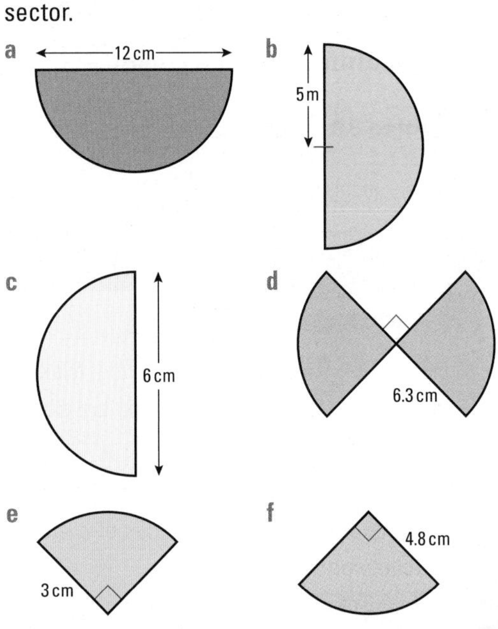

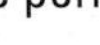

2 A window is in the shape of a rectangle with a semicircle on top.

A02
A03

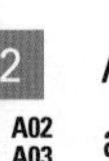

The width of the window is 56 cm.
The height of the rectangular pane of glass is 65 cm.
Calculate the total area of glass in the window.

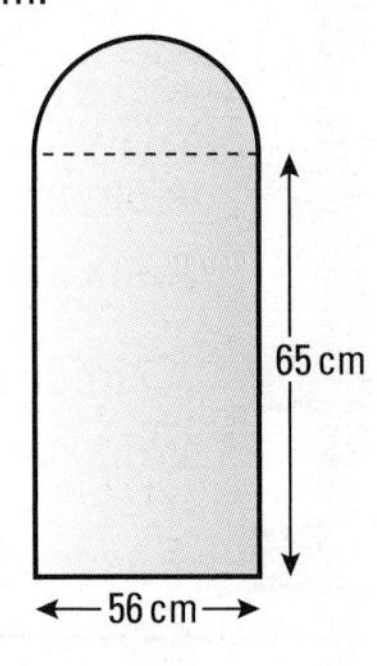

3 This diagram is made up of part A, a right-angled triangle, and part B, a segment. Together they make a quadrant of a circle.

A02

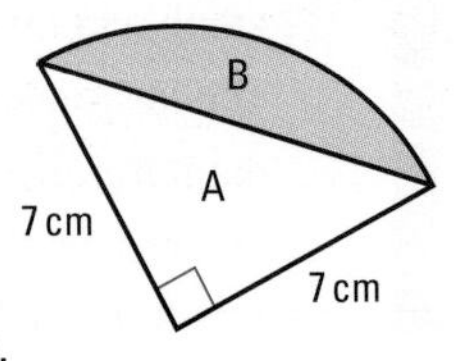

a What is the area of the segment?
b What is the perimeter of the quadrant?

9 Three-dimensional shapes

Key Points

- **net:** a pattern of 2D shapes that can be folded to make a hollow solid shape.
- **plans and elevations:** 2D views of a 3D object drawn from different angles.

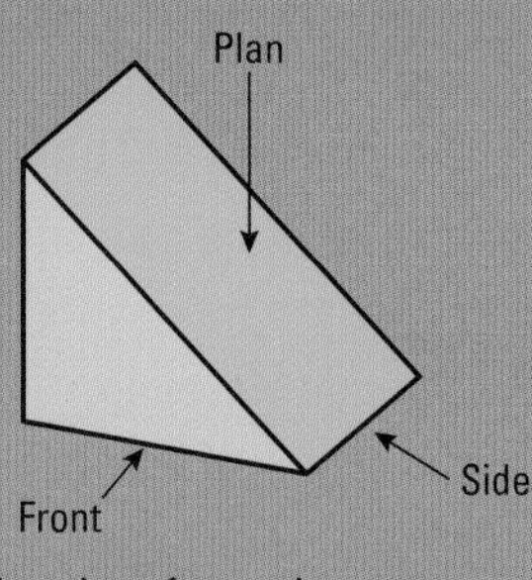

  - **plan:** the view from above.
  - **front elevation:** the view from the front.

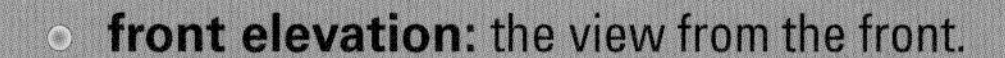

  - **side elevation:** the view from the side.
- **cylinder:** a prism with a circular cross section.
- **volume of a cylinder:** the amount of space the cylinder takes up.
  - volume of a cylinder = the area of the circular cross section × height
  - volume of a cylinder = $\pi r^2 h$
- **surface area of a cylinder:** the area of the net that can be used to build the cylinder.
  - surface area of a cylinder = $2\pi rh + 2\pi r^2$
- **litres:** used to measure capacity or the amount a container can hold.
  - 1 litre (l) = 1000 cm³
  - 1 cm³ = 1 millilitre (ml)
- **finding the perimeter, area and volume of enlarged shapes:**
  - enlarged perimeter = original perimeter × scale factor
  - enlarged area = original area × scale factor²
  - enlarged volume = original volume × scale factor³
- **converting units of measure:** when converting from a large unit to a smaller unit, multiply. When converting from a small unit to a larger unit, divide.

Length	Area
1 cm = 10 mm	1 cm² = 10 × 10 = 100 mm²
1 m = 100 cm	1 m² = 100 × 100 = 10 000 cm²
1 km = 1000 m	1 km² = 1000 × 1000 = 1 000 000 m²

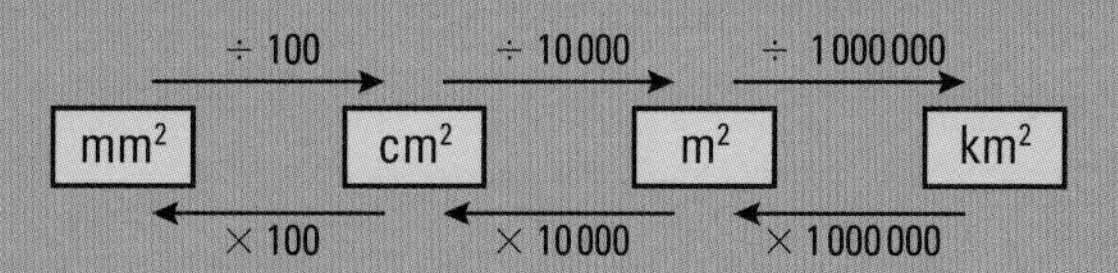

Length	Volume
1 cm = 10 mm	1 cm³ = 10 × 10 × 10 = 1000 mm³
1 m = 100 cm	1 m³ = 100 × 100 × 100 = 1 000 000 cm³
1 km = 1000 m	1 km³ = 1000 × 1000 × 1000 = 1 000 000 000 m³

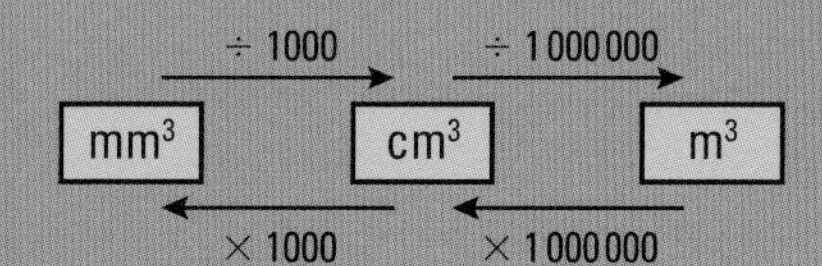

9.1 Nets

Exercise 9A

Questions in this chapter are targeted at the grades indicated.

E

1 The diagrams below show some solids.

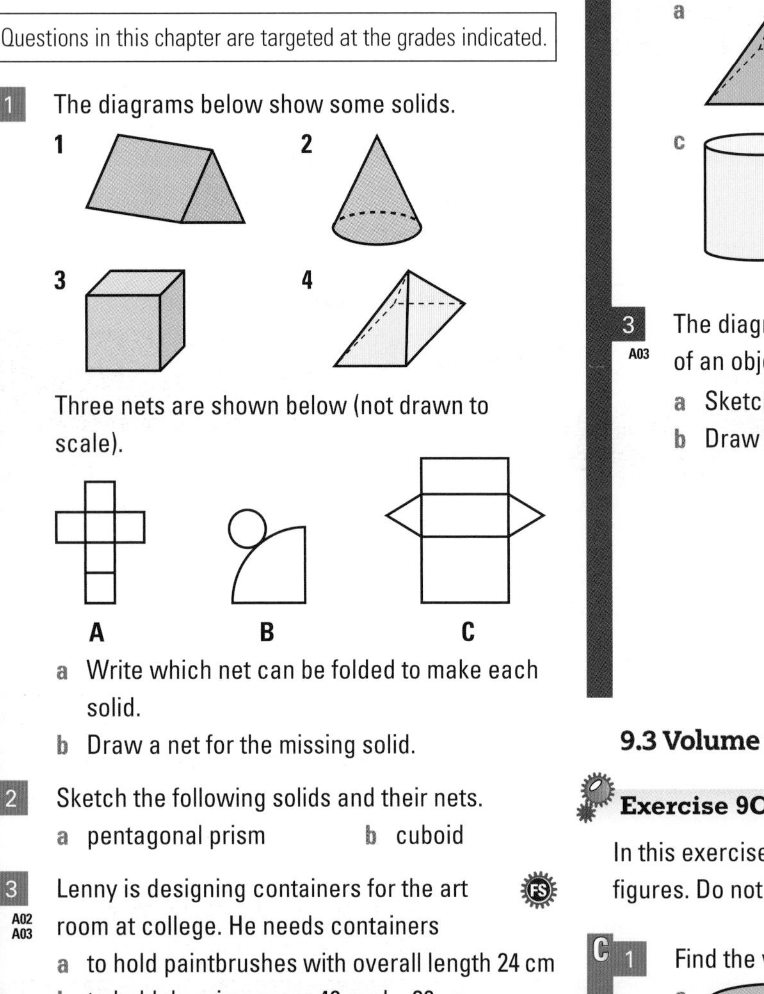

Three nets are shown below (not drawn to scale).

a Write which net can be folded to make each solid.

b Draw a net for the missing solid.

2 Sketch the following solids and their nets.

a pentagonal prism b cuboid

3 Lenny is designing containers for the art room at college. He needs containers

A02 A03

FS

a to hold paintbrushes with overall length 24 cm

b to hold drawing paper 40 cm by 30 cm

Make a sketch of the net of possible containers for a and b.

9.2 Plans and elevations

Exercise 9B

D

1 Use squared paper to make an accurate drawing of the plan, front elevation and side elevation for this cuboid.

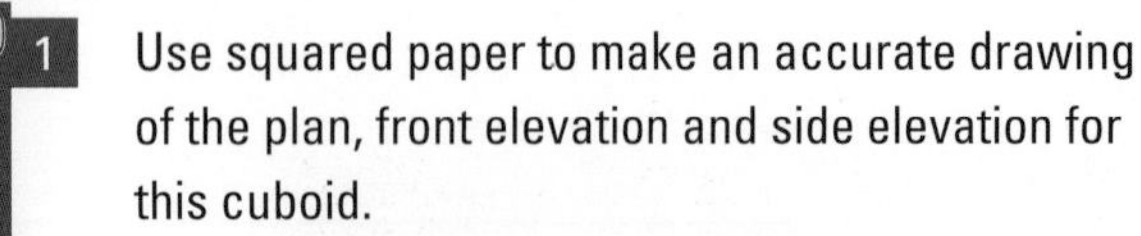

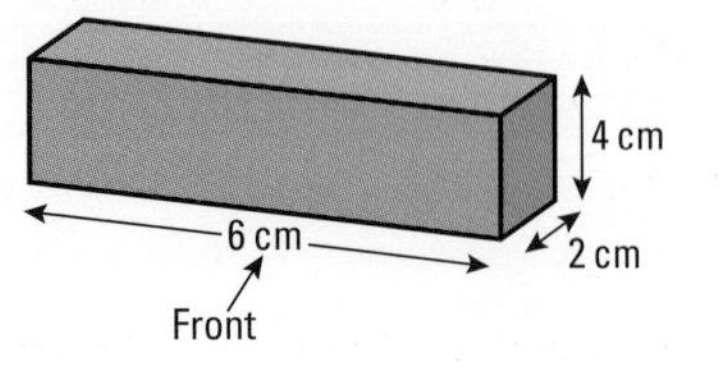

D

2 Sketch the plan, front elevation and side elevation for these shapes.

A03

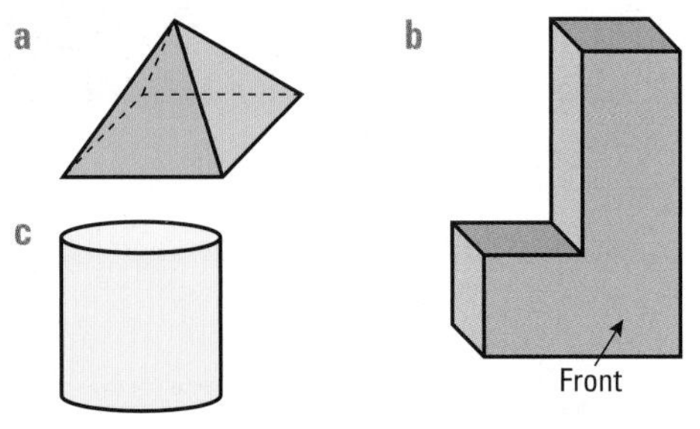

3 The diagram shows the plan and front elevation of an object.

A03

a Sketch the side elevation.

b Draw a 3D sketch of the shape.

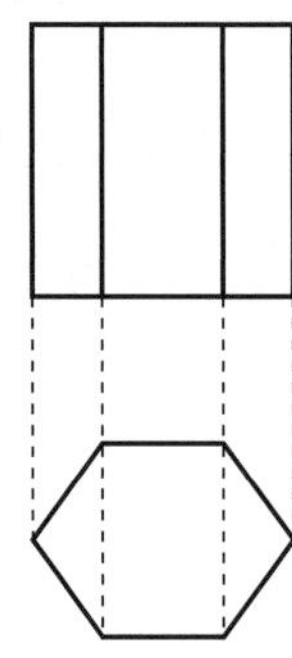

9.3 Volume and surface area of a cylinder

Exercise 9C

In this exercise give your answers to 3 significant figures. Do not forget to give the units.

C

1 Find the volume of these cylinders.

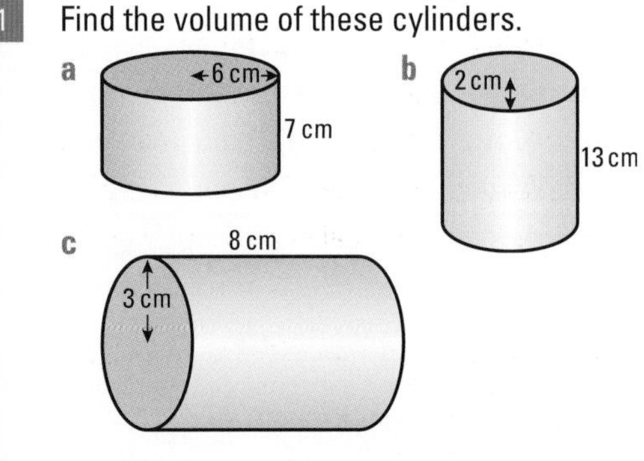

2 A hexagonal prism has a length of 7 cm. If the area of the cross-section is 9 cm^2, calculate its volume.

3 A triangular prism has length 12 cm. The triangular face has base 6 cm and height 7 cm. Calculate the volume of the prism.

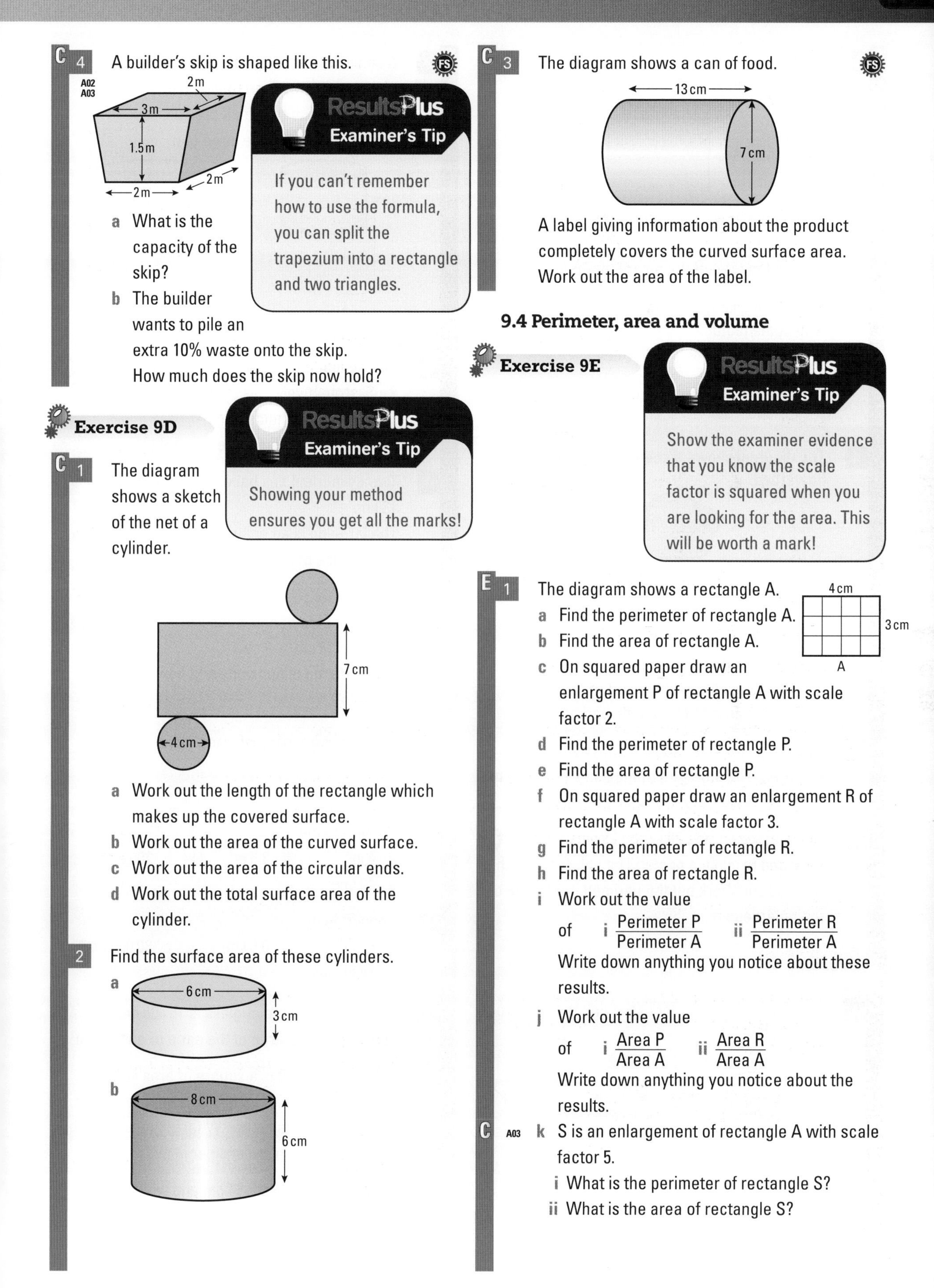

C 4 A builder's skip is shaped like this.

A02
A03

> **ResultsPlus Examiner's Tip**
>
> If you can't remember how to use the formula, you can split the trapezium into a rectangle and two triangles.

a What is the capacity of the skip?

b The builder wants to pile an extra 10% waste onto the skip. How much does the skip now hold?

Exercise 9D

> **ResultsPlus Examiner's Tip**
>
> Showing your method ensures you get all the marks!

C 1 The diagram shows a sketch of the net of a cylinder.

a Work out the length of the rectangle which makes up the covered surface.

b Work out the area of the curved surface.

c Work out the area of the circular ends.

d Work out the total surface area of the cylinder.

2 Find the surface area of these cylinders.

a

b

C 3 The diagram shows a can of food.

A label giving information about the product completely covers the curved surface area. Work out the area of the label.

9.4 Perimeter, area and volume

Exercise 9E

> **ResultsPlus Examiner's Tip**
>
> Show the examiner evidence that you know the scale factor is squared when you are looking for the area. This will be worth a mark!

E 1 The diagram shows a rectangle A.

a Find the perimeter of rectangle A.

b Find the area of rectangle A.

c On squared paper draw an enlargement P of rectangle A with scale factor 2.

d Find the perimeter of rectangle P.

e Find the area of rectangle P.

f On squared paper draw an enlargement R of rectangle A with scale factor 3.

g Find the perimeter of rectangle R.

h Find the area of rectangle R.

i Work out the value of **i** $\frac{\text{Perimeter P}}{\text{Perimeter A}}$ **ii** $\frac{\text{Perimeter R}}{\text{Perimeter A}}$

Write down anything you notice about these results.

j Work out the value of **i** $\frac{\text{Area P}}{\text{Area A}}$ **ii** $\frac{\text{Area R}}{\text{Area A}}$

Write down anything you notice about the results.

C A03 **k** S is an enlargement of rectangle A with scale factor 5.

i What is the perimeter of rectangle S?

ii What is the area of rectangle S?

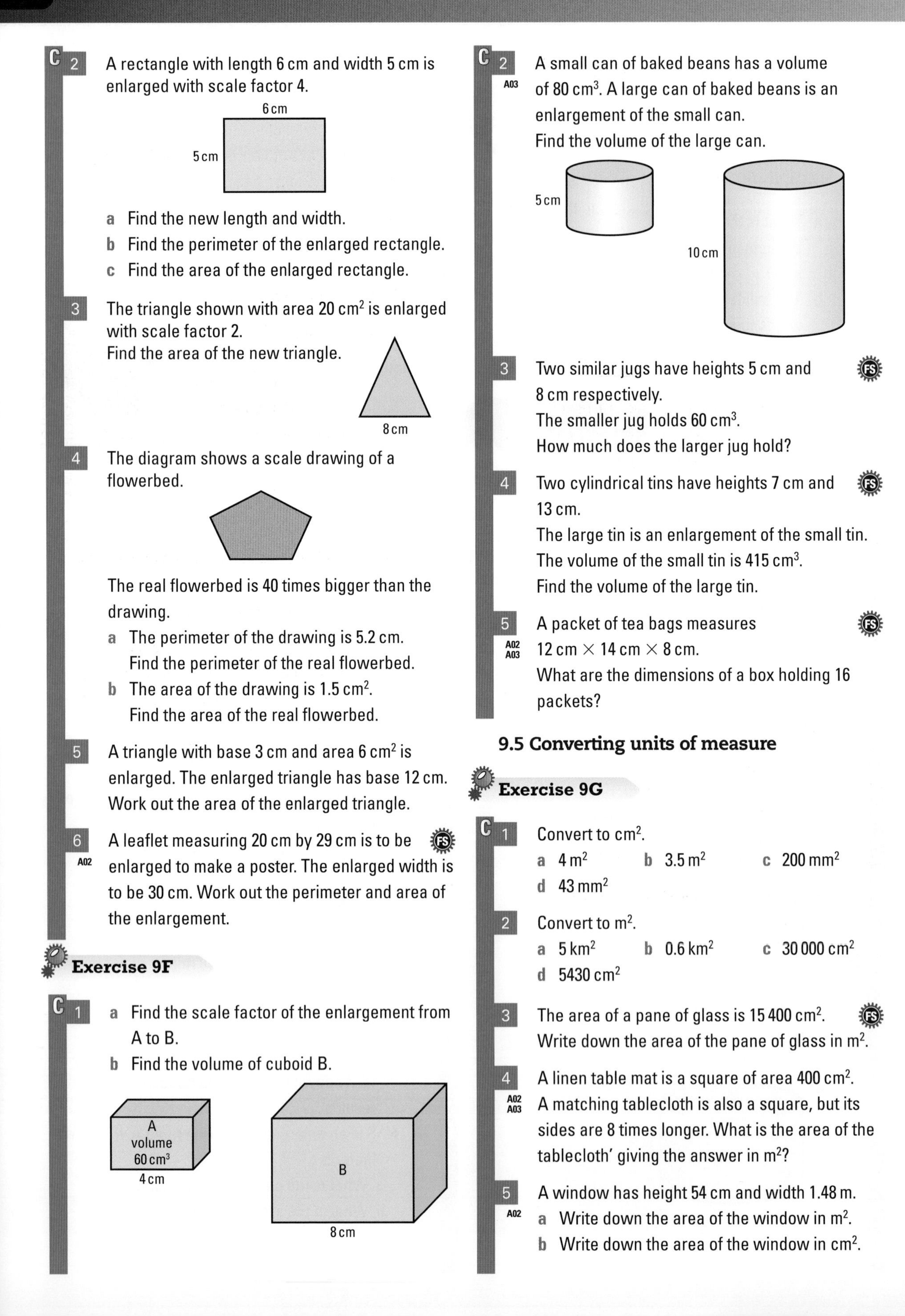

C 2 A rectangle with length 6 cm and width 5 cm is enlarged with scale factor 4.

a Find the new length and width.
b Find the perimeter of the enlarged rectangle.
c Find the area of the enlarged rectangle.

3 The triangle shown with area 20 cm^2 is enlarged with scale factor 2.
Find the area of the new triangle.

4 The diagram shows a scale drawing of a flowerbed.

The real flowerbed is 40 times bigger than the drawing.
a The perimeter of the drawing is 5.2 cm.
Find the perimeter of the real flowerbed.
b The area of the drawing is 1.5 cm^2.
Find the area of the real flowerbed.

5 A triangle with base 3 cm and area 6 cm^2 is enlarged. The enlarged triangle has base 12 cm. Work out the area of the enlarged triangle.

6 A02 A leaflet measuring 20 cm by 29 cm is to be enlarged to make a poster. The enlarged width is to be 30 cm. Work out the perimeter and area of the enlargement. FS

Exercise 9F

C 1 a Find the scale factor of the enlargement from A to B.
b Find the volume of cuboid B.

C 2 A03 A small can of baked beans has a volume of 80 cm^3. A large can of baked beans is an enlargement of the small can.
Find the volume of the large can.

3 Two similar jugs have heights 5 cm and 8 cm respectively. FS
The smaller jug holds 60 cm^3.
How much does the larger jug hold?

4 Two cylindrical tins have heights 7 cm and 13 cm. FS
The large tin is an enlargement of the small tin.
The volume of the small tin is 415 cm^3.
Find the volume of the large tin.

5 A02 A03 A packet of tea bags measures 12 cm × 14 cm × 8 cm. FS
What are the dimensions of a box holding 16 packets?

9.5 Converting units of measure

Exercise 9G

C 1 Convert to cm^2.
a 4 m^2 b 3.5 m^2 c 200 mm^2
d 43 mm^2

2 Convert to m^2.
a 5 km^2 b 0.6 km^2 c 30 000 cm^2
d 5430 cm^2

3 The area of a pane of glass is 15 400 cm^2. FS
Write down the area of the pane of glass in m^2.

4 A02 A03 A linen table mat is a square of area 400 cm^2.
A matching tablecloth is also a square, but its sides are 8 times longer. What is the area of the tablecloth' giving the answer in m^2?

5 A02 A window has height 54 cm and width 1.48 m.
a Write down the area of the window in m^2.
b Write down the area of the window in cm^2.

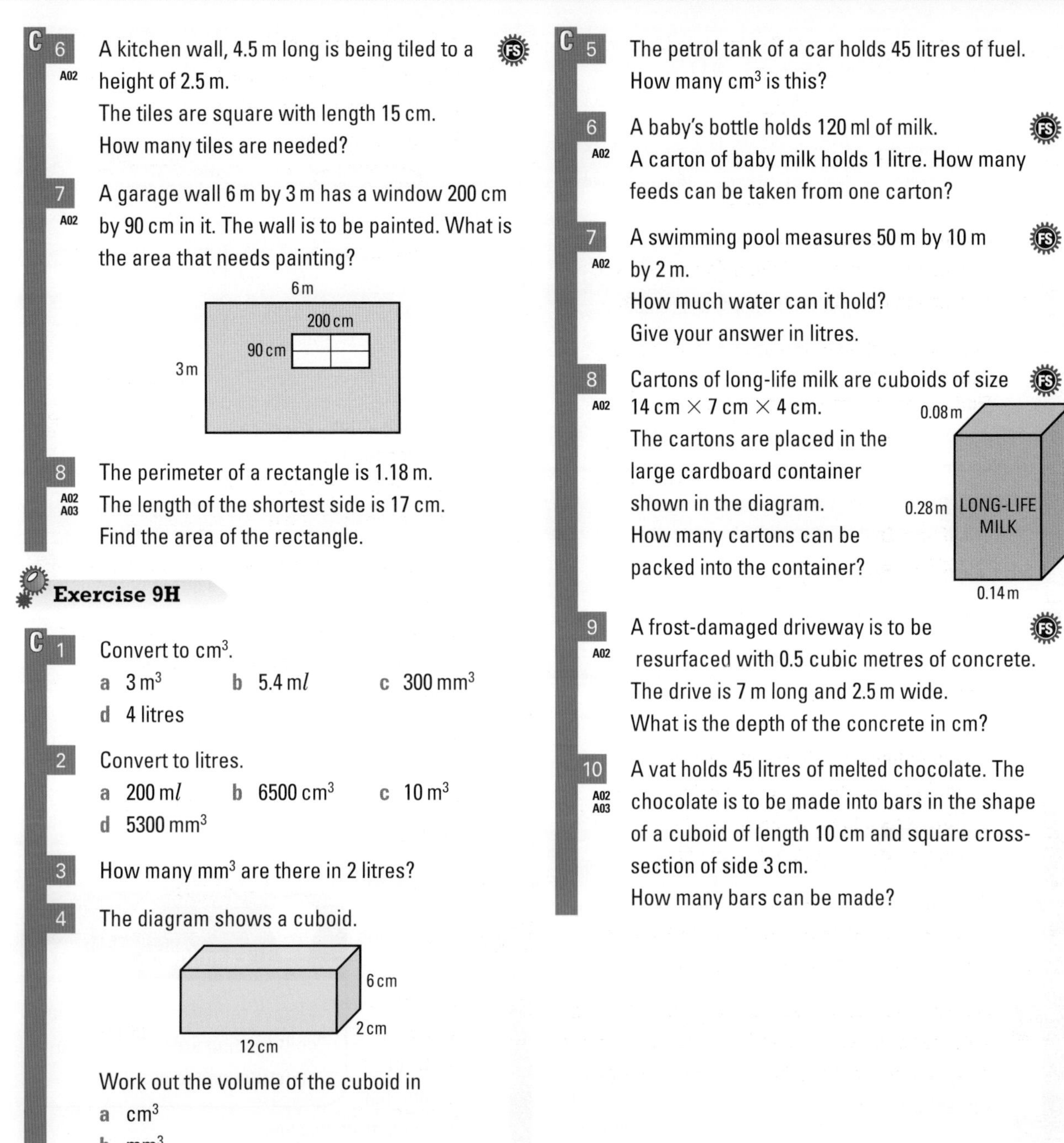

C

6 A kitchen wall, 4.5 m long is being tiled to a height of 2.5 m.
The tiles are square with length 15 cm.
How many tiles are needed?

7 A garage wall 6 m by 3 m has a window 200 cm by 90 cm in it. The wall is to be painted. What is the area that needs painting?

8 The perimeter of a rectangle is 1.18 m.
The length of the shortest side is 17 cm.
Find the area of the rectangle.

Exercise 9H

C

1 Convert to cm^3.
a 3 m^3 b 5.4 ml c 300 mm^3
d 4 litres

2 Convert to litres.
a 200 ml b 6500 cm^3 c 10 m^3
d 5300 mm^3

3 How many mm^3 are there in 2 litres?

4 The diagram shows a cuboid.

Work out the volume of the cuboid in
a cm^3
b mm^3.

C

5 The petrol tank of a car holds 45 litres of fuel.
How many cm^3 is this?

6 A baby's bottle holds 120 ml of milk.
A carton of baby milk holds 1 litre. How many feeds can be taken from one carton?

7 A swimming pool measures 50 m by 10 m by 2 m.
How much water can it hold?
Give your answer in litres.

8 Cartons of long-life milk are cuboids of size 14 cm $\times$ 7 cm $\times$ 4 cm.
The cartons are placed in the large cardboard container shown in the diagram.
How many cartons can be packed into the container?

9 A frost-damaged driveway is to be resurfaced with 0.5 cubic metres of concrete.
The drive is 7 m long and 2.5 m wide.
What is the depth of the concrete in cm?

10 A vat holds 45 litres of melted chocolate. The chocolate is to be made into bars in the shape of a cuboid of length 10 cm and square cross-section of side 3 cm.
How many bars can be made?

10 Constructions and loci

Key Points

- **locus:** a set of points, or a region, obeying a given rule.
- **loci:** the plural of locus.
- **finding the locus of points equidistant from a single point:** draw a circle around the point.
- **finding the locus of points equidistant from two points:** draw a line that is the perpendicular bisector of the line joining the two points.
- **finding the locus of points equidistant from two lines:** draw the bisector of the angle between the two lines.
- **constructing lines and angles:** these must be learned:
  - constructing an angle of 60°
  - bisecting an angle (giving angles of 30°, 45°)
  - bisecting a line
  - constructing a perpendicular line from a point on a given line
  - constructing a perpendicular line from a point to a given line

10.1 Constructions

Exercise 10A

Questions in this chapter are targeted at the grades indicated.

For each of the following questions use one or more of the five constructions.

C

1 Draw lines of the lengths shown. Then bisect each of the lines.
a 10 cm b 5 cm c 6 cm d 8 cm

2 Construct equilateral triangles with sides of the following lengths.
a 4 cm b 7 cm c 6 cm d 5 cm

3 Construct squares with sides of the following lengths.
a 5 cm b 3 cm c 7 cm d 4 cm

4 Draw pairs of parallel lines that have a distance between them of:
a 2.5 cm b 6 cm c 4 cm d 5.5 cm

5 Draw a 6 cm line. Mark a point P on your line. Construct a perpendicular line at point P.

6 Draw a 6 cm line. Mark a point P at least 5 cm above your line. Construct a perpendicular line from point P down to the line.

7 Construct an equilateral triangle of side 5 cm. Construct a perpendicular line from a vertex to the opposite side of the triangle.

C

8 Draw a regular hexagon in a circle of radius 5 cm.

9 Draw a regular octagon in a circle of radius 5 cm.

10 Draw a regular pentagon in a circle of radius 5 cm.

10.2 Loci

Exercise 10B

C

1 A03 Draw a line of length 7 cm. Draw the locus of points that are 3 cm from this line.

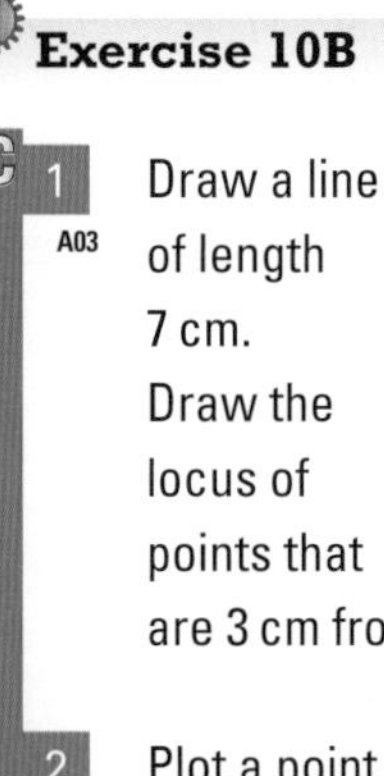

ResultsPlus Examiner's Tip

Always remember to show your construction arcs when using constructions to draw loci.

2 A03 Plot a point P on your page. Draw the locus of points that are 2 cm from point P.

3 A03 Draw two points, P and Q, 10 cm apart.

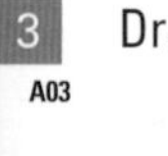

Draw the locus of points that are the same distance from point P and from point Q.

4 A03 Plot a point A on your page. Draw the locus of points that are 5 cm from point A.

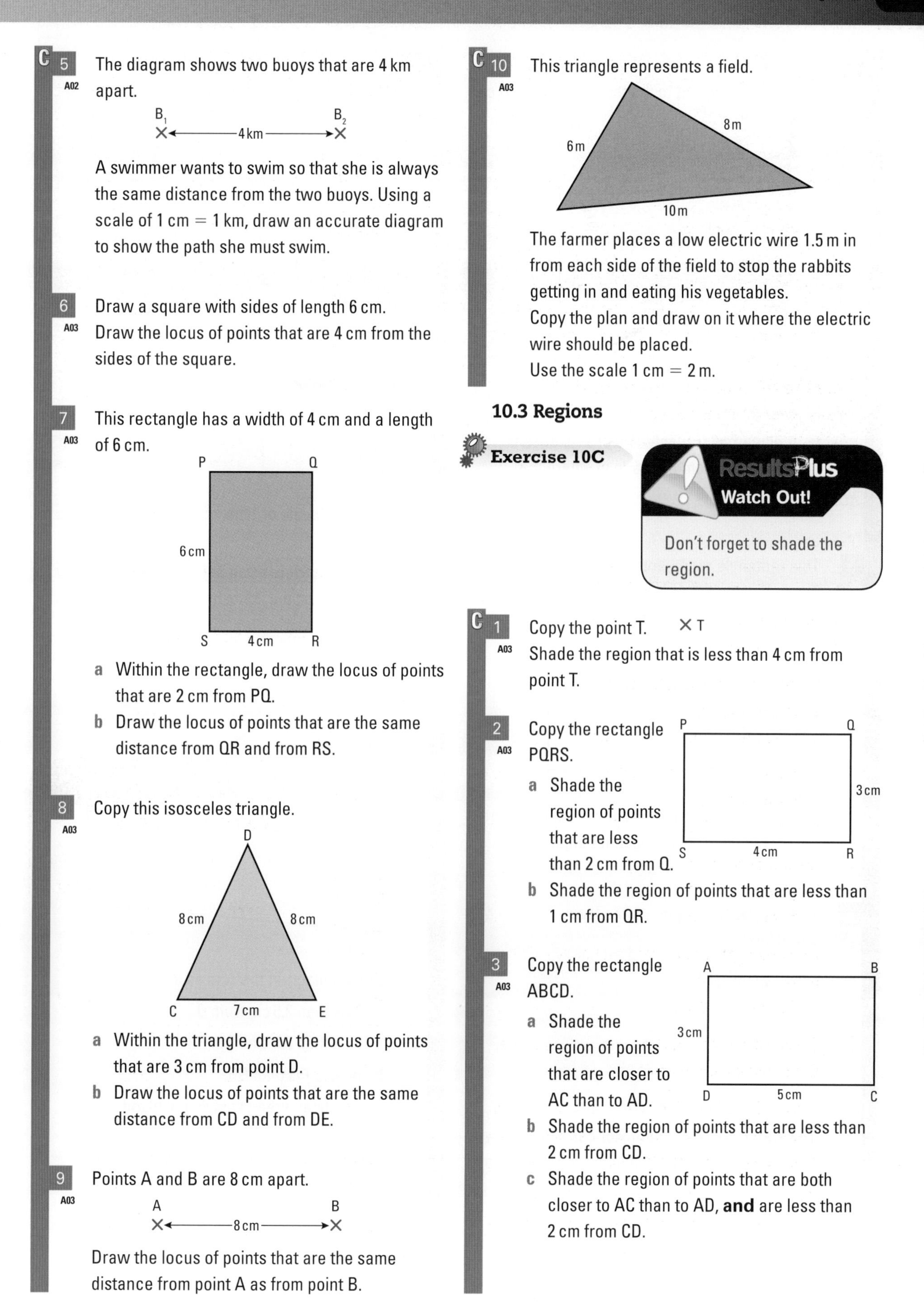

C 5 A02 The diagram shows two buoys that are 4 km apart.

A swimmer wants to swim so that she is always the same distance from the two buoys. Using a scale of 1 cm = 1 km, draw an accurate diagram to show the path she must swim.

6 A03 Draw a square with sides of length 6 cm. Draw the locus of points that are 4 cm from the sides of the square.

7 A03 This rectangle has a width of 4 cm and a length of 6 cm.

a Within the rectangle, draw the locus of points that are 2 cm from PQ.

b Draw the locus of points that are the same distance from QR and from RS.

8 A03 Copy this isosceles triangle.

a Within the triangle, draw the locus of points that are 3 cm from point D.

b Draw the locus of points that are the same distance from CD and from DE.

9 A03 Points A and B are 8 cm apart.

Draw the locus of points that are the same distance from point A as from point B.

C 10 A03 This triangle represents a field.

The farmer places a low electric wire 1.5 m in from each side of the field to stop the rabbits getting in and eating his vegetables.
Copy the plan and draw on it where the electric wire should be placed.
Use the scale 1 cm = 2 m.

10.3 Regions

Exercise 10C

ResultsPlus
Watch Out!
Don't forget to shade the region.

C 1 A03 Copy the point T. × T
Shade the region that is less than 4 cm from point T.

2 A03 Copy the rectangle PQRS.

a Shade the region of points that are less than 2 cm from Q.

b Shade the region of points that are less than 1 cm from QR.

3 A03 Copy the rectangle ABCD.

a Shade the region of points that are closer to AC than to AD.

b Shade the region of points that are less than 2 cm from CD.

c Shade the region of points that are both closer to AC than to AD, **and** are less than 2 cm from CD.

C **4** A03 ABCDEF is a fence around a garden. A hedge is to be planted next to the fence and not more than 1 m away.

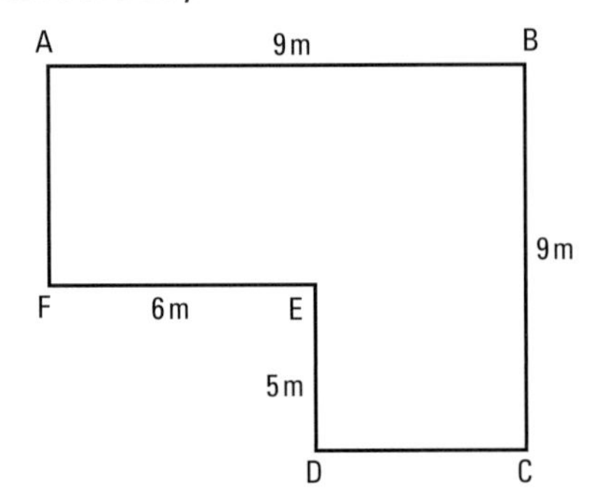

a Copy the diagram and shade in where the hedge will be.

b A flowerbed is planted so that its edge is always 2 m from B. Shade in where the flowerbed will be.

5 A03 This rectangle represents a school library. The half of the library nearest to side LM is reserved for the younger pupils.

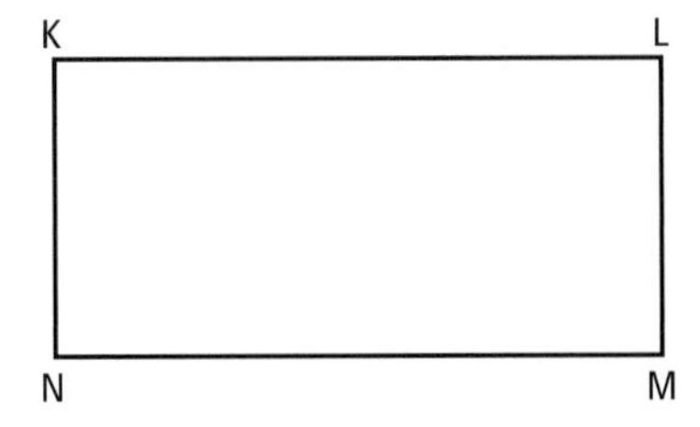

Copy the rectangle and shade the region of the library reserved for the younger pupils.

6 A03 The diagram represents three lawn sprinklers. FS

× ×

The water from each sprinkler can reach 2 m. Copy the diagram and shade the region where water is sprinkled.
Scale: 1 cm = 1 m.

C **7** A03 In the diagram, FG is the wall of a house and L is a security light. The light can shine a distance of 5 m. FS

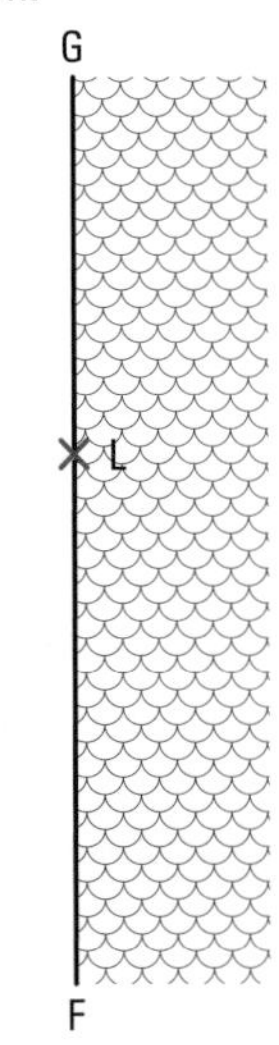

a Use a scale of 1 cm = 2 m to represent this on a plan.

b If lights were fitted at G and F instead, how far would the lights have to shine to cover the whole wall?

8 A03 PQR is a triangle. PQ = 8 cm, QR = 5.5 cm, PR = 10.5 cm.

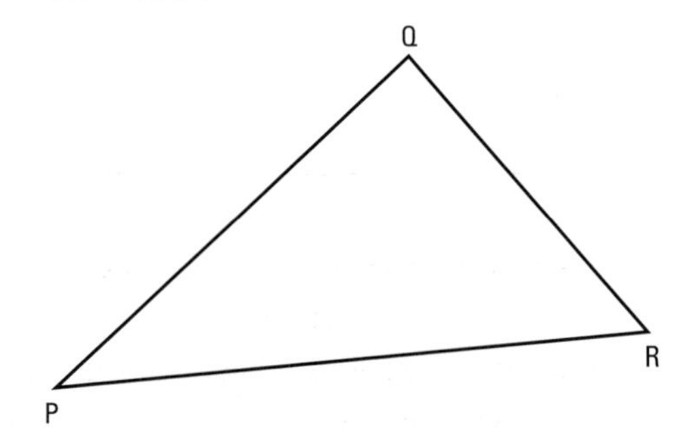

Draw a diagram to show an area

a less than 2.5 cm from Q

b less than 2.5 cm from PR

c nearer PQ than QR.

d Shade an area satisfying **a**, **b** and **c**.

C 9 AO3 Copy this rectangle.

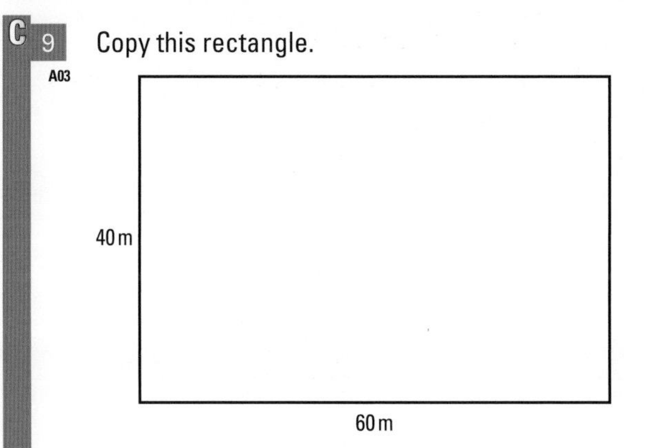

The rectangle represents a plan of a compound which has dimensions of 40 m by 60 m, and a scale of 1 cm = 10 m. Lights are placed on the walls all around the compound. The lights can illuminate a region no more than 10 m from the compound wall, both inside and outside the compound.
Shade on your plan the region illuminated.

10 AO3

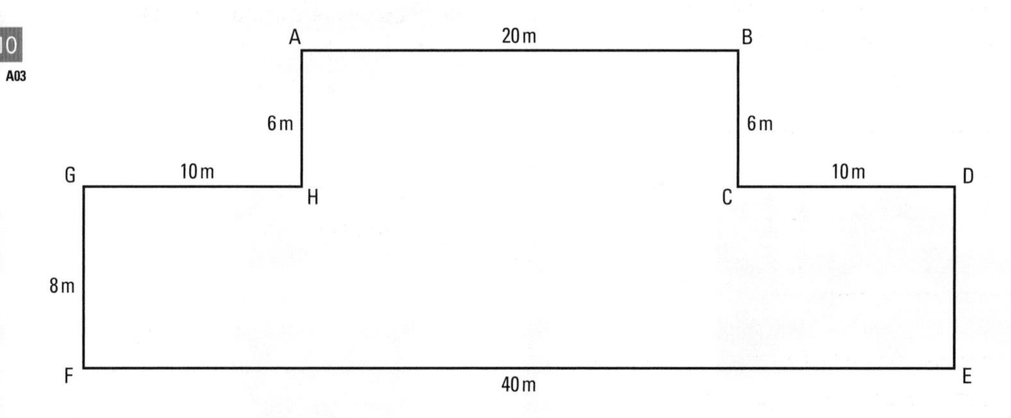

The above is a plan of a hall. Draw an accurate plan, using a scale of 1 cm = 4 m.
Then draw the following regions on the plan.

a The locus of points that are nearer to BC than AB.

b The locus of points that are within 6 m of point H or of point C.

c The locus of points that are more than 2 m from FE.

Indicate clearly any points that are within all of the regions indicated.

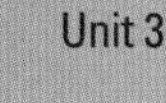

11 Transformations

Key Points

- **translation:** a sliding movement where lengths and angles do not change.
  - **column vector:** a way of describing a translation, e.g. $\begin{pmatrix}3\\2\end{pmatrix}$. The top number describes the movement to the right, and the bottom number describes the movement up.
- **rotation:** turning by a fraction of a turn or by an angle. Lengths and angles do not change.
  - **centre of rotation:** the point about which the shape is turned
- **reflection:** an image of a shape produced by reflecting an object in the line of reflection or mirror line. Lengths and angles do not change.
- **enlargement:** a change in the size of a shape.
  - **scale factor:** the value that multiplies the lengths to give the enlarged image
  - **centre of enlargement:** the point from which the enlarged lengths are measured

11.1 Introduction

Exercise 11A

Questions in this chapter are targeted at the grades indicated.

In each of the following cases, identify which of the four transformations is being shown.

E

1

2

3

E

4

5

6

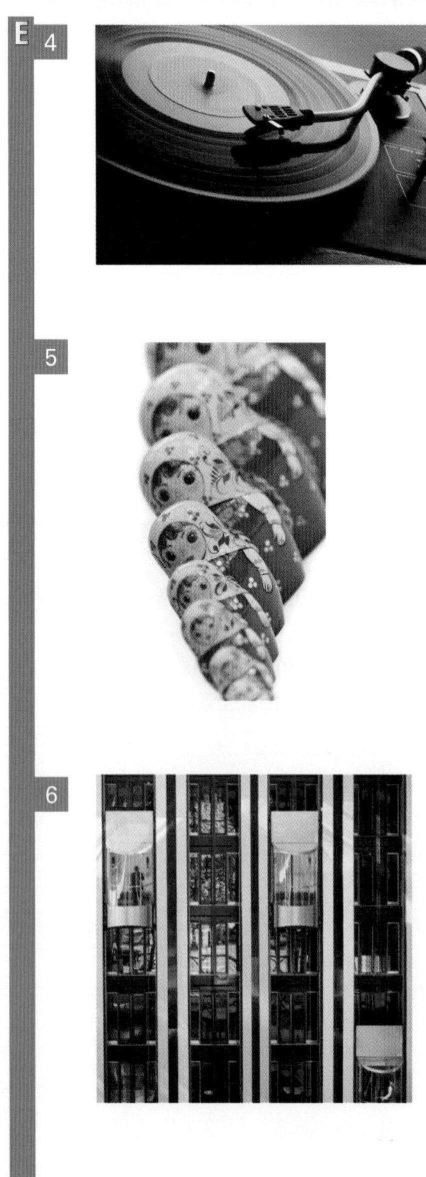

11.2 Translations

Exercise 11B

ResultsPlus
Examiner's Tip

You can use tracing paper to make sure you copy the exact shape.

C 1 Copy each shape and carry out the translation described.

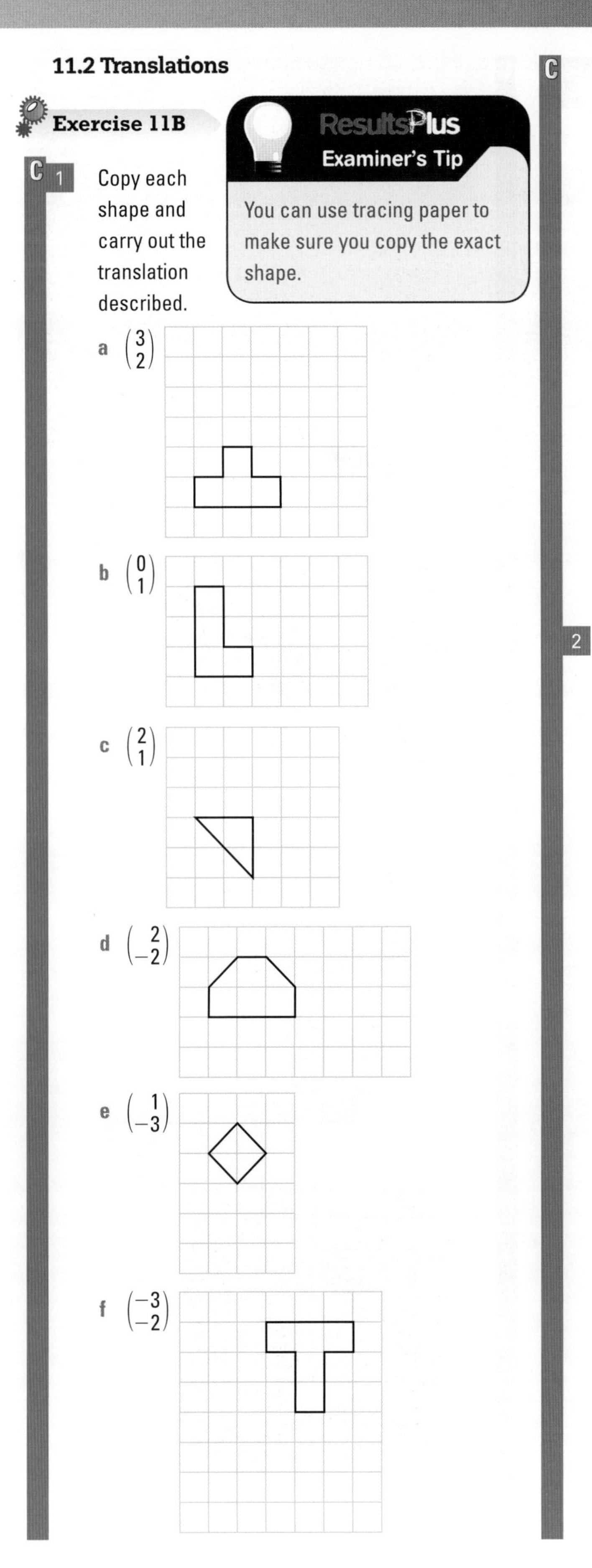

a $\begin{pmatrix} 3 \\ 2 \end{pmatrix}$

b $\begin{pmatrix} 0 \\ 1 \end{pmatrix}$

c $\begin{pmatrix} 2 \\ 1 \end{pmatrix}$

d $\begin{pmatrix} 2 \\ -2 \end{pmatrix}$

e $\begin{pmatrix} 1 \\ -3 \end{pmatrix}$

f $\begin{pmatrix} -3 \\ -2 \end{pmatrix}$

g $\begin{pmatrix} -1 \\ 3 \end{pmatrix}$

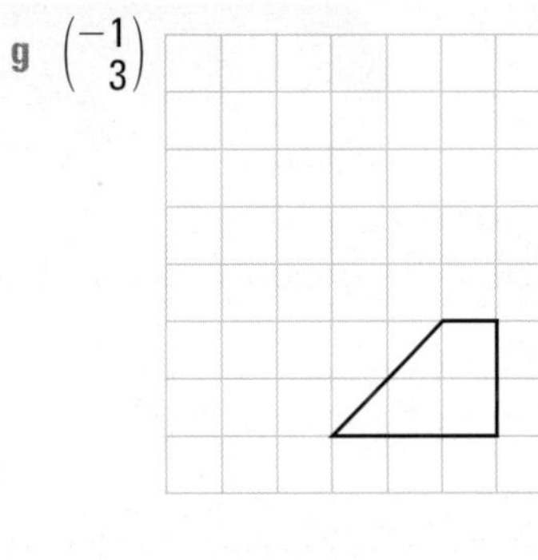

h $\begin{pmatrix} -1 \\ -2 \end{pmatrix}$

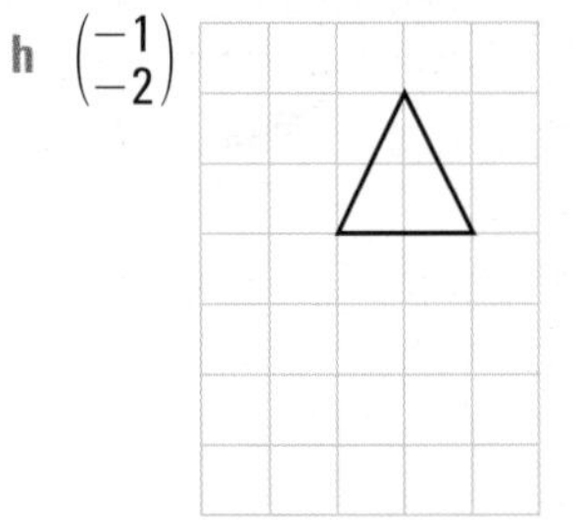

2 Carry out these translations on the shaded shape.

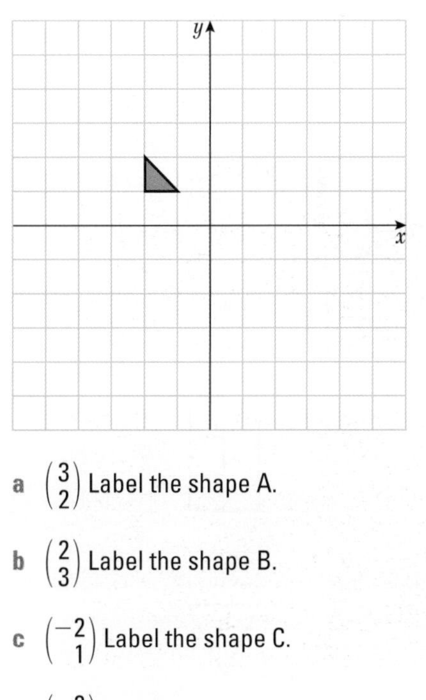

a $\begin{pmatrix} 3 \\ 2 \end{pmatrix}$ Label the shape A.

b $\begin{pmatrix} 2 \\ 3 \end{pmatrix}$ Label the shape B.

c $\begin{pmatrix} -2 \\ 1 \end{pmatrix}$ Label the shape C.

d $\begin{pmatrix} 2 \\ -4 \end{pmatrix}$ Label the shape D.

e $\begin{pmatrix} -2 \\ 0 \end{pmatrix}$ Label the shape E.

f $\begin{pmatrix} 2 \\ -3 \end{pmatrix}$ Label the shape F.

C

3 Describe, as a column vector, each transformation from A to B given below.

> **ResultsPlus**
> **Watch Out!**
> Don't forget the negative signs when writing column vectors.

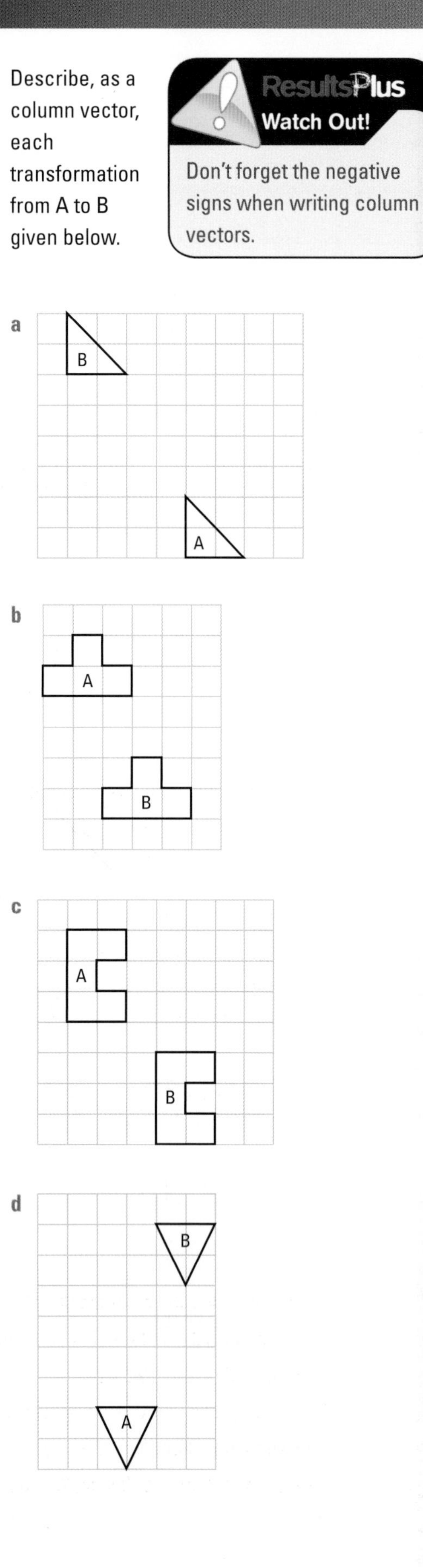

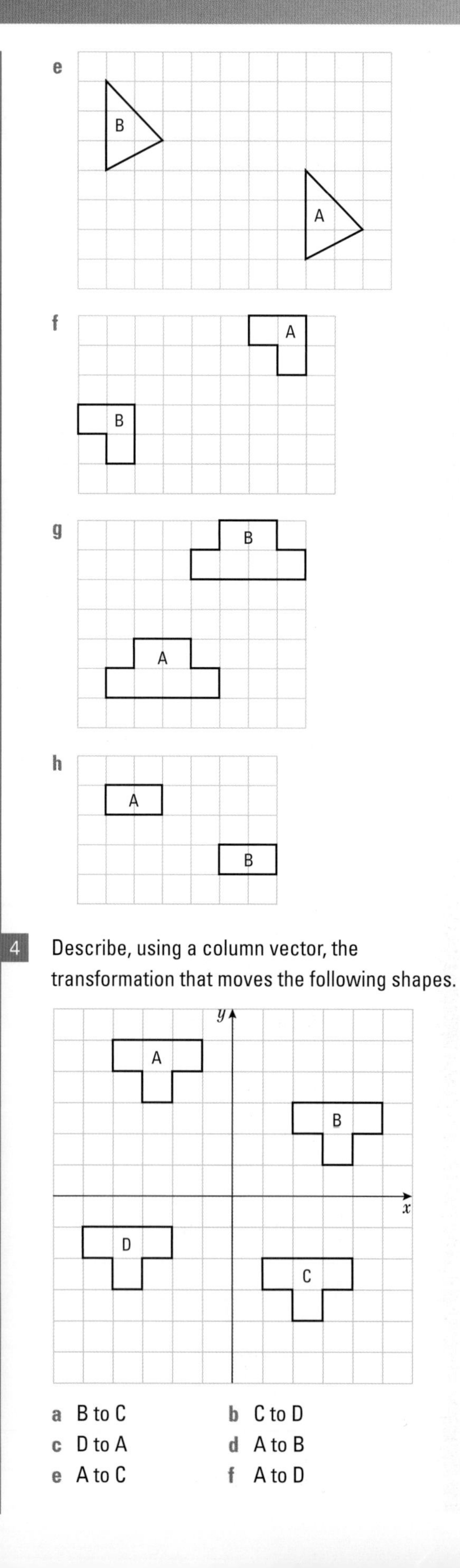

4 Describe, using a column vector, the transformation that moves the following shapes.

a B to C **b** C to D
c D to A **d** A to B
e A to C **f** A to D

11.3 Rotations

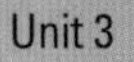

Exercise 11C

D **1** Copy each diagram. Draw the image of each shape after the rotation requested, using the point shown as centre of rotation.

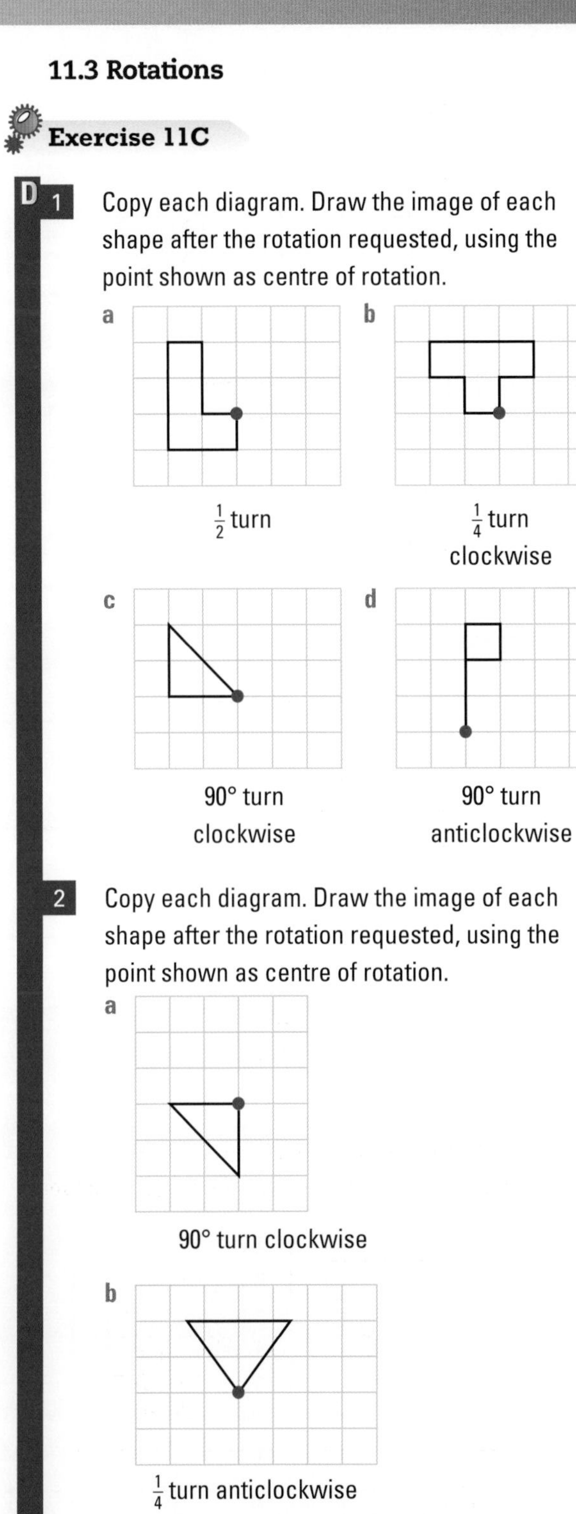

a $\frac{1}{2}$ turn

b $\frac{1}{4}$ turn clockwise

c 90° turn clockwise

d 90° turn anticlockwise

2 Copy each diagram. Draw the image of each shape after the rotation requested, using the point shown as centre of rotation.

a 90° turn clockwise

b $\frac{1}{4}$ turn anticlockwise

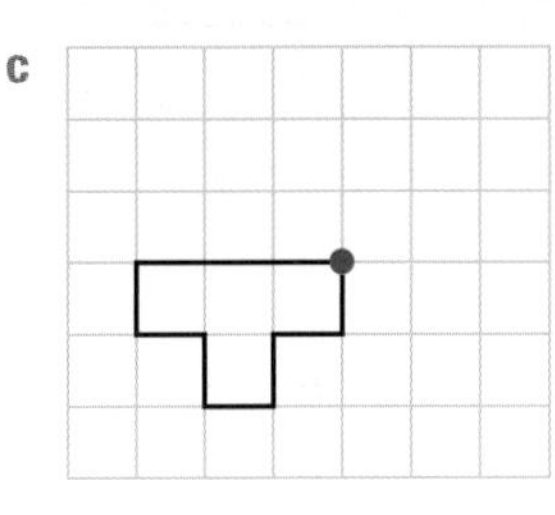

c $\frac{1}{4}$ turn clockwise

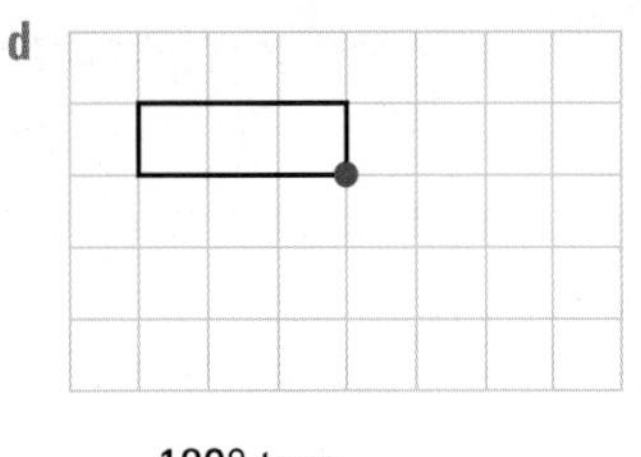

d 180° turn

3 Copy each diagram. Draw separate images for each shape after a rotation of 90° clockwise about each of the centres marked.

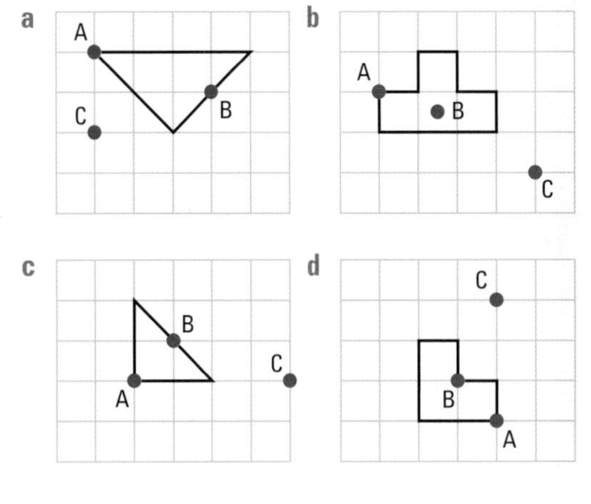

C **4** Copy each diagram. Draw separate images for each shape after the rotation requested, using the given point as centre of rotation.

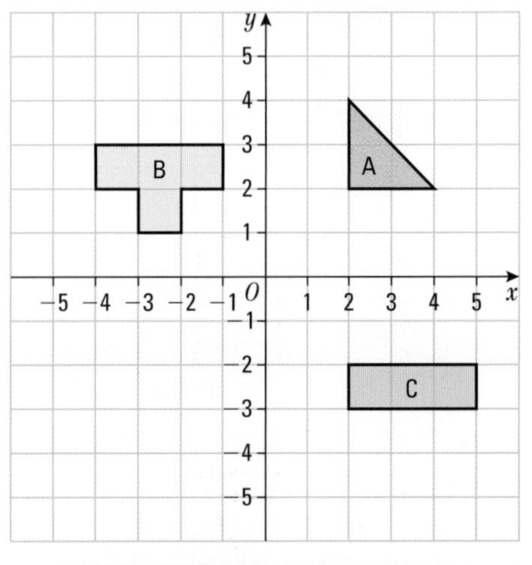

a A rotation of A anticlockwise 90° about (1, 0)

b A rotation of B 90° anticlockwise about (−1, 2)

c A rotation of C 180° about (2, −3)

d A rotation of A clockwise 90° about (1, −1)

e A rotation of B 180° about (−3, 0)

f A rotation of C anticlockwise 90° about (2, 2)

Exercise 11D

D **1** Describe fully the rotation that maps shape B onto shape A.

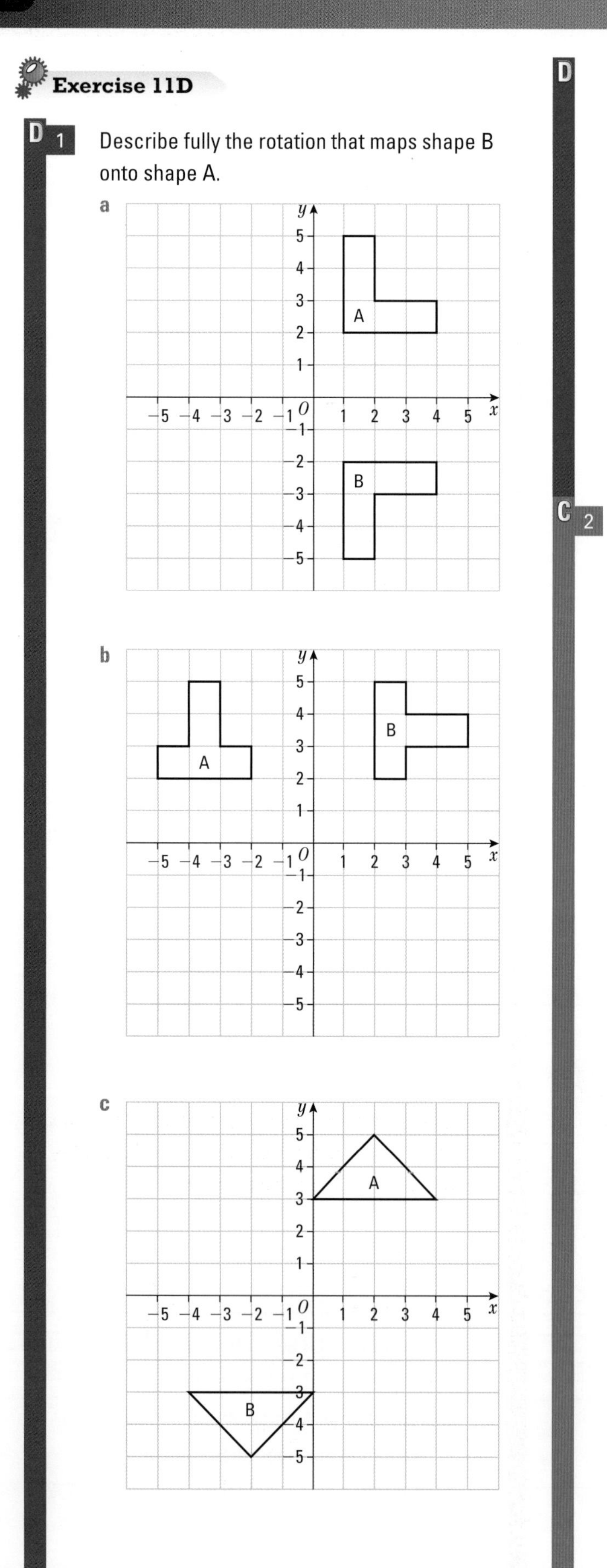

d

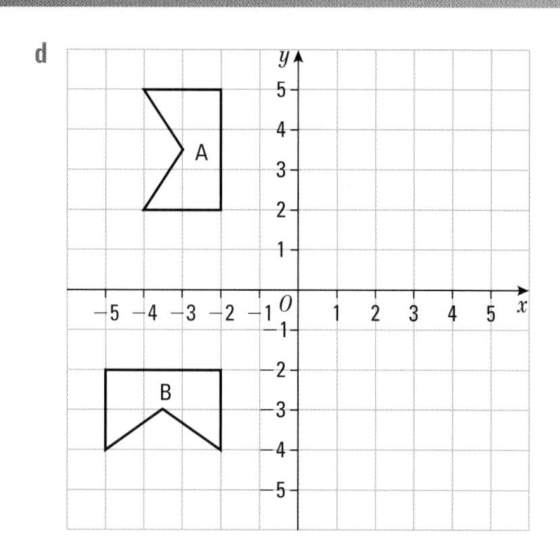

C **2** 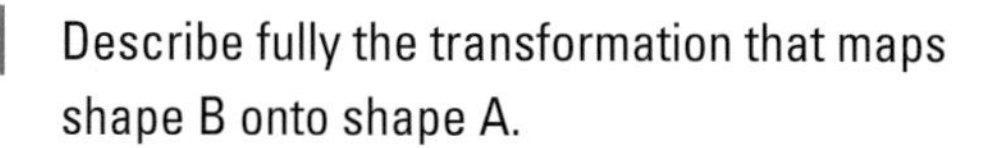

Describe fully the transformation that maps shape B onto shape A.

a

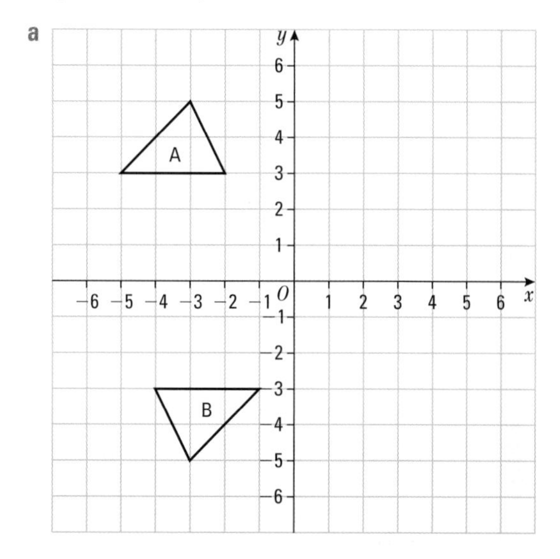

b

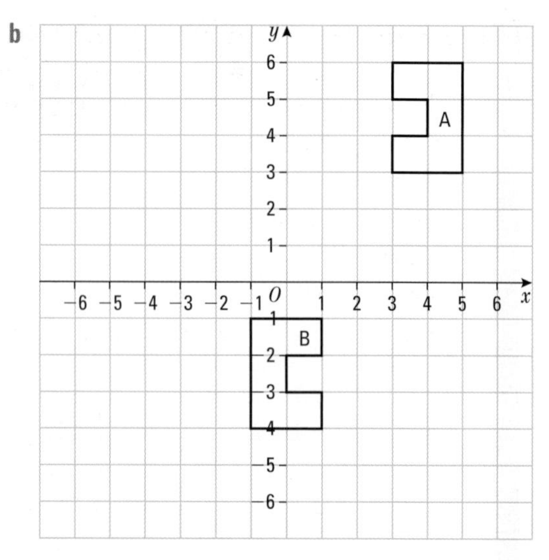

C

c

d

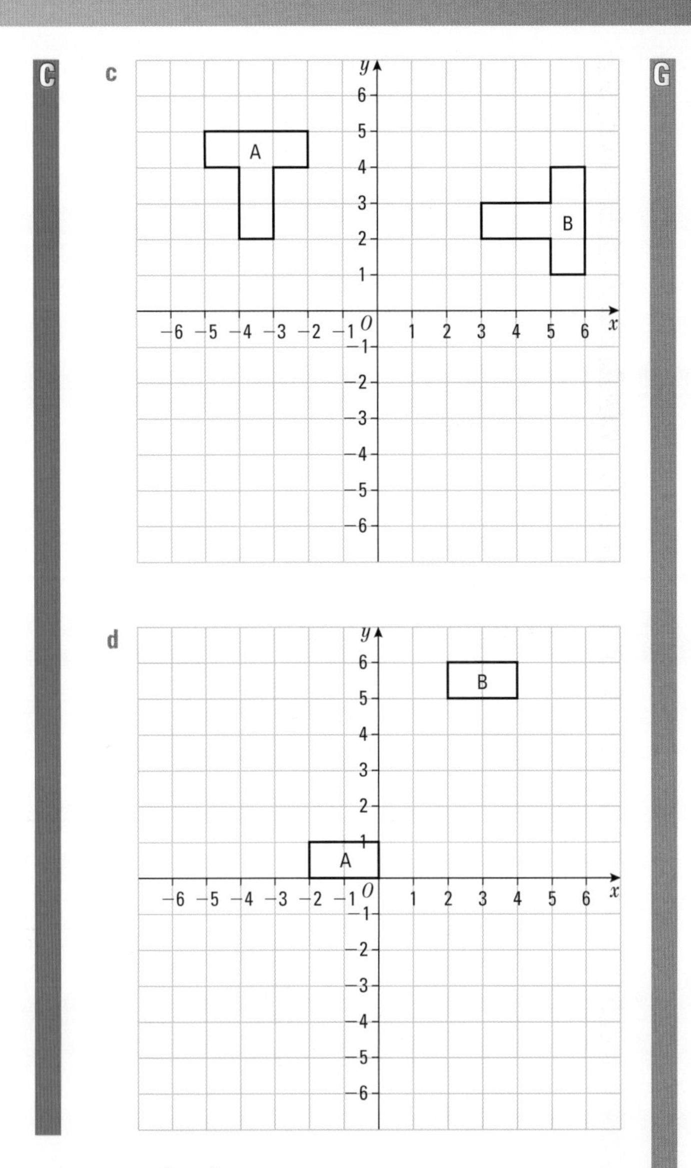

11.4 Reflections

Exercise 11E

Don't forget that the reflection must be the same distance away from the mirror as the original.

G **1** In each of these diagrams the dotted line is a line of reflection. Copy each diagram and draw the reflection of the shape in the line.

a

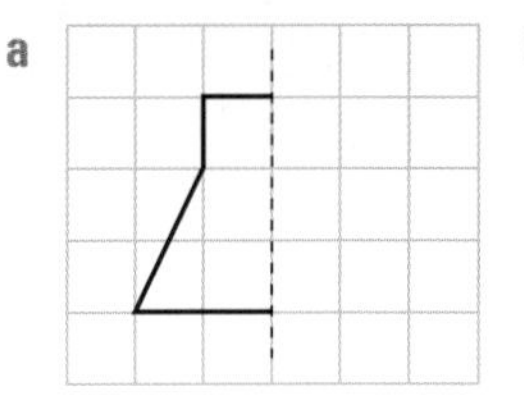

b

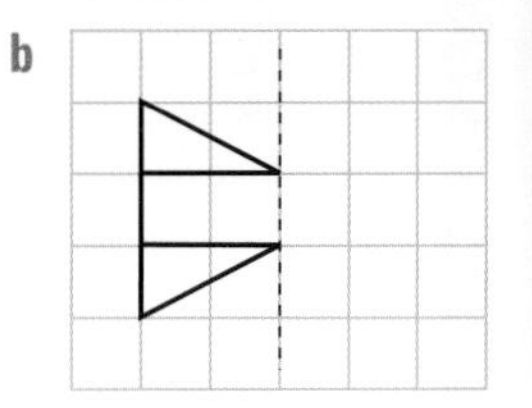

c

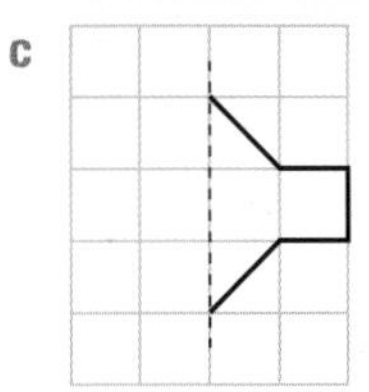

d

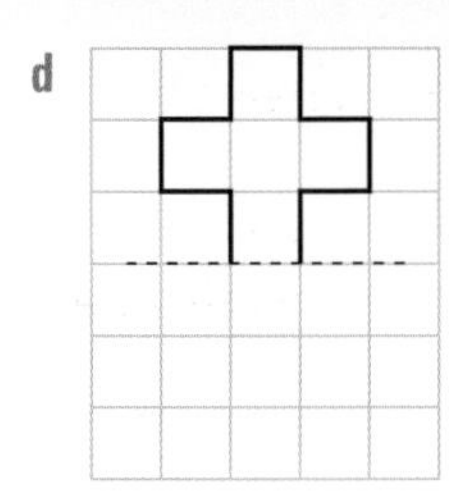

e

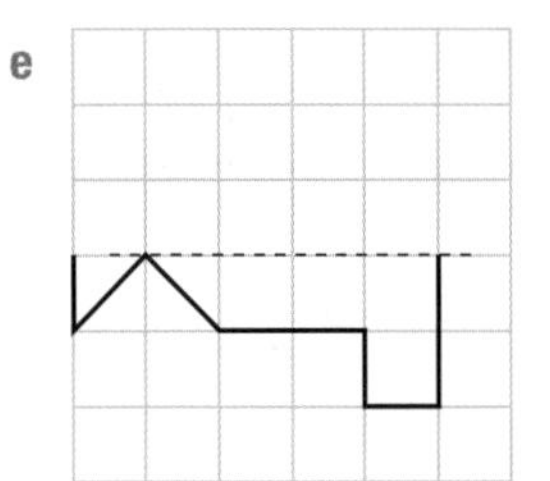

f

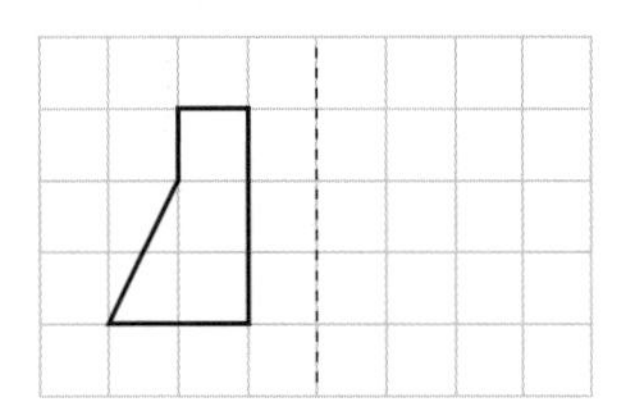

g

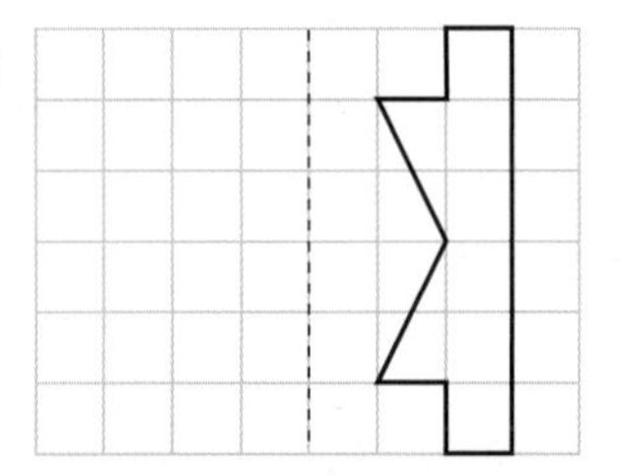

h

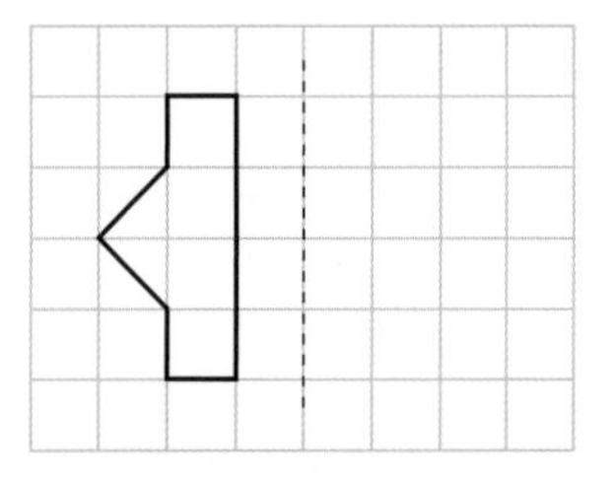

F **2** In each of these diagrams the dotted line is a line of reflection. Copy each diagram and draw the reflection of the shape in the line.

a b

c

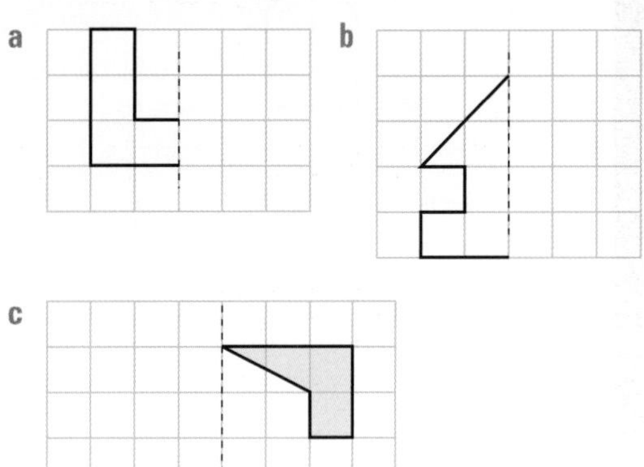

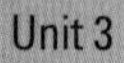

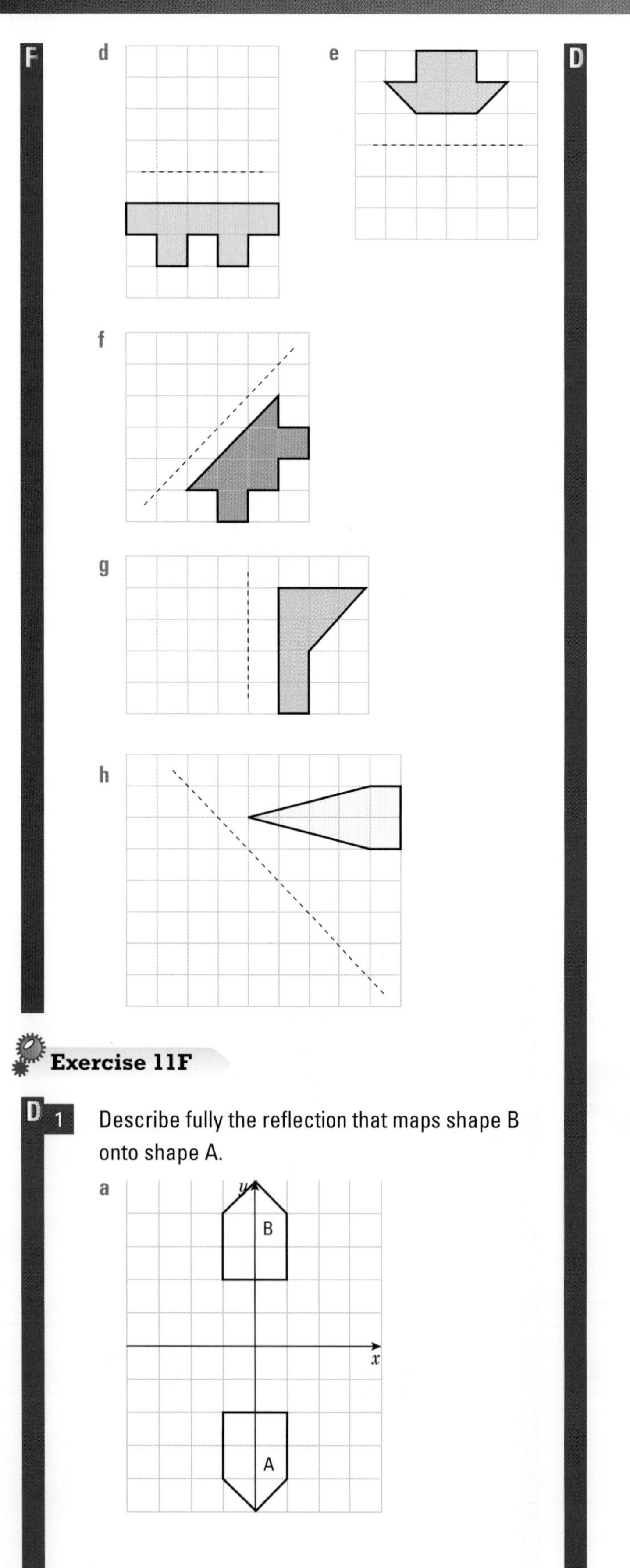

Exercise 11F

D 1 Describe fully the reflection that maps shape B onto shape A.

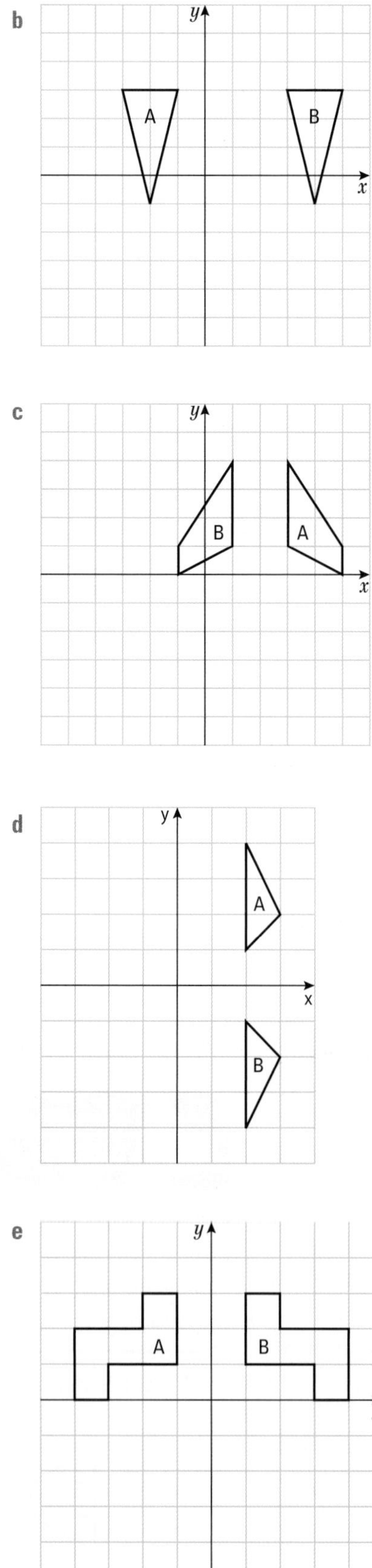

D

f

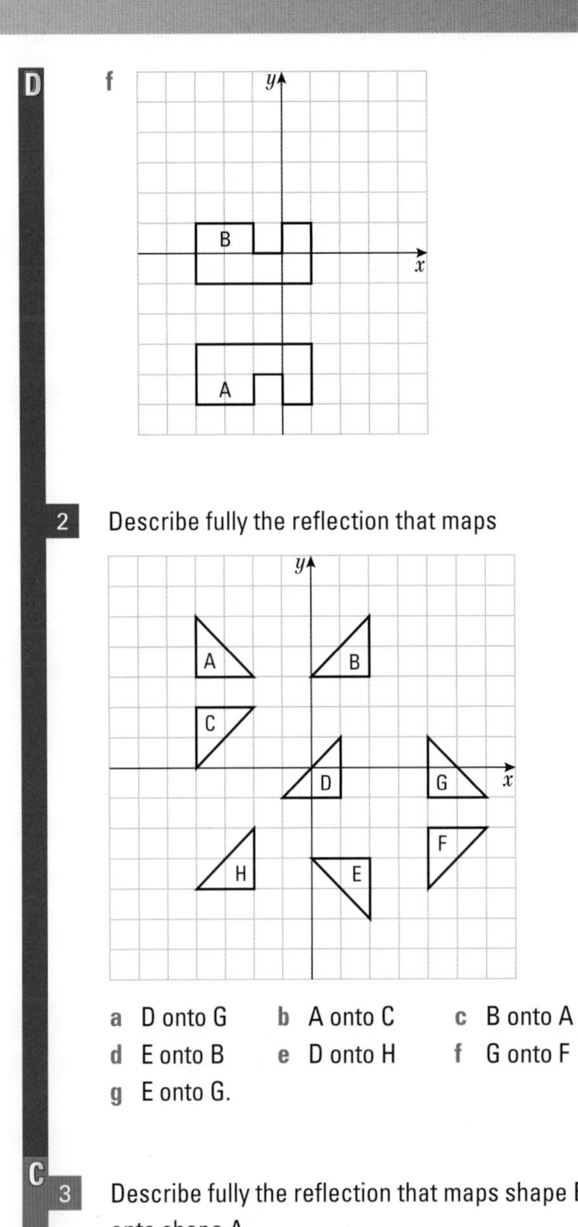

2 Describe fully the reflection that maps

a D onto G b A onto C c B onto A
d E onto B e D onto H f G onto F
g E onto G.

C

3 Describe fully the reflection that maps shape B onto shape A.

a

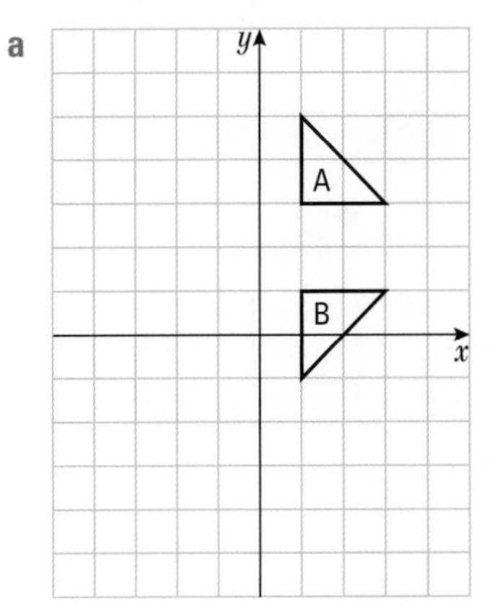

b

c

d

e

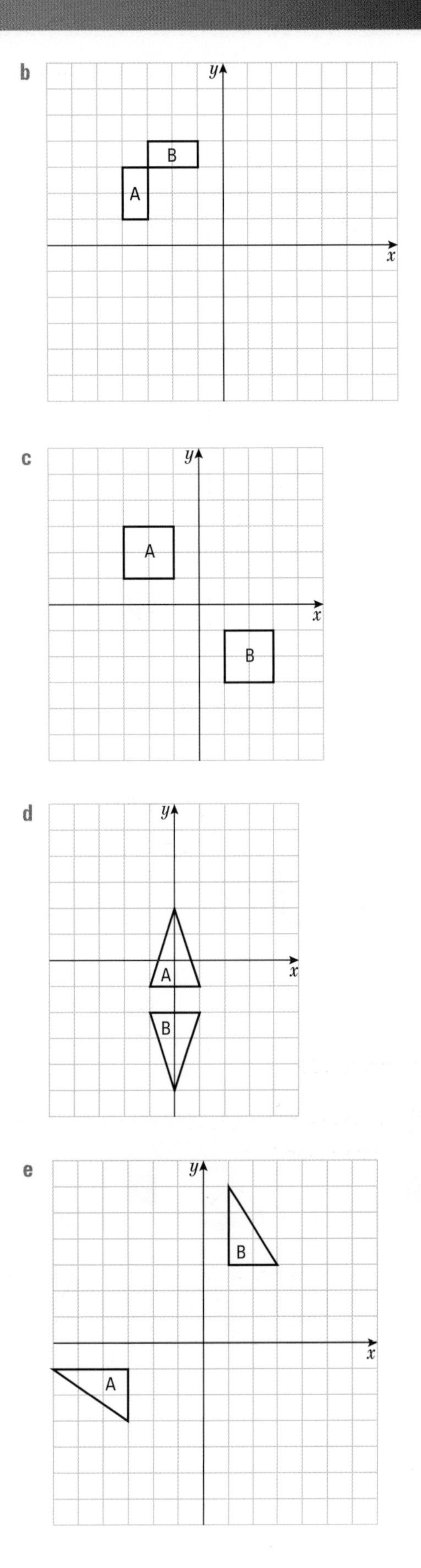

C

f

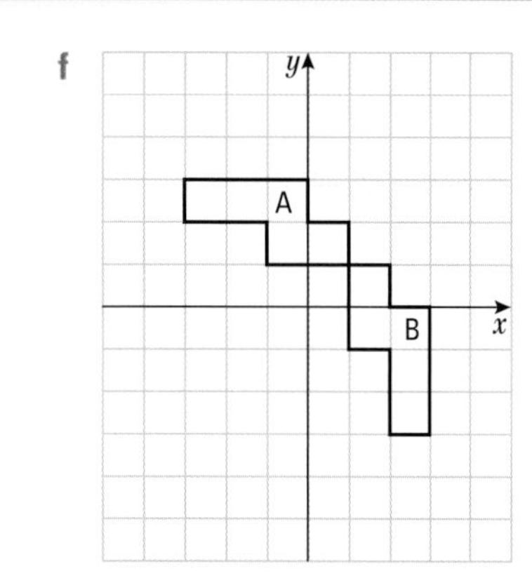

11.5 Enlargement

Exercise 11G

Copy the diagrams and enlarge each of the following shapes by the stated scale factor (sf).

D

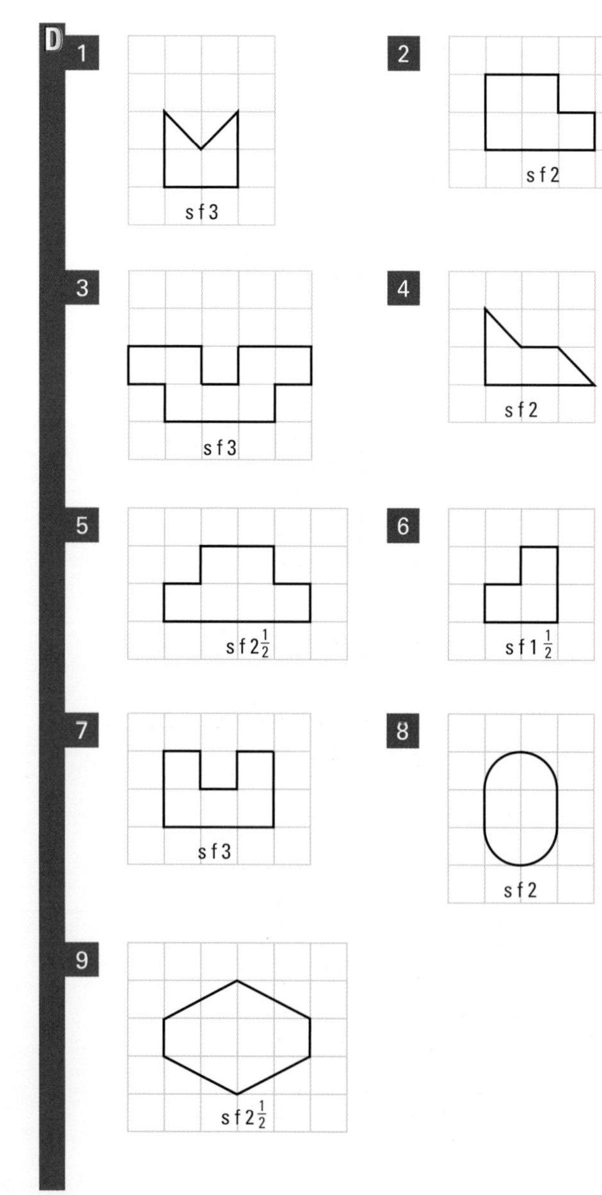

Exercise 11H

D

1 Copy each diagram onto squared paper. Enlarge each of these shapes by the stated scale factor (sf), from the given point of enlargement.

a

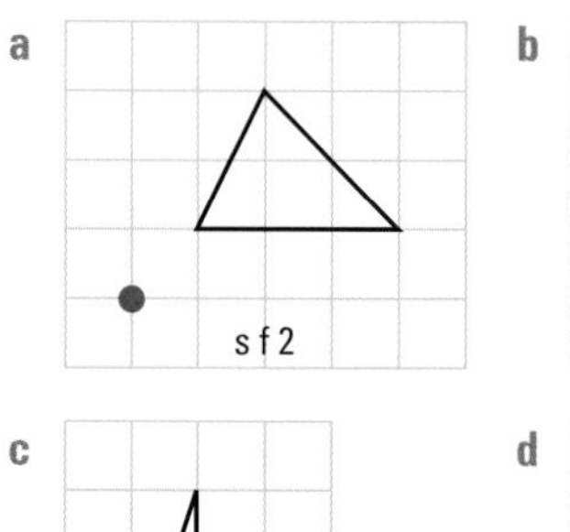

b

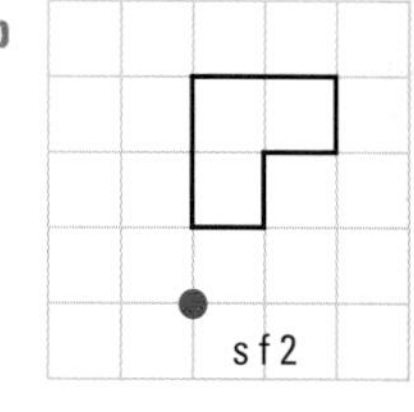

c

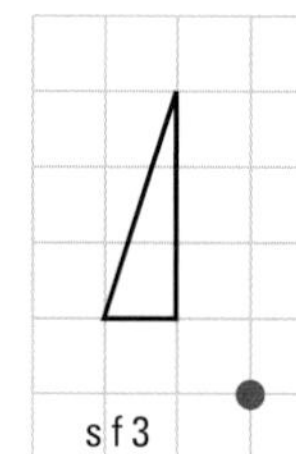

d

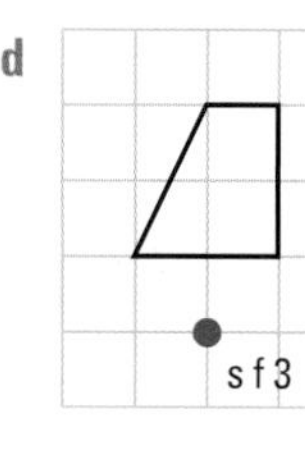

e

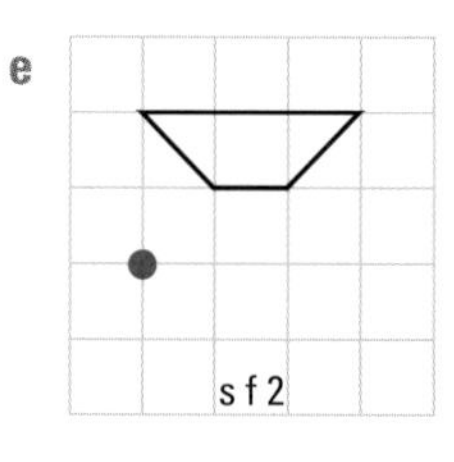

f

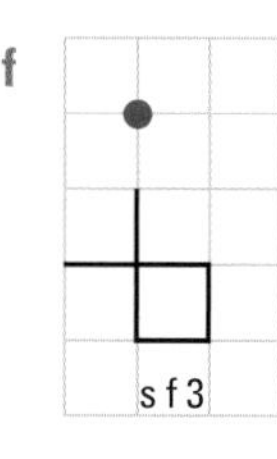

2 Copy each diagram onto squared paper.
For each diagram draw two images, one from each of the points of enlargement given.

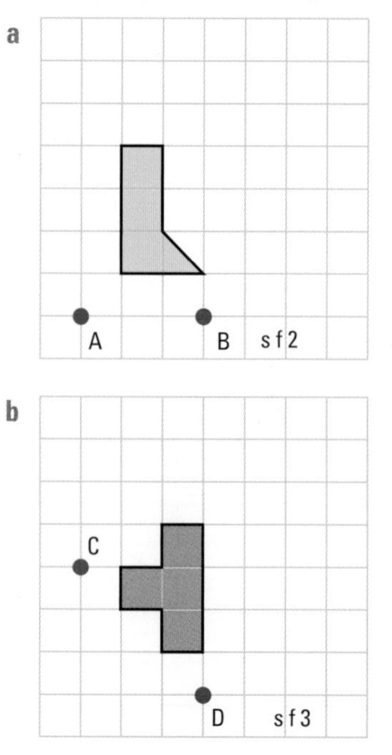

D

c

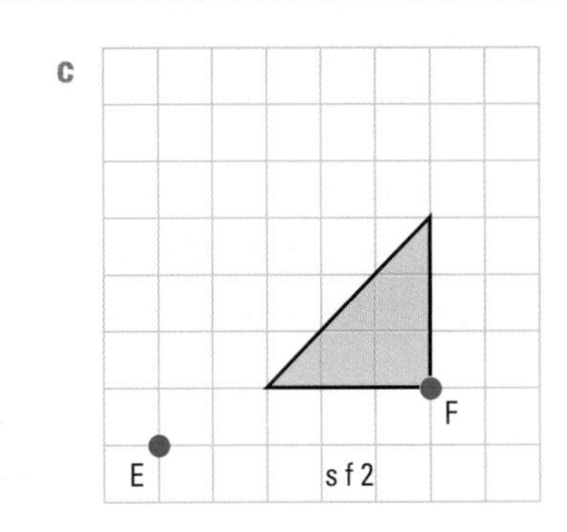

3 Copy each diagram onto squared paper.
For each diagram draw two images, one from each of the points of enlargement given.

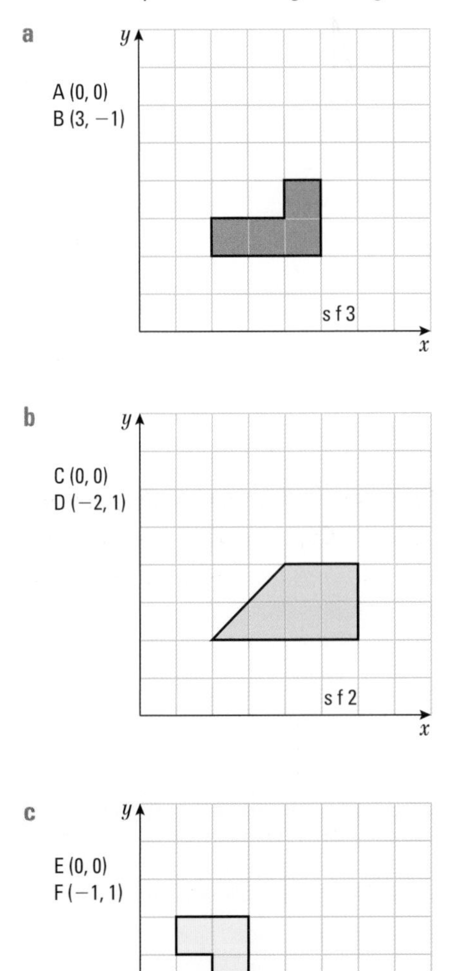

Exercise 11I

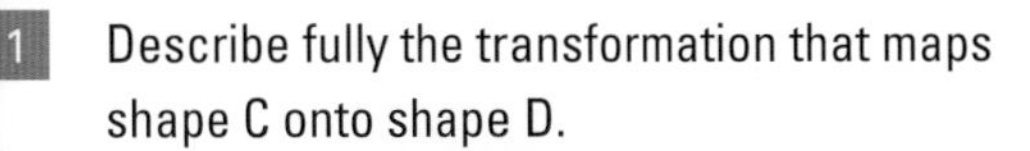

C

1 Describe fully the transformation that maps shape C onto shape D.

a

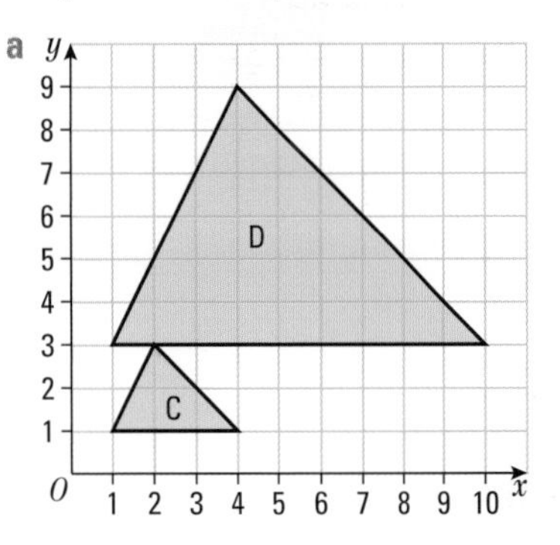

b

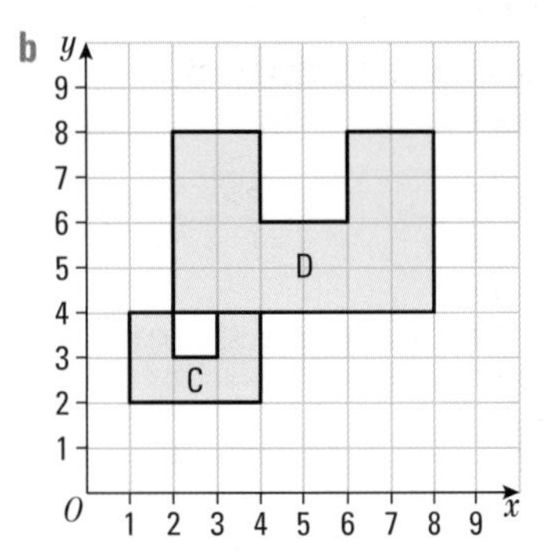

c

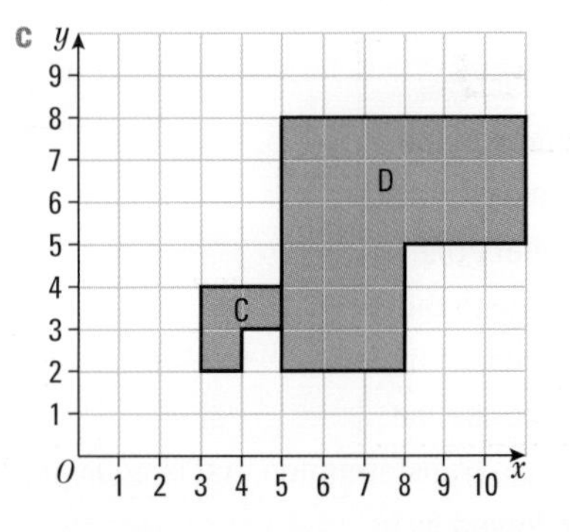

2 Describe fully the transformation that maps shape E onto shape F.

a

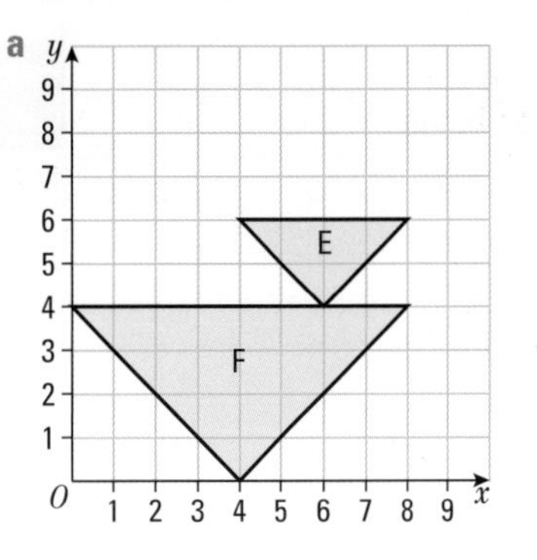

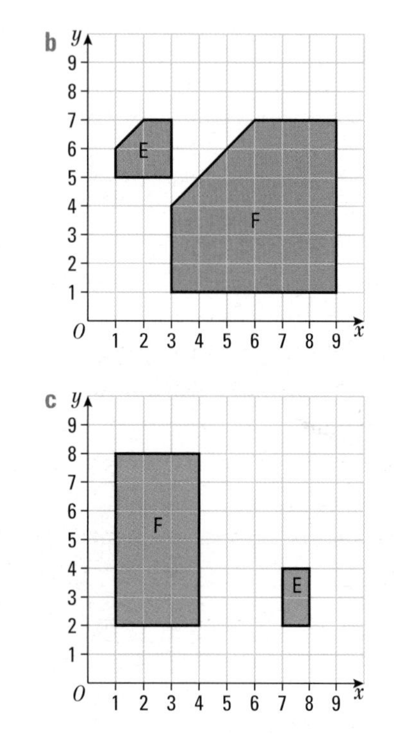

11.6 Combinations of transformations

Exercise 11J

1 Copy the diagram.

a Reflect the shape in the y-axis.

b Reflect the image in the x-axis.

c Describe the single transformation that is equivalent to **a** followed by **b**.

2 Copy the diagram.

a Reflect the shape in the line $y = x$.

b Rotate the image 90° anticlockwise about the origin.

c Describe the single transformation that is equivalent to **a** followed by **b**.

3 Copy the diagram.

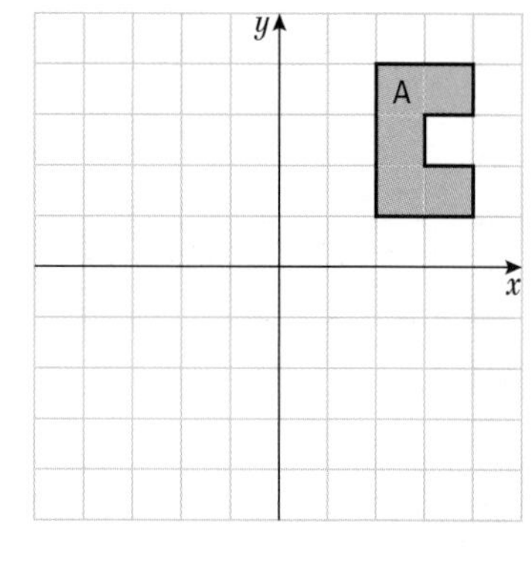

a Reflect the shape A in the x-axis. Label the image B.

b Reflect the image B in the y-axis. Label this image C.

c Describe the single transformation that maps A onto C.

4 Copy the diagram.

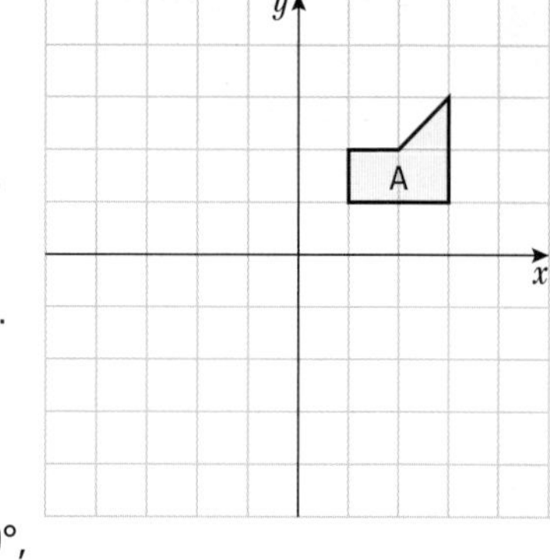

a Rotate the shape A 90° clockwise, centre (3, 3). Label the image B.

b Rotate the image B 180°, centre (3, 3). Label this image C.

c Describe the single transformation that maps A onto C.

12 Pythagoras' Theorem

Key Points

- **Pythagoras' Theorem:** for a right-angled triangle $c^2 = a^2 + b^2$

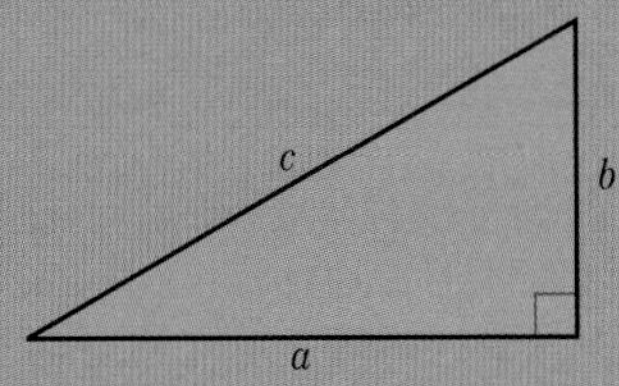

- **hypotenuse:** the longest side of a right angled triangle (c).
- **finding the hypotenuse:** add the square of each of the other sides, and square root the answer. $c = \sqrt{(a^2 + b^2)}$
- **finding the length of a shorter side:** subtract the square of the other short side from the square of the hypotenuse, and square root the answer. E.g. $a = \sqrt{(c^2 - b^2)}$
- **checking if a triangle is right-angled:** see if the square of the longest side equals the sum of the squares of the other two sides.
- **finding the distance between two points on a coordinate grid:**
  - find the difference between the x-coordinates and square
  - find the difference between the y-coordinates and square
  - add the results
  - square root the answer

12.1 Finding the length of the hypotenuse of a right-angled triangle

Exercise 12A

Questions in this chapter are targeted at the grades indicated.

C **1** Calculate the length of the hypotenuse in these right-angled triangles.

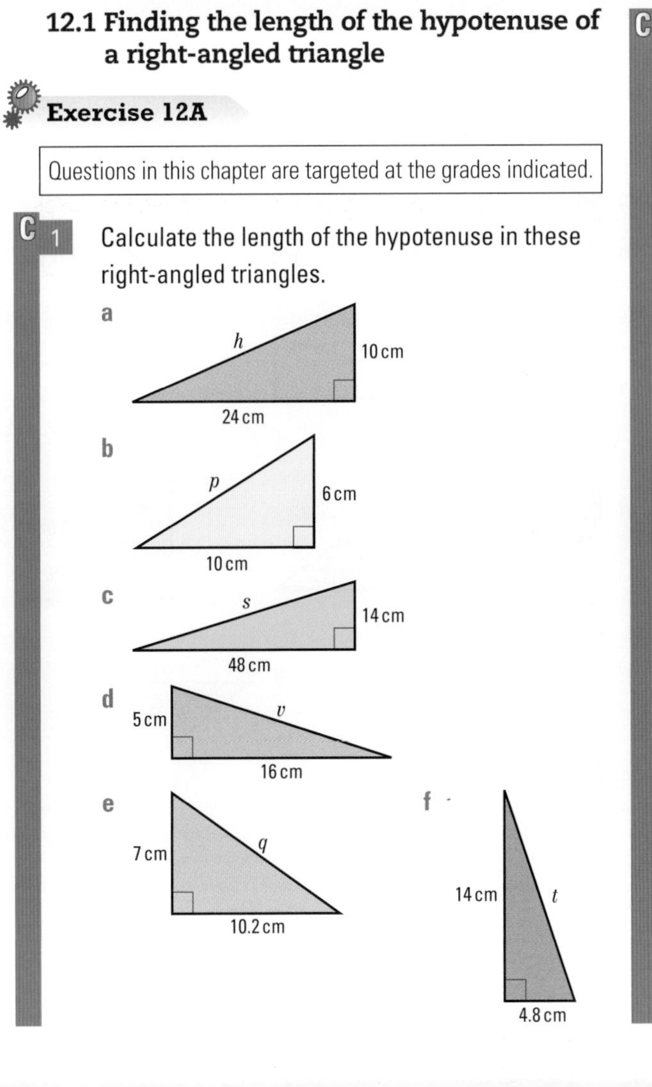

C

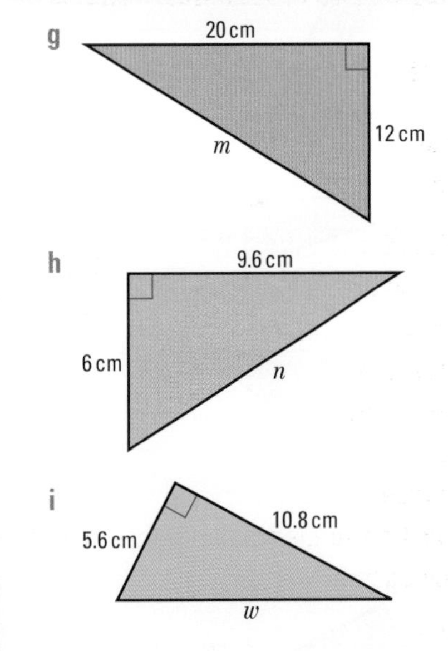

2 Calculate the length of the missing side in these right-angled triangles.

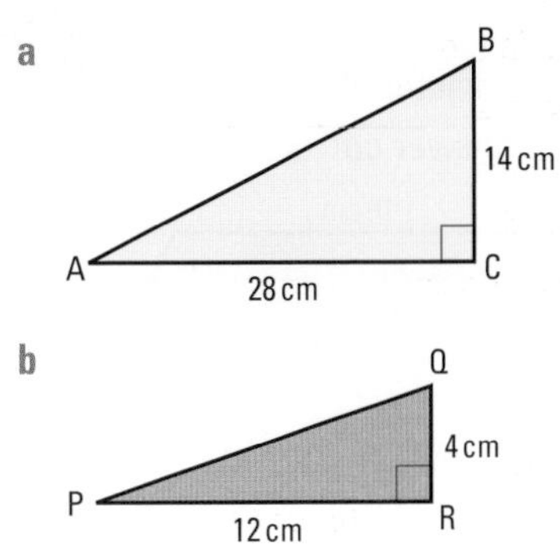

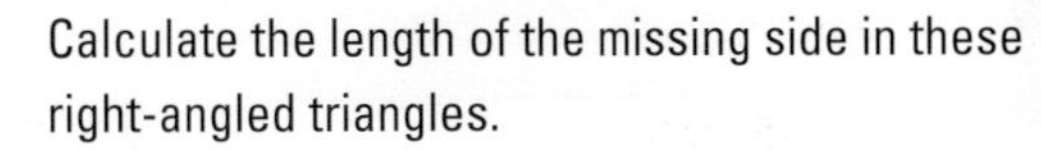

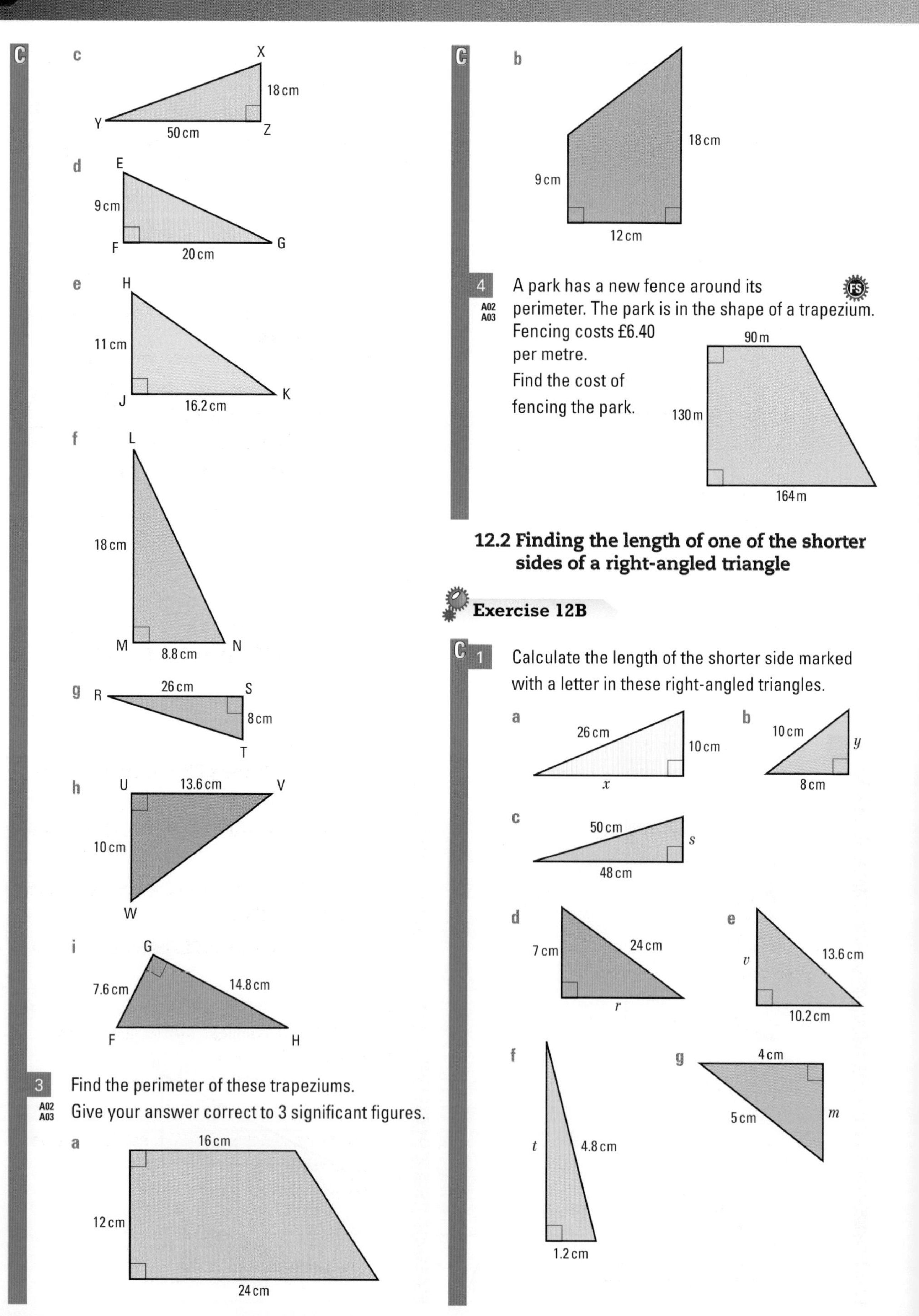

3 (AO2, AO3) Find the perimeter of these trapeziums.
Give your answer correct to 3 significant figures.

4 (AO2, AO3) A park has a new fence around its perimeter. The park is in the shape of a trapezium.
Fencing costs £6.40 per metre.
Find the cost of fencing the park.

12.2 Finding the length of one of the shorter sides of a right-angled triangle

Exercise 12B

1 Calculate the length of the shorter side marked with a letter in these right-angled triangles.

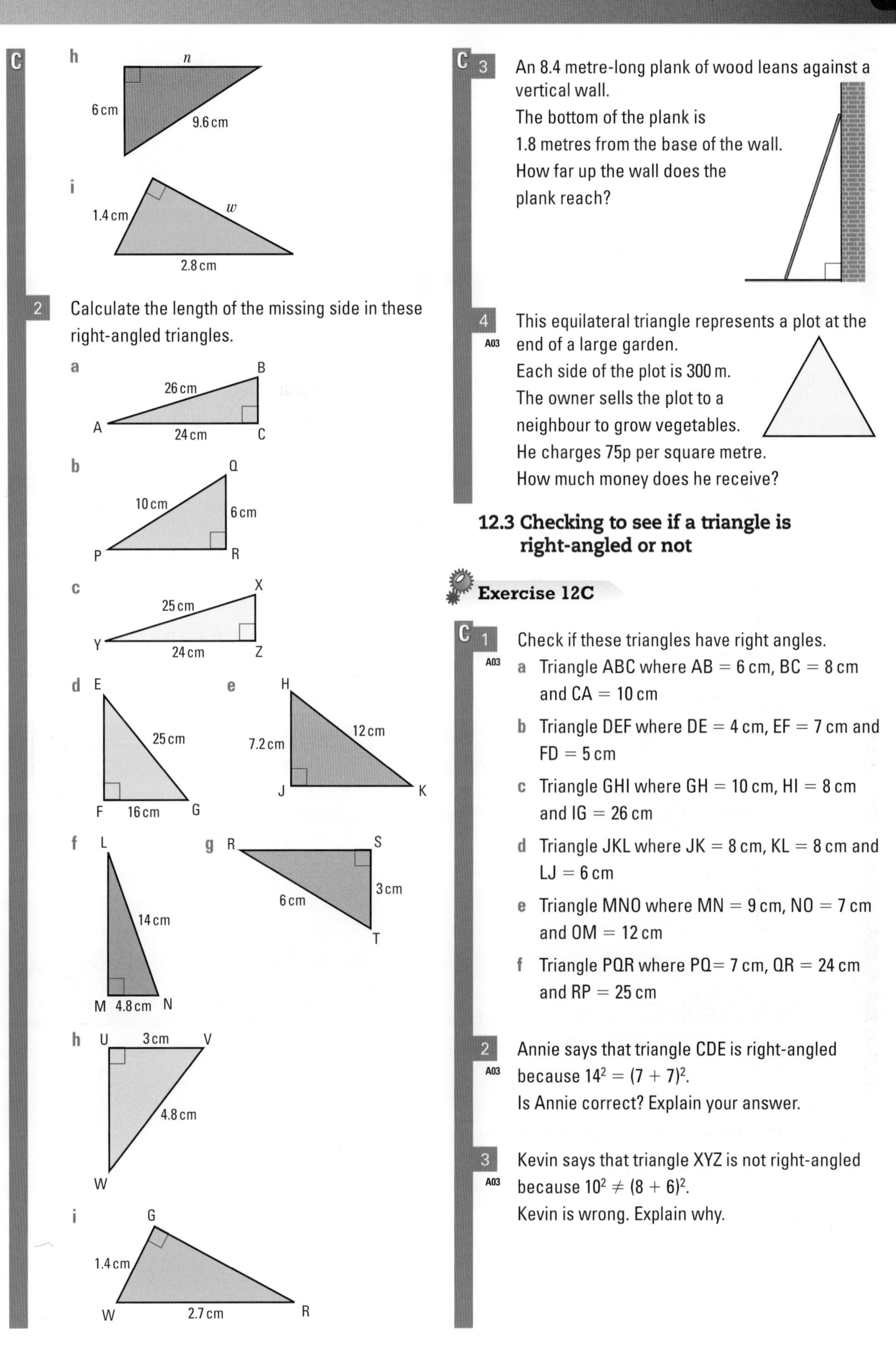

C

h

i

2 Calculate the length of the missing side in these right-angled triangles.

a

b

c

d

e

f

g

h

i

C 3 An 8.4 metre-long plank of wood leans against a vertical wall.
The bottom of the plank is 1.8 metres from the base of the wall.
How far up the wall does the plank reach?

4 This equilateral triangle represents a plot at the
A03 end of a large garden.
Each side of the plot is 300 m.
The owner sells the plot to a neighbour to grow vegetables.
He charges 75p per square metre.
How much money does he receive?

12.3 Checking to see if a triangle is right-angled or not

Exercise 12C

C 1 Check if these triangles have right angles.

A03

a Triangle ABC where AB = 6 cm, BC = 8 cm and CA = 10 cm

b Triangle DEF where DE = 4 cm, EF = 7 cm and FD = 5 cm

c Triangle GHI where GH = 10 cm, HI = 8 cm and IG = 26 cm

d Triangle JKL where JK = 8 cm, KL = 8 cm and LJ = 6 cm

e Triangle MNO where MN = 9 cm, NO = 7 cm and OM = 12 cm

f Triangle PQR where PQ= 7 cm, QR = 24 cm and RP = 25 cm

2 Annie says that triangle CDE is right-angled
A03 because $14^2 = (7 + 7)^2$.
Is Annie correct? Explain your answer.

3 Kevin says that triangle XYZ is not right-angled
A03 because $10^2 \neq (8 + 6)^2$.
Kevin is wrong. Explain why.

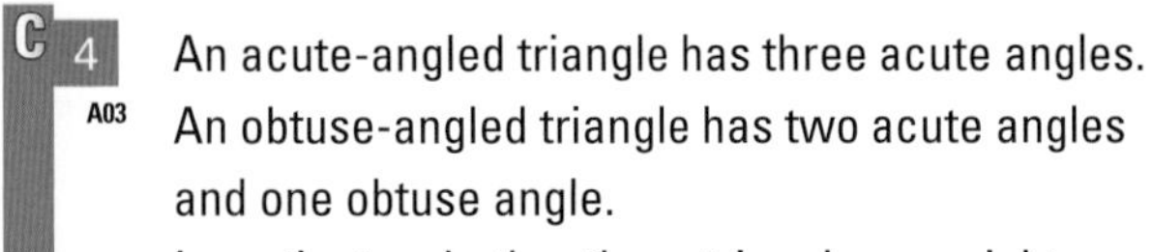

C **4** A03 An acute-angled triangle has three acute angles.
An obtuse-angled triangle has two acute angles and one obtuse angle.
Investigate whether these triangles are right-angled, acute-angled or obtuse-angled.

- **a** Triangle ABC where AB = 12 mm, BC = 35 mm and CA = 37 mm
- **b** Triangle DEF where DE = 8 cm, EF = 14 cm and FD = 10 cm
- **c** Triangle GHI where GH = 4 cm, HI = 4 cm and IG = 3 cm
- **d** Triangle JKL where JK = 30 cm, KL = 40 cm and LJ = 50 cm
- **e** Triangle MNO where MN = 4 cm, NO = 3 cm and OM = 6.5 cm
- **f** Triangle PQR where PQ = 4 mm, QR = 4 mm and RP = 4 mm

12.4 Finding the length of a line segment

Exercise 12D

C **1** Work out the length of each of the line segments shown on the grid.

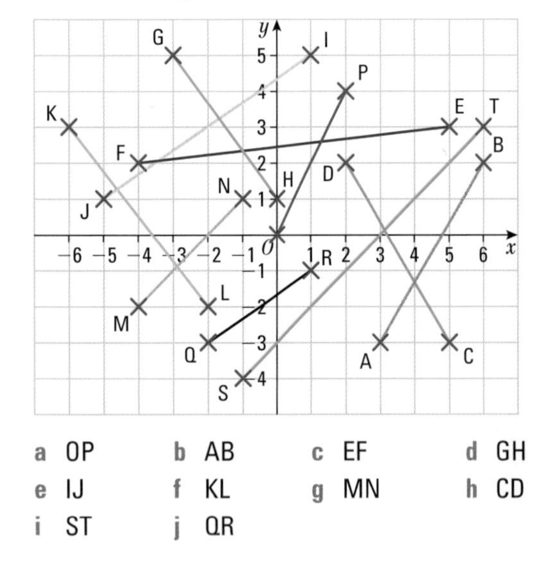

a OP	**b** AB	**c** EF	**d** GH
e IJ	**f** KL	**g** MN	**h** CD
i ST	**j** QR		

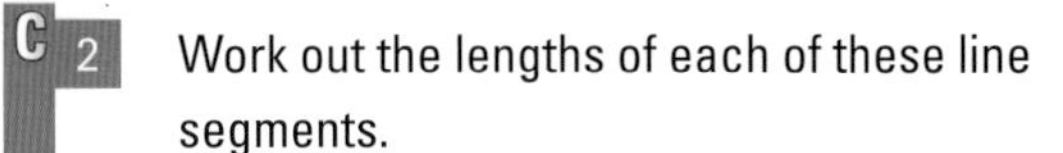

C **2** Work out the lengths of each of these line segments.

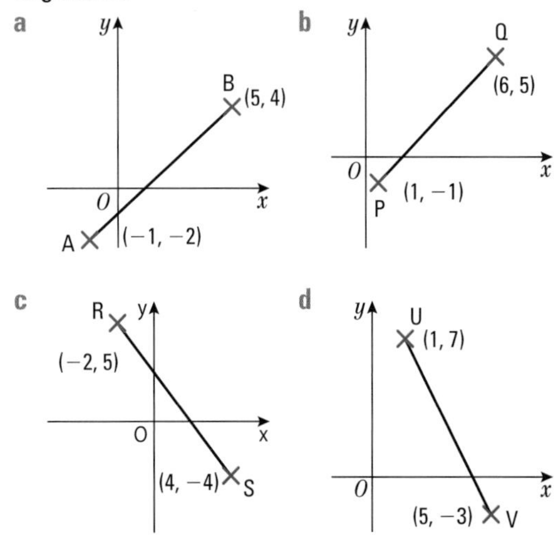

3 Work out the lengths of each of these line segments.

- **a** AB when A is (−1, −1) and B is (7, 7)
- **b** PQ when P is (2, −4) and Q is (−5, 8)
- **c** ST when S is (5, −8) and T is (−3, 2)
- **d** CD when C is (1, 7) and D is (−5, 3)
- **e** UV when U is (−2, 3) and V is (7, −9)
- **f** GH when G is (−2, −6) and H is (6, 4)